The Joan Palevsky Imprint in Classical Literature

In honor of beloved Virgil—

"O degli altri poeti onore e lume . . ."

—Dante, *Inferno*

The publisher gratefully acknowledges the generous
support of the Classical Literature Endowment
Fund of the University of California Press Foundation,
which was established by a major gift from Joan Palevsky.

SATHER CLASSICAL LECTURES

Volume Seventy-Two

Greek Gods Abroad

Greek Gods Abroad

Greek Gods Abroad

Names, Natures, and Transformations

Robert Parker

UNIVERSITY OF CALIFORNIA PRESS

University of California Press, one of the most distinguished university presses in the United States, enriches lives around the world by advancing scholarship in the humanities, social sciences, and natural sciences. Its activities are supported by the UC Press Foundation and by philanthropic contributions from individuals and institutions. For more information, visit www.ucpress.edu.

University of California Press
Oakland, California

Library of Congress Cataloging-in-Publication Data

Names: Parker, Robert, author.
Title: Greek gods abroad : names, natures, and transformations / Robert Parker.
Other titles: Sather classical lectures ; v. 72.
Description: Oakland, California : University of California Press, [2017] | Series: Sather classical lectures ; volume 72 | Includes bibliographical references and index.
Identifiers: LCCN 2016049350 (print) | LCCN 2016052477 (ebook) | ISBN 9780520293946 (cloth : alk. paper) | ISBN 9780520967250 ()
Subjects: LCSH: Gods, Greek. | Civilization, Classical. | Mediterranean Region mCivilization. | Polytheism.
Classification: LCC BL783 .P365 2017 (print) | LCC BL783 (ebook) | DDC 292.2/11 mdc23
LC record available at https://lccn.loc.gov/2016049350

Manufactured in the United States of America

24 23 22 21 20 19 18 17
10 9 8 7 6 5 4 3 2 1

CONTENTS

PREFACE

This book has its origin in three of the six Sather lectures that I had the honour of delivering in the spring of 2013. Those three lectures form the core of what are now chapters 2 and 3 and appendix H. Lectures that sought to be accessible have expanded into a somewhat intricate treatise, and I am grateful to the Sather Lectures committee and the press for accepting this hybrid product in the Sather Classical Lectures series.

Any reader of Sather prefaces has heard of the fabled friendliness and hospitality with which the professor is received; one could give a graduate student interested in rhetoric the assignment of tracing the development of the variations on the theme over the years. In an idle moment I investigated its origin, and discovered the first occurrence in the third of the published volumes, T. R Glover's *Herodotus* of 1924: this book, Glover writes, "will recall pleasant days spent in a community of ten thousand students, with men who added to the ties of kindred studies those of a signal kindness." I can warmly echo and expand Glover's remarks in recording my gratitude to both men and women who did indeed add to the ties of kindred studies those of a signal kindness; and so I can confirm, what classicists sometimes doubt, that some commonplaces are common because they are true. I hope it is not invidious to single out for thanks the late Tony Bulloch, Susanna Elm, Mark Griffith, Eric Gruen, Tony Long, Emily Mackil, Donald Mastronade, Trevor Murphy, Kim Shelton, and especially John Ferrari, chair of the department during my stay, and Ron and Conny Stroud; from outside the department, Robert Goldman and Martin Schwartz; and for much practical assistance, Nancy Lichtenstein. I remember with affection the participants in my graduate seminar. I am also very grateful for warm hospitality and helpful suggestions to hosts and audiences at the University of California Davis, Los Angeles, and Santa Barbara, Seattle University, Stanford, and Yale. I owe advice on particular problems to Jim Adams, John Baines, John Barton, John Healey, Charlotte Potts, Jo Quinn, Beth Richert, Richard Stoneman, and Elisabeth Tucker. At the press I thank my editor, Eric Schmidt, his assistant, Maeve Cornell-Taylor, and project editor Cindy Fulton for friendly guidance, and, for her meticulous copy-editing, Marian Rogers. I am indebted, too, to the University of Oxford for granting me the leave that allowed me to visit Berkeley, and more especially for the academic home that (with two of its colleges, Oriel and New) it has provided me with for almost fifty years.

1

Names and Epithets

By the third century B.C., gods bearing Greek names were worshipped from Spain to Afghanistan. This book is about aspects of that divine diaspora. Its particular emphasis is on names and naming, one of the two principal aspects (iconography is the other) under which that diffusion is revealed to us. Later chapters will enquire how ways of addressing and referring to gods in Greek developed outside old Greece and over time, two processes that can scarcely be kept separate. But first some account of those naming practices in the period when they were relatively isolated from such external contact is needed. The isolation was only relative because contacts always existed, but a real change occurred when use of the Greek language spread and non-Greek gods had to be named in Greek. This chapter will attempt an outline of the status quo ante.

The obvious starting point is the names of the gods themselves, but first the concept of "naming the gods" must be complicated a little. In studying divine names, one needs, at a minimum, to distinguish ways of referring to the gods from ways of addressing them.[1] Direct address brings respect and the desire to conciliate into play, often in very high degree; it may lead to avoidance of proper names in favour of respectful titles or at least the addition of such titles. But it was often necessary to refer to the gods, usually in relation to their shrines or property or priests, in a less charged but accurate manner in the third person. A dedication is perhaps halfway between these two registers: it is a direct address to the god, but there may also be concern to identify the addressee accurately. Respectful forms of naming

1. Cf. Dickey, *Greek Forms of Address*, 9–12, "Lexical versus Address Meaning"; on address itself, F. Braun, *Terms of Address* (Berlin, 1988).

can also spill over from direct address into referential naming. In many ancient Near Eastern cultures, periphrastic avoidance of the actual god's name occurred in all contexts, not just in prayers, so that, for instance, we are uncertain what the real name, if she had one, of the figure referred to as "the lady of Byblos" may have been.[2] In Greece we find certain euphemisms similarly applied almost invariably: Demeter's daughter Persephone is Kore, "Maiden," in inventories as well as in invocations,[3] and certain gods of mysteries are never referred to except by titles.

Another distinction is that between prose and verse. Many epithets are given to gods in poetry but not in cult; there are also cases such as that of Agesilas/Agesilaos ("Leader of the People"), an alternative name, probably euphemistic, for Hades which had a long life in poetry[4] but is unattested in prose. The distinction is not between "literature" and "real belief," because, for instance, the periphrastic impulse present in Agesilaos is certainly an expression of religious feeling. It is a difference rather of register, between language that an ordinary Greek would recognise and respond to and that which he would actually use.

As a rule, gods had one name each. Though it was a title of honour for a god to be "many-named," what was meant by this was in fact "many-epitheted"; the multiple variant names of Akkadian gods lack a Greek equivalent. Phoibos and Pallas are not so much alternative as additional names for Apollo and Athena, regularly used in conjunction with the main name.[5] Kore is not a second name for Persephone of equivalent standing, but, as just noted, a euphemistic alternative. The same is probably true of Plouton (deriving from πλοῦτος, "wealth") as an alternative name for Hades.

In contrast to some ancient polytheisms where gods are named from their functions,[6] the names of the familiar Olympian gods and goddesses are opaque;

2. E.g., *KAI* 10; on figures to whom she was assimilated, see Bonnet, *Enfants de Cadmos,* 168.

3. So, e.g., in *IG* II² 1424a.241 and repeatedly; the ascription of the same objects there listed to "Demeter and Pherrephatte" (an Attic variant for Persephone) in *IG* II² 1437 58 is a rare exception. On Kore for Persephone, see K. Clinton, *OpAth* 16 (1986): 44; for exceptions, R. Parker, *Greece and Rome* 38 (1991): 15 n. 22. Ko-wa = Kore in Mycenaean is very controversial: Rougemont, "Noms des dieux en linéaire B," 332.

4. Aesch. fr. 406 Radt; Callim. *Hymn* 5. 130; *Anth. Pal.* 7. 545.4 (Hegesippus V in Gow-Page, *HE*); Nicander fr. 74.72; *Suppl. Hell.* 990.9; *IC* I.xxii.58.5; *IC* II.v.49.2. Note too Hesych. α 495, Ἀγήσανδρος· ὁ Ἅιδης.

5. Epithets standing in for theonyms, especially in poetry (e.g., Eriounios for Hermes), are a different case. Deo for Demeter is probably a by-form of the same name (for the possibilities, see the note of Richardson on *Hymn. Hom. Dem.* 47).

6. S.A. Geller, in *One God or Many?* 339: "Canaanite gods are famous for being what they are called—so the god of death is Mot, that is 'death', and the sea-god is Yam, 'sea'. They don't have personal names other than that of the thing they represent. In Phoenicia also, the gods are all things; the name of it is what it is." But similar claims for other ancient polytheisms have failed to distinguish etymology from continuing semantic force: contrast S. Morenz, *Egyptian Religion,* trans. A Keep (London, 1973), 21 ("Almost all the gods' names can be translated and as a rule denote a characteristic feature of

none bears a speaking name with unmistakable meaning. The names of various Peloponnesian and Aeginetan figures—certainly or plausibly independent goddesses without mythological connections with the Olympian family but, within their own orbit, as important as any other—are equally obscure: Alea, Orthaia, Aphaia, Mnia, Auzesia.[7] It is true that Greeks tried to extract meaning from the names of some Olympians by popular etymology: they heard *aphros*, "foam," in Aphrodite, and related it to her birth from the sea; they connected Apollo, sometimes a death-dealing god, with the verb *apollumi*, "destroy"; Demeter could be analysed as *Ge meter*, "earth mother," while the accusative case of Zeus, Dia, indicated that it was *dia*, "because of," him that most processes in the world occurred.[8] Aristotle casually alludes to use of the name "in praises of the gods"; he is presumably referring to such attempts to infer the powers of gods from their names.[9] But these were occasional interpretations; no automatic and transparent meaning attached to any of these names. It is true also that the euphemistic names just noted for the lords of the underworld, Kore and Plouton, have transparent meanings, but their true names were obscure like any other. Sun (Helios), Earth (Ge, Gaia), and Hearth (Hestia) were physical entities as well as deities, and therefore had speaking names, but their importance in cult was modest. The one major figure with a transparent name was Mother (Meter), and even she was sometimes

their nature or function") with J. Baines, in *One God or Many?* 29 ("Much effort has gone into the search for etymologies of divine names, but while plausible origins in nouns or domains of action can be proposed for a number of major deities, in synchronic perspective they possess proper names, and their sphere of action is not limited to what these might imply: there is altogether more to a deity than a name might encompass").

7. I give the names of the two last as they appear in the fifth-century inventory *IG* IV² 2.787 (the *e* of Auzesia is short), not Damie and Auxēsie as in literature (e.g., Hdt. 5.82–83). Forms in Azesi- (with short or long *e*) and Azosi- are also attested as theonyms or in related epithets or month names (see Polinskaya, *Local History*, 274–78; particularly important is the Attic Demeter Azesia, *Agora* XIX H 16, 4th c. B.C.). Auxesie has usually been accepted as the proper form and as a speaking name, "Increaser" (so, e.g., Chantraine, *Dictionnaire étymologique*, s.v. αὔξω); Polinskaya considers the possibility that Auxesia is contrasted with an Azosia/Azosia deriving from ἀζαίνω (so Hesych. s.v. Ἀζήσια), "I wither," thus a double-sided pair of deities with speaking names. But I am more inclined to see a single name underlying the *aux*- and *az*- forms. Other possible local gods: Φειλάτα, *IG* IX.I² 3.663; Πρεινάτις, *IG* IX.I² 4.1730.

8. Aphrodite: Hes. *Theog.* 197; or she could be linked with *aphrosyne*, "folly": Eur. *Tro.* 989–90. Apollo: Macrob. *Sat.* 1.17.9–10, citing Archilochus fr. 26.5–6; Eur. fr. 781.11–12 (*Phaethon*). Demeter: Eur. *Bacchae* 275–76, with Dodds's note. Zeus: perhaps Hes. *Op.* 3, and certainly passages cited by West in his note ad loc. "The earliest object of fundamental theological speculation is in all seeming the name of the god," B. Gladigow, "Götternamen und Name Gottes," in H. v. Stietencron, ed., *Der Name Gottes* (Düsseldorf, 1975), 13–32, at 24; for Hesiod, D. Arnould, *REG* 122 (2009): 1–14; still in the much later Orphic Hymns, A. F. Morand, in A. Bernabé et al., eds., *Orfeo y el Orfismo: Nuevas perspectivas*, www.cervantesvirtual.com, 157–76.

9. *Rhet.* 2.23, 1400b19; cf. Pernot, "Lieu du nom," 30–34, with examples such as plays on Isis and *isos*, "equal."

identified with the opaquely named Rhea (or Cybele). Most major heroes too have meaningless names such as Theseus or Achilles or Herakles. Gods and heroes were thus marked off from mortals, who in the historical period typically had names compounded from ordinary Greek words and thus readily comprehensible. It has indeed been argued that divine names with recognisable meanings were dissimilated from their etymological origins to render them opaque.[10] And, though mortals regularly bore theophoric names based on those of a wide range of gods, they never bore divine names unadjusted (with the possible exception of Artemis) until quite a late date, even if the difference was no more than a single letter (as in Dionysios from Dionysus).[11]

At a lower level in the divine hierarchy, speaking names do appear. The Nymphs as a group are a collectivity of brides, *nymphai,* and many female figures of that level of power have transparent names: the Praxidikai, "Justice-Exacters," for instance, or Kourotrophos, "Child-Nurturer," sometimes a minor independent goddess, sometimes an epithet of some larger figure. A minority of cult heroes too have names indicating a function: Matton, "Kneader," and Keraon, "Mixer," culinary heroes in Sparta; the Attic Sosineos, "Save Ship"; the Thessalian Poliphylax, "City Guard" (honoured by human "city guards"); and others.[12] There is also a great swarm of what we would call personified abstractions, figures such as Eros, "Sexual Desire"; Pheme, "Rumour"; Phobos, "Fear"; and many others. Some of these existed only as figures of speech or iconography, while others actually received cult, but there was no sharp division between the two groups: a personification who had only existed at a verbal or pictorial level could easily cross over to become a recipient of cult. Of personifications that received cult, a large number were closely linked with major deities whose power they expressed in some way; such

10. Burkert, *Greek Religion,* 182 (but the examples depend on etymologies not universally accepted); F. Graf, "Namen von Göttern im klassischen Altertum," in *Namenforschung,* 2:1823–37, at 1826.

11. See R. Parker, "Theophoric Names and the History of Greek Religion," in S. Hornblower and E. Matthews, eds., *Greek Personal Names: Their Value as Evidence* (Oxford, 2000), 53–80, at 57–59. I know from conversation that the late Anna Morpurgo Davies doubted Masson's interpretation, on which the case for the name Artemis as an exception depended (although the case was very limited, because the name was not borne by citizen Greek women). The frequency of Dionysos, Souchos, Sarapis, and Helios as personal names in Egypt is drastically reduced by W. Clarysse, *ZPE* 186 (2013): 259–66; and G. Jennes, ibid., 267–69.

12. Polemon fr. 40 Preller, ap. Ath. 2.9, 39C-D, where Deipneus, "Diner," of Achaia, and Akratopotes, "Drinker of Unmixed Wine," of Mounichia are also mentioned. Sosineos: *SEG* XXXIII 147.50; Poliphylax: *SEG* XXVII 205. To others listed by L. R. Farnell, *Greek Hero-Cults and Ideas of Immortality* (Oxford, 1921), 418–20, add, e.g., from Attica Heudanemos, "Sleep Wind" (Arrian, *Anab.* 3.16.8), Kalamites ἥρως, "Hero of the Stalk" (Dem. 18.129; Clinton, *Myth and Cult,* 106 n. 6: presumably the stalk of the growing corn); Kuamites, "(Hero) of the Bean" (Paus. 1.37.4, though Kearns, *Heroes of Attica,* 180, thinks he may have been named from the nearby bean-market); from Eretria Naustolos, "Ship Sender," *IG* XII.9.256.

was the case of two who were linked with Aphrodite, Peitho, "Persuasion," and Eros, "Love," and of Athena's associate Nike, "Victory." But a minority had a free-standing existence in cult, most notably Nemesis, "righteous outrage";[13] there were also groups such as the Graces (Charites), the Seasons (Horai), and the Muses, who might be broken down into individuals who bore speaking names in turn. There is again here a difference, though of a different kind, between divine and human naming. Whereas most human names had semantic content, no one will have expected the conduct of a Philodemos, "People-Lover," to be governed by his name; but we presume that a worshipper of "Save Ship" will normally have looked to him to do just that.[14]

A majority of Greek divine figures have names of one of these two types: they are either opaque, or relate directly to their powers or functions. A minority—goddesses more commonly than gods—are normally referred to by titles or by adjectival descriptions of some kind. Such replacement of name by title may once have been commoner than it became. "Mistress (Potnia) of the Labyrinth" is one of numerous Mycenaean usages of Potnia in lieu of a name (whether Wanax and Wanassa, "Lord" and "Lady," are similarly used is controversial).[15] At Perge in Pamphylia the goddess later familiar as Artemis of Perge is named on earlier coins and inscriptions simply as "Wanassa" of Perge, while Aphrodite on Cyprus is initially plain "Wanassa" of Paphos or of Golgoi.[16] Somewhat similar is Alcaeus's remarkable address to what must be Hera as "glorious Aeolian goddess, source of all things."[17]

These usages occur either very early or in places on the fringes of the Greek world where older naming conventions may have survived (though external influence is also possible). Later on the mainland, titles or adjectival descriptions are usually found in relation to mystery cults or gods who inspire fear or invite euphemism in other ways;[18] the two factors, unwilligness to name mystery gods directly

13. On all this, see Parker, *Athenian Religion,* 228–37.

14. But name is not a wholly reliable guide to function: Eunostos, "Fair Return," is attested by lexicographers (Hesych.; Phot. s.v. Eunostos) as a "spirit (*daimon*) of the mill," but no trace of that function remains in the story of the Tanagraean hero Eunostos in Plut. *Quaest. Graec.* 40, 300D-301A.

15. See Rougemont, "Noms des dieux en linéaire B," 344–60; on Wanax and Wanassa, J. T. Killen, in Killen and A. M. Davies, eds., *Documents in Mycenaean Greek,* 3rd ed. (Cambridge, forthcoming).

16. *IPerge* 1; cf. C. Brixhe, *Le dialecte grec de Pamphylie: Documents et grammaire* (Paris, 1976), 140, 160–61, and his no. 3, line 29; Brixhe, in M. Fritz and S. Zeilfelder, eds., *Novalis indogermanica* (Graz, 2002), 50–58; on the chronology of the coins, cf. J. Nollé, *Chiron* 44 (2014): 292. (For a possible Pamphylian goddess ΔιϜια, see Brixhe, *Dialecte grec,* 139 on no. 3, line 1.) Cyprus: M. Egetmeyer, "'Sprechen sie Golgisch?'," in P. Carlier et al., *Études mycéniennes 2010* (Pisa/Rome, 2012), 427–34.

17. Αἰολήιαν κυδαλίμαν θέον/πάντων γενέθλαν, fr. 129.6 Voigt; that this is Hera is fairly clear from the parallel with Sappho fr. 17 Voigt; the goddess's importance in Lesbos (cf. D. L. Page, *Sappho and Alcaeus,* Oxford 1955, 168) has been underlined by the new "brothers' song" of Sappho with its prayer to "queen Hera" for salvation at sea (cf. V. Pirenne-Delforge and G. Pironti, *ZPE* 191, 2014: 27–31).

18. Plouton is identified as a euphemism for Hades inspired by fear in Pl. *Crat.* 403A.

and fear of the underworld, often coincided, as in relation to the gods of Eleusis. Despoina, "Mistress," of Lycosura, the goddess most revered by the Arcadians according to Pausanias, presided over mysteries.[19] Megaloi Theoi, "Great Gods," are found in separate mystery cults at Andania in Messenia, and in the Aegean. There are mysteries of Megalai Theai, "Great Goddesses," in Arcadia,[20] while the Eumenides, "Friendly Ones," were those whom in Sophocles' words "we are afraid to name, and whom we pass without looking, without sound, without speech, moving our lips in respectful silence"; they were also called Semnai Theai, "Reverend Goddesses."[21] A sanctuary on a hill at Pallantion in Arcadia, at which "oaths on the most important matters were sworn," was identified simply as belonging to the "pure ones" (*katharoi*): either "they do not know the names of the gods or are unwilling to reveal them," says our sole source, Pausanias (8.44.5–6).[22]

Often the gods addressed by titles also had names known to their worshippers (Pausanias declines to tell the true name of Despoina to the uninitiated),[23] but there was always an impulse respectfully to avoid direct naming when dealing with such powers. The Erinyes/Eumenides are even described by Euripides as "the nameless goddesses," and it is apparently this degree zero of naming that we encounter in the Eleusinian pair "God and Goddess" (presumably Hades and Persephone).[24] Individual gods could be known under many different periphrases. The "Reverend Goddess" (*hagnē theos*) of a fifth-century curse tablet from Selinus and a fourth-century Attic calendar is probably a further name for Persephone (we have already noted Kore and plain Goddess); she also appears in dedications (in northern Greece) as "Only Child" (Μουνογόνη, Μωνογενίη) and "Bride" (Nymphe), while her husband is Despotes ("Lord"), Basileus ("King"), Klymenos

19. *LSCG* 68; *SEG* XLI 332.7; Paus. 8.37.1–10; she was for the Arcadians a daughter of Demeter and Poseidon (Paus. 8.37.9; cf. 8.25.7, Thelpousa; 8.42.1, Phigaleia), but symbolically dominated her mother (Paus. 8.37.4, a work of Damophon showing Demeter with a torch, Despoina with a sceptre); the "Despoinai" who receive *IG* V.2.525 (Hadrianic?) should be her and Demeter; Despoinai again at Olympia, Paus. 5.15.4, and joint cult of Demeter and Despoina on the Arcadian-Messenian border 8.35.2. *IG* V.2.524 is a dedication from Lycosura by "king Julius Epiphanes Philopappos" to Despoina and Soteira; Jost, *Arcadie*, 335, sees the latter as Artemis.

20. See p. 14n55.

21. Soph. *OC* 129–33, trans. Lloyd-Jones; Aesch. *Eum.* 1041, with A. Sommerstein's note ad loc.

22. For the excavated remains possibly associable with them, see the commentary of M. Moggi and M. Osanna on Paus. 8.44, their lines 35–46; on the nature of the gods (unknowable), Jost, *Arcadie*, 590–91.

23. Paus. 8.37.9, with comment also on Kore-Persephone; similarly at Thelpousa, 8.25.7.

24. Nameless goddesses: Eur. *IT* 944 (i.e., the Erinyes, so named in 941, 963, 970); Eur. fr. 494.18: cf. A. Henrichs, "Anonymity and Polarity: Unknown Gods and Nameless Altars at the Areopagos," *Illinois Classical Studies* 19 (1994): 27–58, at 37. Different are the famous unknown gods, ingeniously singularized into an "unknown god" by the author of Acts 17.22–23 (n. 36 below); cf. Henrichs, 29–35. "God and goddess": see Parker, *Polytheism and Society*, 335–36.

("Famed One") (Paus. 2.35.9–10), and much else besides.[25] The resident of Colonus near Athens in Sophocles' play, asked the "solemn name" of the "dread goddesses, daughters of Earth and Darkness" who inhabit the grove there, answers: "The people here would call them the all-seeing Eumenides; but different names are favoured in different places" (*OC* 42–43):[26] the title used, then, was seen as a locally variable euphemism for an invariable essence. Nonetheless, the use of titles could apparently lead to uncertainty and variation when the attempt was made to identify the power concerned: Pausanias has to explain that Klymenos at Hermione in the Argolid is (supposedly) a "king under the earth," not a mortal Argive, while in the mystery cult at Andania in the southern Peloponnese the apparent sex-change of the chief honorands between the first century B.C. and the time of Pausanias from Great Gods to Great Goddesses is a standing conundrum.[27]

About some other deities known by adjective or title we know too little to interpret with confidence. Several distinct goddesses called Parthenos, "Maiden," are known, one from the Tauric Chersonese on the northern coast of the Black Sea, one from Leros and adjacent regions of Caria, one attested across a swath of northern Greece from Epirus to Neapolis (modern Kavalla) in Thrace.[28] The title may be a euphemism—among Greeks, at least, the Parthenos of the Chersonese had a dire reputation, as a supposed recipient of human sacrifice; it may (for we are outside

25. "Reverend goddess": *IGDS* 38; *SEG* LIV 214.23, 26; cf. Paus. 4.33.4. Only Child, Nymphe, Despotes: see Parker, "Theonyms in Northern Greece." On the mysterious Τελεσσαι, "Complete Ones, Completers" (?) of Cyrene (*LSS* 116.12), perhaps Demeter and Kore, cf. *BÉ* 2014, no. 538. Basileus: *SEG* LVII 405 (Olympia). Klymenos: Lasus of Hermione fr. 1, *PMG* 702, ap. Ath. 14, 624E-F; Callim. fr. 285 with Pfeiffer's note ad loc.; Paus. 2.35.9–10; *IG* IV.686–91. "Temenios," "He of the precinct," is probably again Hades in *IG* V.I.497, 589, 608. For further mostly poetic terms for Hades, see E. Rohde, *Psyche*, trans. W. B. Hillis (London, 1925), 183 n. 6. I do not open here the difficult questions of Zeus Chthonios and (Zeus) Eubouleus. When an Athenian in Pl. *Leg.* 796b speaks of Athena as "our maiden (*kore*) and mistress (*despoina*)" he seems to be playfully juxtaposing terms contrasted in themselves and normally applied to gods other than Athena.

26. Line 41, where Oedipus speaks of their σεμνὸν ὄνομα, hints at one of the different names, Semnai.

27. Paus. 2.35.9; sex change: p. 142n55 below.

28. Chersonese: Hdt. 4.103.1–2 (said by Hdt. to be identified by the natives with Iphigeneia); *IOSPE* I² index p. 553; note esp. *IOSPE* I² 352.23 (*Syll.*³ 352): ἁ διὰ παντὸς Χερσονασιτᾶν προστατοῦσα [Πα]ρθένος, *IOSPE* I² 344 = *FGrH* 807 T 1 (her ἐπιφάνειαι); in imperial time she becomes θεὰ βασίλισσα Παρθένος (*IOSPE* I² index p. 553, first in no. 359. 20–21, apparently because she regularly served in the magistracy of *basileus*: *IOSPE* I² index p. 552 s.v. βασιλεύουσα Παρθένος); Strabo 7.4.2, 308 (described as δαίμων τις!). Leros: *ASAtene* n.s. 25–26 (1963–64): 304–10, nos. 2–5; Klytos of Miletos, *FGrH* 490 F 1 ap. Ath. 14, 655C (interpreted as Artemis in Ant. Lib. 2.6 and *FGrH* 475 F 2), *IG* XII.3.440 (dedication to Parthenos Leria, from Thera). Halicarnassus: *SEG* XLIII 713.2; *SGDI* 5733; note too Diod. 5.62.4 (Bybassos). Sosibios's reference to an enkomion of the Parthenos in Sparta is isolated: *FGrH* 595 F 6 ap. Ath. 646; ἡ πάτριος ἁγνὴ Παρθένος in *IAssos* 26.20 (37 A.D.) is probably a late designation for Athena. Northern Greece: Parker, "Theonyms in Northern Greece."

the Greek heartland) translate an indigenous term, or serve to label an originally anonymous goddess. Another anonymous figure from a border zone is Polystephanos, "Many-Crowned (Goddess)," worshipped at a spring near Butera in Sicily.[29] The "Great God," or perhaps "Great God of the Odesitai," known from Hellenistic coins of Odessos on the west coast of the Black Sea, reappears in later inscriptions as "Great God Derzelas/Darzalas": such had presumably always been his name, but one initially shunned in favour of periphrasis by the Greek settlers.[30] Gods in Greece itself who were recurrently addressed not by a proper name were a Basileia, "Queen," in West Locri (there was also a less prominent, perhaps euphemistic, Basile in Attica),[31] and Kalliste, "Most Beautiful One (feminine)," sometimes paired with Ariste, "Best," in Attica;[32] there are also isolated occurrences of "Beautiful Goddess," "Good God/s," "Good Goddess" (this last possibly in one instance identical with Kalliste).[33] Agathos Daimon, "Good Power," "Good Destiny," grew from a figure toasted at symposia into a popular domestic god of Roman Egypt, often

29. *SEG* XVI 593; but her cult (as Polystephanos Soteira) is also attested by a domestic (?) altar from Acragas, *Kokalos* 13 (1967): 202–4 (*ThesCRA* V p. 237, no. 528).

30. See Chiekova, *Pont gauche,* 179–200: for "great god" (*theos megas*) without the proper name still in the Roman period, see *IGB* I² 150, 186ter; III 1855; for plain Darzalas, *IGB* II 768 (cf. probably 770, Derzes). It is ambiguous whether the genitive Ὀδεσιτῶν of coins attaches to the god's name (so Chiekova), or just indicates the issuing polis (Mihailov in *IGB* I² p. 92). On this figure, cf. p. 142n56.

31. Basileia: see Lerat, *Locriens de l' Ouest,* 2:158–61, on *IG* IX.I² 3.659, 685, 715, arguing that she had a cult on a hilltop and might be Hera (or Artemis); on the Attic figure (figures?) linked with (a) Neleus and Kodros and (b) Zeuxippos, see H.A. Shapiro, *ZPE* 63 (1986): 134–36; Kearns, *Heroes of Attica,* 151; note too a *thea* Basileia on Thera (*IG* XII.3.416: 1st c. B.C. or later) apparently possessing a temple. Basileia is also (see Lerat, *Locriens de l'Ouest,* 2:158–61) an epithet of Hera (as in the new Sappho poem, *ZPE* 189, 2014: 32–49), Persephone (?) (cf. *SEG* XLIV 910, Mylasa), Aphrodite, and, questionably (see Lerat, on Hdt. 4.33.5), Artemis. On an apparently freestanding Basileia at Iasos, see M. Nafissi, *SCO* 61 (2015): 123–24, who sees her as the Mother of the Gods.

32. Kalliste: *IG* II² 788–89, 4665, 4667–68 (all speaking simply of Kalliste); Paus. 1.29.2 (who adds Ariste and identifies the titles as epithets of Artemis); *IG* II² 1298, found near the dedications 4665, 4667–68, is to be displayed in a sanctuary of Artemis. Note too n. 59 on Hagemone at Lykosoura and Messene.

33. "Beautiful goddess" (one each in Macedonia and Alexandria): see Parker, "Theonyms in Northern Greece." Good gods (with Zeus Meilichios): *IG* IX.I² 3.693; cf. Lerat, *Locriens de l' Ouest,* 2:147–49. Good god: two of the Tegean "standing stelai" (Gaifman, *Aniconism,* 211–22), *IG* V.2.60, 67 (note too 59 to Agathos Daimon); votive stelai from Larisa, sometimes in association with Agathe Tyche: *SEG* XLIII 285 (2nd c. A.D.?), XLV 618 (2nd c. B.C.?); Heinz, *Thessalische Votivstelen* K 267; *IG* IV² 1.394 (186 A.D.), 406 (224 A.D., showing a bearded figure holding sceptre and cornucopia, traversed by a snake: J. Harrison, *Themis,* Cambridge 1912, 285, fig. 75). Good goddess: *IG* II² 4589, Piraeus, ca. 300: she holds a cornucopia and is approached by two worshippers, above whom is shown a leg; *SEG* LVI 203 (3rd c. B.C.), a *synodos* dedicated to her in Athens; possible link with Kalliste: R. Parker, in J. Dijkstra et al., eds., *Myths, Martyrs, and Modernity* (Leiden, 2010), 207–8. The male "valorous, excellent ones" (Ἀριστεῦσι, *SEG* XXIX 519, 1st/2nd c.) who receive a dedication at Gonnoi must be quite different.

associated with Agathe Tyche, "Good Luck."[34] Anake (dual), the title by which the Dioscuri (or figures assimilated to them) were known in Attica and Argos, is a mysterious fossil: Greeks probably heard in it a variant on Anakte, "Lords," so a title of respect.[35] The explanation for these expressions may vary from case to case: euphemism; replacement of the theonym by an honorific accompaniment that has become fixed; emphasis on divine attributes particularly desirable in a particular case ("beauty," in patronesses of young girls; goodness, in the sense of friendliness to man); uncertainty about the identity of the power addressed; in the case of Anake, habitual usage.

A minor category is that of gods whose names were not yet known, acknowledged by those altars addressed to "unknown gods" out of which the apostle Paul seems to have created his altar to an "unknown god."[36]

THE CULTIC DOUBLE NAME

In narrative, a god is normally designated simply by a single name. It is unusual when Apollonius (*Argon.* 2.2–3) speaks of a Nymph who slept with Poseidon Genethlios ("of Begetting"), as opposed to plain Poseidon. But, in cult, divine names were typically accompanied by an epithet, to give what is called the cultic double name;[37] this makes the god what has been called "declinable" ("dieu decliné"). The specification added by an epithet is so important that, it has been claimed, a god's name taken alone reveals nothing about the function a god performs in a particular context, with the single exception of Asclepius, always

34. See F. Dunand, in *LIMC* s.v. Agathodaimon.

35. B. Hemberg, *Anax, Anassa und Anakes als Götternamen unter besonderer Berücksichtigung der attischen Kulte* (Uppsala, 1955). Note too the Anaktes Paides at Amphissa (Paus. 10.38.7), variously identified according to Pausanias as Dioskouroi, Kouretes, Kabeiroi. Who the Anaktores of *IEph* 719.8 (Trajanic) are is uncertain (Hemberg, ibid., 24). Equally obscure are the "First Gods" of *ID* 2310, Semitic according to Plassart, Kabeiroi according to Vallois (see the commentary in *ID* ad loc.; but though the Kabeiroi were certainly viewed as primeval gods, they never receive this title); the statement in Paus. 8.31.3 that in Megalopolis the Horai, Pan, and Apollo are described in an epigram as "among the First Gods" is equally enigmatic, but probably unconnected (Jost, *Arcadie,* 475). The "thirty-day goddess" of a late inscription from Epidaurus (*IG* IV² 1.532, θεὰ τριακονθάμερος) is an isolated and mysterious figure.

36. Paus. 1.1.4, 5.14.8; Hesych. α 682; Philostr. *Ap. Ty.* 6.3.5; Tertullian, *Ad nat.* 2.9; *Adv. Marc.* 1.9.2; Paul was corrected by Jerome, *Ad Tit.* 1. 12 (*Commentarius in Epistulas Pauli Apostoli ad Titum et Philemonem,* ed. F. Bucchi, CCSL LXXVII C, Turnhout 2002, p. 30, lines 666–68): "inscriptio autem arae non ita erat, ut Paulus asseruit, ignoto deo, sed ita, diis Asiae et Europae et Africae, diis ignotis et peregrinis." But Bickerman, *Studies,* 2:615–17, postulates a private altar set up by a pagan god-fearer to the Jewish god. The issue was famously discussed by E. Norden, *Agnostos Theos* (Leipzig, 1913). Just what inscription the ἀνώνυμοι βωμοί supposedly set up at Athens by Epimenides (Diog. Laert.1. 110) carried is uncertain; the reader for the press suggests they had none, that is, they could be used for any god.

37. For a searchable database of cult epithets, not yet complete, go to http://ntarcheo2.univ-rennes1.fr/epicleses/accueil.php (quondam http://www.sites.univ-rennes2.fr/lahm/crescam/).

associated with healing.[38] (Triple and quadruple names, i.e., theonym plus two or even three epithets, are a rarity before the Hellenistic period;[39] they become common later when the line between cult epithets *stricto sensu* and epithets of more celebratory type becomes blurred.) As one example out of thousands of its use one might take a dedication by an important Greek living in Egypt in the third century B.C., Apollonios the *dioikētēs* (i.e., head of the civil administration), to "Apollo Hylates ("of Hylai"), Artemis Phosphoros ("Light-Bringer"), Artemis Enodios ("in the Road"), Leto Euteknos ("of Fair Children"), Herakles Kallinikos ("of Fair Victory"): it neatly illustrates a feeling on Apollonios's part that every god should be accorded an epithet. Heroes by contrast normally lacked epithets; the two who acquired a good set, Herakles and Asclepius, were the two who became functionally equivalent to gods.

The concept of cult epithet was already familiar in antiquity,[40] but defining it is difficult, and establishing fixed boundaries between what is one and what is not is

38. Polinskaya, *Local History,* 105. "Dieu décliné": J.P. Albert et al., "Conclusions," in *Dieux des autres,* 239–51, at 247. The "epithet" is occasionally not adjectival in form, and in such cases the syntactical relation between the two parts is an unexplored problem. That two names could be juxtaposed to give, e.g., Aphrodite Peitho or Zeus Trophonios is clear as a fact, if linguistically odd. But what other juxtapositions were possible? *IG* V.2.288 (5th c. B.C., Mantinea) gives Zeus Keraunos, Zeus presumably being identified with his manifestation, as in Latin *Iovi fulmini fulguri, ILS* 3052–53 (but M.L. West, *Indo-European Poetry and Myth,* Oxford 2007, 243–44, takes Keraunos as an old theonym; a new Arcadian inscription of ca. 500 does indeed list an offering to plain Keraunos; see J. Clackson and J.M. Carbon, *Kernos* 29, 2016: 131, who cite also *SEG* XL 1457). On Apollo Korax at Cyrene, see n. 100 below. P. Perlman, *Arethusa* 22 (1989): 127–30, interprets Artemis's Cyrenean title Καταγωγίς from the use of καταγωγίς for a kind of tunic (Sappho fr. 22 Voigt). But the analogy she cites, Artemis Κιθώνη, is imperfect, Κιθώνη being an adjectival formation. Pausanias 4.23.10 and 6.21.4 takes the shrines of "Herakles Mantiklos" and "Asclepius Demainetos" to have been founded respectively by a Mantiklos and a Demainetos, which is linguistically bizarre: for the view that both are ordinary, though unexplained, epithets, see Jessen, in *RE* s.v. Demainetos; Kruse and Ehrenberg, in *RE* s.v. Mantiklos 1 and 2.

39. I have noted Apollo Didymaios Milesios (*SEG* XXXVI 694B, graffito, Berezan, late 6th c.?); Zeus Milichios Panphylos at Megara (*JHS* 18, 1898: 332, 5th c.: a special case, Panphylos here standing in for the genitive normally found in cults of Zeus Meilichios); Ennodia Stropika Patroa (*SEG* LIV 561, Larisa, 5th c.; cf. *BÉ* 2013, no. 238); Nike Athana Polias (Soph. *Phil.* 134); Hera Argeia Heleia Basileia (*IG* XII.4.274 [*LSCG* 151 B] 5, Cos, 4th c.); Athenaia Ergane Polias and Zeus Epiteleios Philios (*IG* II^2 4318 and 4627, Athens, 4th c.?); Poseidon Kranaios Pylaios (*SEG* XXXV 590, Thessaly, *stoichedon,* so probably not later than the early 3rd c.); Demeter Krisaia Epidamos (*IG* VII 3213, Koroneia?, 3rd c.? [*LGPN*]: Schachter, *Cults of Boiotia,* 1:155); Hermes Pylios Harmateus (*IErythrai* 201 d 31, early 3rd c.); Zeus Aristaios Ikmios (Callim. fr. 75.33–34); whether Demeter Erinys Tilphossaia (Schachter, *Cults of Boiotia,* 1:164) ever received all three names at once is uncertain. The reader for the press adds: "*IG* II^2 5012 Ἀρτέμιδος Ὀρθωσίας Ἡγεμόνης, with the interesting echo of the Tenian 2nd c. BC *IG* XII.5.894 [Ἀρ]τέμιδος Ἁγεμονείας Ὀρθωσίας."

40. See, e.g., Paus. 7.21.7, where it is distinguished from the poetic epithet. I seek here to complement, but only partially repeat, my article "The Problem of the Greek Cult Epithet," *OpAth* 28 (2003): 173–83. Apollonios's dedication: *OGIS* 53 = Bernand, *Portes du désert,* 47.

impossible. It is normal and correct to distinguish between poetic or honorific epithets and true cult epithets, though there was certainly overlap and possibility of crossover between the two classes. The cult epithet is perhaps best defined as one used in prayers and appeals to the god in prose, in dedications, and in indirect references to the god, and usually following the god's name. One cannot simply make it "an epithet used in a cult context," because hymns performed in cult often contained ornamental and honorific epithets borrowed from the poetic tradition; "in prose" is added in the definition above to exclude such cases.[41] "Usually following the god's name" is added to exclude titles of respect such as *anax, potnia, despoina,* and *kurios* (all roughly meaning "master/mistress"):[42] these are common in prayers, but are not found in calendars of sacrifices, for instance; they do not individualise the god in the way that is here taken as a necessary characteristic of the cult epithet. A little different again are "acclamatory" epithets such as *megas, epēkoos, epiphanes,* and *sōtēr* ("great," "who gives heed," "manifest," and "saviour") which celebrated the power of a god in hopes of assistance or, very often, in gratitude for assistance received. But the dividing line between acclamatory and true cult epithets is again a porous one, *sōtēr*, for instance, being frequent in both roles.

As a general rule, the "respect" epithets do not appear in dedications (though they do in prayers), and the acclamatory epithets do not appear at all, before the Hellenistic period. This absence connects with a point about the actual use of cult epithets, which in the classical period was not primarily one of glorification. Arrian's account of the explorations of Alexander in the region of Nysa (supposedly a foundation of Dionysus) in India includes a striking moment (5.2.6): visiting Mount Meros ("thigh," a name evoking the myth of Dionysus's birth from the thigh of Zeus), the troops of Alexander were delighted to see ivy once again after a long interval, and at once made themselves ivy garlands; as they did so they sang hymns to Dionysus and "called on his various names (ἐπωνυμίαι)." Ovid similarly (*Met.* 4.11–17) has a scene in which the women of Thebes call on Dionysus by fourteen different names and "all the very many other names you have, Liber, through the peoples of Greece"; in the *Greek Anthology* (9.524) there is also an ingenious "Hymn to Dionysus" which goes through the alphabet including four epithets beginning with each letter, one letter per hexameter line. But these examples illustrate exceptions and not the rule: the early Greek cult epithet was not typically a tessera in such a mosaic of praise. Possibly this use was distinctive of the cult of Dionysus:[43] at all events the remains of early cult poetry contain nothing comparable to the strings of epithets found in the late Orphic Hymns or the Egyptian and

41. Cf. Z. Stewart, *JRS* 50 (1960): 40: "Poetic epithets were avoided with what appears meticulous care in dedicatory prose inscriptions." These distinctions become blurred in late antiquity.

42. Respectful and acclamatory epithets are discussed in chapter 5 below.

43. His epithets are discussed by Diod. Sic. 4.5.1–2, as if of special interest.

Babylonian honorific listings of the names of Amun and Marduk.[44] Where an accumulation is found, it is of poetic and not cult epithets,[45] and never of great length. Having many names was, it is true, a mark of a god's standing, a proof that he or she was worshipped far and wide under many aspects. A poet or orator could express doubt about which of a god's many names it was appropriate to use on a given occasion. Callimachus gives the baby Artemis a precocious awareness of the prestige of polyonymy when he shows her on her father's knee asking that she may have more names than her brother Apollo.[46] But this does not entail that Artemis was imagining a cult hymn celebrating her under all her epithets. In early dedications the norm is to have just one, if any.

The cult epithet, therefore, was not primarily honorific. It was a way of addressing or referring to the god, not a form of praise. Nor did it normally designate a particular iconographic type. The distinction is blurred (and was so perhaps in some degree for the ancients)[47] because a cult statue could have an epithet which was formally indistinguishable from what is here defined as a cult epithet (so, e.g., Athena Promachos, "Frontline Fighter"). But in Greece (in contrast to what is apparently the case in some ancient Near Eastern cultures) cult epithets did not normally refer to statues, nor were statues visual embodiments of cult epithets: Hippia, "of Horses," was a cult epithet of Athena with no corresponding visual image, Promachos a statue type with no corresponding cult. (But some god-epithet combinations such as Zeus Meilichios, "of Propitiation (?)," did have dis-

44. Hornung, *The One and the Many,* 90: "In Egypt, the cultic naming of the deity was the original form of hymnic praise"; cf. Assmann, *Search for God,* 84. Contrast too the Islamic practice (Dhikr) of reciting the names of God. It is not significant for early usage that Aristides ends his *Hymn to Zeus* with an epithet section (XLIII 30). The fragment of Orpheus's *Hymns* quoted in the Derveni papyrus (col. XXII 12) lists the names, not epithets, of different goddesses in asyndeton. An oracle addressed to the people of Tralleis probably after an earthquake urged them to call on Poseidon under five epithets (*SGO* 02/02/01, 200–250 A.D.): here, as in the Orphic Hymns, we seem to have a new application of the epithet.

45. See, e.g., *Hymn. Hom.* 2.492; 18.12; 19.1–2; 23.1–2; 27.1–3. The practice of Callimachus can be somewhat different, e.g., in *Hymn* 3.225–36; and cf. from an actual dedication, *IG* XII.I.914 (Rhodes, "3rd c BC"), cited on p. 148n94 below. Different again are the accumulations of epithets in Lycophron's *Alexandra* (excellently discussed, with much material of wider import, by Hornblower, "Cult Epithets in the *Alexandra*"): many of these are true cult epithets, but used in a distinctive, cryptographic manner.

46. *Hymn* 3.6–7; on the epithet πολυώνυμος, L. Bricault, in C. Berger et al., eds., *Hommages à Jean Leclant* (Cairo, 1994), 3:69–70; Hornblower, "Cult Epithets in the *Alexandra*," 100 n. 42; Chaniotis, "Megatheism," 132 n. 84. Orator: Menander Rhetor 2.445.28–446.9 Russell-Wilson; cf. already Callim. *Hymn* 2.69–71. For the use of epithets in oratory, see Pernot, "Lieu du nom," 34–37.

47. Cf. *OpAth* 28 (2003): 174 n. 7; Pausanias does not sharply distinguish the two types of epithet. Just why different statues of the same deity within a single sanctuary (Paus. 9.16.3: Aphrodite Ourania, Pandemos, Apostrophia; 9.2.7: Hera Teleia and Numpheuomene) were separately named (because of different ritual functions? different iconography?) is not clear. The statue of "gaping Apollo" seen by Polemo (fr. 71 M ap. Clem. Al. *Protr.* 2.38.4) was presumably given a nickname from the depiction. For Roman "iconographic epithets," see Carter, *De deorum romanorum cognominibus quaestiones,* 9.

tinctive representations; thus when Artemidorus in his dream book talks of the significance of dreaming of gods under particular epithets, he sometimes has visual images in mind, sometimes not.[48])

Though at first sight Greek cult epithets fall into numerous distinct classes,[49] their main functions can be reduced to two. One is to distinguish the god worshipped in one place from the same god worshipped in another. This was often done by simply adding a place-name or other local description in adjectival form (Apollo of Amyclae; Artemis Epipyrgidia, "on the Bastion"; Zeus Alseios, "of the Grove"), but many other epithets achieve the same effect more indirectly. Hera was identified as Aigophagos, "Goat-Eating," at a sanctuary in Sparta (Paus. 3.15.9) not in order to convey a general truth about the goddess, but because goat sacrifice was a distinctive trait of the cult at this particular shrine. Apollo was Spodios, "Ashy," where his altar was made from the ashes of sacrificial victims (Paus. 9.11.7); Artemis was Philomeirax, "Lover of Boys," where her shrine abutted a gymnasium (Paus. 6.23.8). Or a singularity of the image of a god (Artemis Lygodesma, "Bound in Withies") might give its name to the sanctuary that contained it (Paus. 3.16.11).[50] Such epithets functioned in a sense as simple addresses; they were, among other things, a practical necessity, needed to distinguish one cult site from another. So the three Poseidons who appear in the accounts of the Treasurers of the Other Gods at Athens[51] are carefully distinguished as "at Sounion," "of Kalaureia," and "Hippios" ("of Horses").

Cult epithets that apparently have very different origins and relations to the god all fulfill this function equally well: whether they refer to the authority that established the cult (Pythochrestos, "Decreed by Apollo"), to the funding that supports it (Demoteles, "Publicly Financed"), to a main festival celebrated at the sanctuary (Demeter Thesmophoros, referring to the festival Thesmophoria), to rites performed at such a festival

48. 2.34–40. Artemis "Agrotera and Elaphebolos is always better for activities than one fashioned in any other way," while "for those who have chosen a more dignified life, one more restrained in deportment is better such as the Ephesian and the Pergaian and the one among the Lycians called Eleuthera" (p. 159.19–24 Pack). Here the reference is clearly to iconographic types, as also in a discussion of different forms of Hermes (pp. 170.26–171.4). But "Apollo Delphinios," who "normally indicates travel and movement," must have been recognised in some other way (p. 160.11–12), since no iconographic type of Delphinios is known. Again, it is not clear iconographically how one distinguished Aphrodite Pandemos from Aphrodite Pelagia (pp. 171.13–172.5), or recognised Hekate Chthonia (p. 158.6; cf. p. 175.19). Different is the appeal to an epithet to explain an interpretation: Artemis is good for parturient women since she is called Locheia (p. 159.15–16). Here there is no suggestion that the epithet is visible.

49. See, for instance, the eight classes distinguished by Gladigow, "Gottesnamen," 1231–32. In *OpAth* 28 (2003): 178 n. 43, I queried whether epithets were ever formed from the profession of a worshipping group, but Poseidon Nauklarios of *ID* 2483 looks like an example: any exclusion is rash.

50. Cf. *OpAth* 28 (2003): 174 n. 7; Apollo of the *petasos* (Petasites, from *petasos,* a type of cap: *ICam.* 132) looks like a case.

51. *IG* I³ 369 and 383; so too with other gods in these lists who had multiple cults, and in the Erythrai priesthood sales list (*IErythrai* 201; Parker, *On Greek Religion,* 100–102).

(Apollo Karneiodromos, "Karneia Runner") or good things eaten at it (Demeter Megalartos, "of Great Loaves"), or are completely opaque (Athena Hellotis)—and there are many other possibilities—their primary role is differentiation, a truth that is obscured by elaborate division into different categories. Other factors too could certainly influence the choice of a name: to call Apollo Archegetes, "Leader," for instance, commemorated his role in the early history of a colonial foundation. Epithets might or might not also convey something to the worshipper about the powers or nature of the god; writers in later antiquity sometimes made an artificial collage of those that did to create a composite portrait of a deity.[52] But an epithet did not need to say anything important about the god to individuate the cult-place in question. This practical role is nicely illustrated by the list in a Roman legal text of nine gods who could be named as heirs under Roman law: each receives a local epithet, because the potential beneficiary was not the god at large, an impossibility, but a particular sanctuary.[53] We might call this the bureaucratic or administrative function of the epithet.

The second broad function of the epithet was to provide focus, to pick out one aspect or power amid the many of a god of broad powers: Poseidon of Horses, Zeus of the Oath, Hermes of Competitions. The epithet related to a particular need of the worshipper (e.g., Iatros, "Doctor"), a particular attitude the worshipper wished the god to adopt (e.g. Soter, "Saviour") or not to adopt (Maimaktes, "Raging"), or a particular occasion on which the god was addressed (as Zeus in Attica became "Zeus Heraios" once a year at a particular festival which celebrated his marriage to Hera[54]). Apollonius Rhodios, for instance, tells of the offerings made by the Argonauts at the appropriate moments to Apollo Embasios and Ekbasios ("of Embarkation" and "of Disembarkation"). At Stymphalos in Arcadia, the mythical founder Temenos supposedly divided the stages of Hera's life, and so perhaps the stages through which Hera could guide mortal women, into three, each embodied in a separate sanctuary: child (παῖς) Hera, mature or fulfilled (τέλεια) Hera, widow (χήρα) Hera; this remarkable arrangement unfortunately was revealed to Pausanias only by local say-so, nothing of it surviving in his day.[55] Some epithets contained a

52. See D. A. Russell, *Dio Chrysostom, Orations VII, XII, XXXVI* (Cambridge, 1992), 206; cf., e.g., Plut. *Ant.* 24.4–5 on the different aspects of Dionysus/Antony, as Charidotes and Meilichios opposed to Omestes and Agrionios; Plut. *E. Delph.* 2, 385B, on epithets of Apollo.

53. *FIRA* II, p. 285, *Tituli ex corpore Ulpiani XXII.6, Iovem Tarpeium, Apollinem Didymaeum Mileti, Martem in Gallia, Minervam Iliensem, Herculem Gaditanum, Dianam Ephesiam, Matrem Deorum Sipylenen, Nemesim quae Smyrnae colitur, et Caelestem Salinensem Carthagini.* Or there are so-called festival coinages bearing name and epithet (Athena Ilias, Artemis Pergaia: Thonemann, *Maeander Valley,* 117–19).

54. Parker, *Polytheism and Society,* 42 n. 20.

55. Ap. Rhod. *Argon.* 1.359 (cf. 421, embarkation offerings), 966, 1186; Paus. 8.22.2; cf. V. Pirenne-Delforge and G. Pironti, "La féminité des déesses à l'épreuve des épiclèses: Le cas d' Héra," in L. Bodiou and V. Mehl, *La religion des femmes en Grèce ancienne* (Rennes, 2009), 95–109.

functional specification within their own meaning ("Doctor," "of Horses"); we assume that others also specialized the god in some way that was conventionally understood (Apollo Delphinios, for instance, whose activities have little relation to the dolphins his epithet evokes).[56] The same focusing effect was sometimes achieved by juxtaposition of an ordinary divine name with a deified abstraction: Athena Nike, "Victory"; Aphrodite Peitho, "Persuasion."[57] A function could be very precise, as with Zeus Kataibates ("Who Comes Down"—i.e., the thunderbolt) or Apollo Parnopios ("of Locusts"), or so general that most major gods could discharge it: thus epithets such as Polieus ("of the City"), Soter ("Saviour"), Hegemon ("Leader"), and (eventually) Ouranios ("Heavenly")[58] came to be shared by many of the greater gods.[59] A different form of sharing was where a pair of gods who were often worshipped together shared an epithet at a particular site (as, for instance, Zeus Phemios and Athena Phemia, "of [Verbal] Omens," at Erythrai in Ionia[60]).

56. But on this epithet, see the doubts of Polinskaya, *Local History*, 222–23.

57. Νυκτὶ Ἀρτέμιδι in *SEG* LVI 601 (Kallipolis in Aetolia, 4th/3rd c.) would be a strange example (both in the combination and the order); the editor may be right to take them as separate powers.

58. See appendix A.

59. For Polieus (and related terms), see U. Brackertz, "Zum Problem der Schutzgottheiten griechischer Städte" (PhD diss., Berlin, 1976), table 3; and Chiai, "Medien religiöser Kommunikation," 74 n. 49; Soter will be discussed in a forthcoming study by T. S. F. Jim. As for Hegemon, Apollo is Hegemon in Phasis in the fifth century (*SEG* L 1383; unless N. Ehrhardt is right, *ZPE* 56, 1984: 156–57, that the silver phiale in question is a forgery), and Hagetor in Laconia (*IG* V.I.977, 3rd c. B.C.?); Aristophanes seems to know Hegemonios as a title of Hermes (*Plut.* 1159; cf. later *IG* II^2 1496.84–85; *SEG* XXIII 547.53, and the Anubis Hegemon of *Syll.*^{3} 1129 [*ID* 1253], where Anubis may be assimilated to Hermes); Xenophon consults Herakles Hegemon (*Anab.* 6.2.15; cf. 4.8.25) and envisages Zeus Hegemon as a password (*Cyrop.* 3.3.58 and 7.1.10; cf. *Anab.* 4.8.25), while Spartan campaigns begin with sacrifice to Zeus Hagetor (Xen. *Rep. Lac.* 13.2); in the third century appear Artemis Hegemone (perhaps Callim. *Hymn.* 3.227, and certainly often later: see the summary in *SEG* LVIII 745) and Aphrodite Hegemone (τοῦ δήμου): *IG* II^2 2798; *IRhamnous* 32.33, 35.8–9; in the second century, Dionysus Kathegemon (Ohlemutz, *Kulte in Pergamon*, 90–122). The meaning seems to develop from a literal "leader" to "patron"; eventually προκαθηγεμών (τῆς πόλεως) becomes a general title meaning "chief divine patron" of a city (L. Robert, *Études Anatoliennes*, Paris 1937, 27: e.g., Apollo: *LSAM* 53.6–7, Miletus; *ASAten* 22–23, 1944–45: 165, no. 145, Kalymnos). The newly attested "goddess Hagemona" of *SEG* XLI 332.18 (Lykosoura) is presumably Artemis Hegemone (Paus. 8.37.1); the preference for title is perhaps influenced by the proximity at Lykosoura of Despoina (mentioned in the same decree, *SEG* XLI 332.7). The same title appears in *SEG* XLI 352 (Messene) and has hitherto been related to Messene because her father Triopas apparently shared the monument (see, e.g., P. Themelis, *Praktika* 1989, 110–12).

60. *IErythrai* 201 c 47–48: the same list of priesthoods offers Zeus Apotropaios and Athena Apotropaia b 8–9, Apollo Kaukaseus and Artemis Kaukasia c 40–41. On shared epithets, see Gladigow, "Gottesnamen," 1229; P. Brulé, *Kernos* 11 (1998): 30–31. As well as those shared by many gods (previous note), a few indicated a more specialised function shared, if in different ways, by few (Hippios, shared by Poseidon and Athena; Lysios, shared by Dionysus and Artemis, *IG* IV^2 1.162, 275). Some related to features of a sanctuary (e.g., location on a high place, Akraios; in a grove, Alseios) that might be common to many cults without connecting the gods concerned. About Koria, supposedly an Arcadian epithet of both Artemis and Athena (Jost, *Arcadie*, 389–90), nothing secure can be said.

The distinction between the two functions of the epithet—to identify sites on earth, to focus divine powers[61]—is often blurred. A focusing epithet can, it is true, be used without any reference to a place on earth. In a famous passage of Herodotus, Croesus reproaches Zeus in three different aspects in each of which, the king claims, the god has let him down: as Zeus Katharsios, "of Purification"; as Zeus Epistios, "of the Hearth"; and as Zeus Hetaireios, "of companionship" (1.44).[62] Aeneas Tacticus speaks of watchwords to be used in different circumstances: for hunting it will be Artemis Agrotera, "Huntress"; for trickery, Hermes Dolios, "Tricksy" Hermes (24.15). To judge from Aristophanes, appeals to figures such as Hermes Agoraios, "of the Marketplace," or Apollo Apotropaios, "Averter (of Evil)," were common in everyday speech. Evening, says a speaker in Plutarch, belongs to Lysios ("Releaser") Dionysus, morning to Ergane ("Worker") Athena and Hermes Agoraios (*Quaest. conv.* 3.6.4, 654F). But a particular sanctuary could also belong to a god bearing a focusing epithet: Poseidon Hippios, "of Horses," at Colonus, as it might be, or Poseidon Asphaleios, "of Safety," in many places; gods received such epithets in the prayers of worshippers, wherever they might be, but they were also inscribed on particular altars. So even a focusing epithet could discharge the bureaucratic role, as the identifier or address of a particular cult. And, as we have seen, some epithets that formally count as focusing are so vague that in effect they are little more than identifiers.

Conversely, one has only to contemplate the figure of Artemis of Ephesus to appreciate that an epithet that is formally topographic can also identify what was for the worshipper a distinctive form or manifestation of the god. Within topographic epithets, one might distinguish between a bureaucratic or practical function and an ontological one that classifies that form of the god as a distinct existent entity. Perhaps some topographic or equivalent epithets never got beyond the bureaucratic role: it may be that worshippers never invoked or made dedications to, say, "Apollo of Saberidai" or "Goat-Eating Hera" under those names, rather than as plain Apollo or Hera. But it was extremely common for a god invoked under a particular topographic epithet to acquire a distinctive identity. That is why, on the one hand, worshippers might use the topographic epithet where it was

61. At Lousoi both were in use: the same god was known topographically as Lousiatis and functionally as Hemera, "Gentle"—or just as Artemis (*IG* V.2.397–403). According to Pausanias (8.23.6–7), at Kaphyae an unfortunate event caused an epithet of Artemis to change from a toponym, Kondyleatis, to a speaking epithet, ἀπαγχομένη, "being strangled." That epithet is related to small images of gods hung in trees by Jost, *Arcadie,* 400–402, an explanation which does not fully account for the choice of verb; and to the sensation of strangulation supposedly experienced by *parthenoi* suffering trouble at menarche by H. King (in A. Cameron and A. Kuhrt, *Images of Women in Antiquity,* London 1983, 113–20), an explanation which would be easier if comparable mirrorings of human sufferings in divine epithets were attested.

62. Cf. Ap. Rhod. *Argon.* 2.1123–33, an appeal in the name of Zeus Epopsios, Xeinios, and Hikesios.

unnecessary, addressing, for instance, Artemis as Artemis Patmia even when bringing an offering to her on Patmos (*Syll.*³ 1152),[63] or could drop the god's own name and address him or her by the local epithet alone, so that Apollo of Amyclae could become "the Amyclaean"; it is also why a god plus toponym combination could be exported far from the original place that gave the name, why Delian Apollo or Ephesian Artemis or Cyprian Aphrodite could become ubiquitous presences in the Mediterranean world. (This did not, however, lead to the kind of double toponym occasionally found for Hittite gods, such as "Ḫebat of Aleppo of Ḫattuša," i.e., the cult at Ḫattuša of "Ḫebat at Aleppo.") From all this (this and the multiplication of functional epithets) arises the issue so troublesome to us, and so straightforward, or merely uninteresting, to the ancients, of determining whether a theonym associable with, say, ten epithets describes one god or ten.[64]

Epithets that fulfilled one of these two functions were so common that a certain sense developed that in a cult context the divine name was incomplete without one (though there were always exceptions of cults without epithets). Ariston, who was sent by Ptolemy to explore the coast of Arabia, founded an altar to Poseidon Pelagios, "of the Sea," at the southern tip of the peninsula of Sinai (Diod. Sic. 3.42.1). There was no need to distinguish this sanctuary of Poseidon from any other in this remote region, nor to specify that the Poseidon here in question was he of the sea. But the epithet added dignity. We noted earlier the dedication by Apollonios the *dioikētēs* to five gods with five epithets. This dignifying role needs to be added as a modest supplement to the two main functions discussed above. Some of the epithets we meet in Pausanias may have been "filled in" by antiquaries for sanctuaries which lacked, or had lost, one handed down by tradition.[65]

63. For similar cases, see *IG* V.2.397, 399, dedications from Lousoi to Artemis Lousiatis or at Lousoi; *IG* I³ 985 from Brauron, to Artemis Brauronia; *IG* IX.1²4.822–83, 826, from Korkyra, to Apollo Korkyraios; *SEG* XXXV 371, presumably from Stymphalos, to Demeter at Stymphalos; *OGIS* 164–66, dedications from Paphos to Aphrodite Paphia, by the Cypriot *koinon* (164–65) and the polis of Paphos (166); cf. *IG* XII.8.361, a *horos* of Zeus Agoraios Thasios from Thasos. (On Thasian Herakles, see appendix E.) According to H. Seyrig, *BCH* 51 (1927): 369–70, such toponymic epithets originate with visitors to the sanctuary from abroad, and are taken up by locals by a "bounce back" ("choc en retour"); the distinction cannot unfortunately be checked in the cases just cited except the Paphian, but for locals using the local epithet in relation to their own deity, see Robert, *Hellenica* IX, 21–22. The crew of the warship Demeter, landed on Paros, where Demeter had a celebrated cult (*Hymn. Hom. Dem.* 491; *RE* s.v. Demeter, 2723), took the opportunity to dedicate to Demeter Paria (*SEG* XXXIII 684). But Apollo becomes Aeginetan only outside Aegina: Polinskaya, *Local History,* 207.

64. See, above all, Versnel, *Coping with the Gods,* 60–87, 517–26, whose answer is "both—in different contexts." Many texts bring out the paradox, e.g., *SEG* LIX 1418 (nr. Hadrianoi, in Mysia, 2nd/3rd c. A.D.): "Zeus Anabatenos dedicated this to Zeus Kersoullos." Allen, *Splintered Divine,* studies the same problem for the ancient Near East, and argues resolutely that different epithets create different gods. Hittite double toponym: Allen, 206.

65. Cf. Schachter, *Cults of Boiotia,* 2:12–13, on the Herakles Hippodetes of the Teneric plain and Herakles Rhinokoloustes of Thebes.

SONDERGÖTTER AND SEMIAUTONOMOUS EPITHETS

The great student of Greek divine names, Hermann Usener, famously argued that most of the focusing epithets had originally been independent deities, what he called *Sondergötter,* "special gods."[66] According to Usener, primitive man's first gods had all been *Sondergötter,* that is, gods with closely circumscribed functions indicated by speaking names. Gods with personalities and broad functions were a secondary development that only became possible once the meaning of originally transparent speaking names was no longer understood; these great gods then recovered contact with specific functions by capturing *Sondergötter* as epithets. There had originally, for instance, been a goddess Korotrophos, "Child-Nurturer," whose name revealed her powers; later she declined into an epithet for which "a whole series of goddesses"[67] competed. The theory might be correct in a few cases. The reduction (occasional or permanent) of a name to an epithet is a well-attested phenomenon,[68] and though the clearest cases involve coalescence of two equally opaque theonyms (e.g., Enyalios and Ares), the process postulated by Usener of an obscure theonym attracting a speaking name is actually easier to envisage. Korotrophos is just one of a number of figures who are apparently attested both on their own and as epithets of major deities (others include Ennodia, "[Goddess] in the Road"; Pasikrata, "Ruler over All" (feminine); Tychon, "Lucky Strike"), and it is quite possible that some of these began as independent figures. But the direction of travel was not necessarily the one that Usener supposed: Ennodia, for instance, though functioning as an independent goddess with her own iconography in Thessaly and Macedonia,[69] is adjectival in formation and appears early as an epithet of other deities; she might rather be an epithet that acquired independence. As for many other adjectival formations claimed by Usener[70] as quondam *Sondergötter,* such as Megalartos, "Big Loaves" (epithet of Demeter), or Panoptes, "All-Seeing" (of Zeus), there is no reason to think that they had ever had an independent existence. More generally, Usener's evolutionary approach has long been

66. On Usener, see *Hermann Usener filologo della religione;* M. Espagne and P. Rabault-Feuerhahn, eds., *Hermann Usener und die Metamorphosen der Philologie* (Wiesbaden, 2011) (*non vidi*); Konaris, *Greek Gods in Modern Scholarship,* 180–94; on reactions to *Götternamen,* see p. 173n1 below.

67. Usener, *Götternamen,* 122–28. But the case as presented fits the theory imperfectly, since on Usener's own showing the main claimant to Korotrophos as an epithet was Earth, Ge, a goddess with a still transparent name.

68. See p. 21 below.

69. Cf. Parker, "Theonyms in Northern Greece" (also ibid. on Pasikrata and Tychon). Erwin Rohde in a letter thanking Usener for the gift of *Götternamen* (1/6/1896, cited by Mette, "Usener und seine Schule," 88–90) already insisted that epithets could become *Sondergötter,* as well as vice versa; for possible Roman instances, see Carter, *De deorum romanorum cognominibus quaestiones,* 11–15; for the reverse process, 26.

70. Usener, *Götternamen,* 218.

deeply out of fashion. In its own terms, the attempt to build a general theory of "religious concept formation" (*Versuch einer Lehre von der religiösen Begriffsbildung,* subtitle of his book) on three Indo-European case studies would have been overoptimistic even if they had proved less open to criticism than they did;[71] Usener seems never to have explained why he ignored the Egyptian documentation that was already available to him, and great amounts of evidence for early divine name-giving have become available since his time, from Sumerian, Akkadian, Hittite (which is Indo-European), and many less-known ancient languages and dialects besides.

But it has been pointed out in relation to Roman religion that if one abandons Usener's evolutionary and generalizing perspective, his ideas can be usefully redeployed;[72] and this is also true of Greece, in several ways. Usener postulated a historical progression from innumerable special gods with transparent names to a small number of major gods with opaque names. But one can apply the contrast between opaque and transparent not in terms of evolutionary progress but of hierarchy. As we have seen, at the top level, the level of greatest power, are gods with obscure names. The more one moves downward in terms of power, the greater is the frequency of transparency, whether of nymphs and heroes with speaking names, of personified abstractions, or of cult epithets. The cult epithets belong to gods, the personified abstractions are often very closely attached to them: the god is thus broken down into smaller, more specific, more comprehensible elements. This is the power in terms of religious psychology of the cult-epithet system, the way in which it associates the great high god, the figure of broad power, a reality not an abstraction, with something more down to earth and local, perceptibly close to the worshipper's specific need.

A second, related value of Usener's perspective is his theory of the original autonomy of the epithet. Again one needs to substitute for a postulated historical process a lasting condition, the permanent tendency of the epithet to achieve at least quasi autonomy. Worshippers constantly chose to address a god by epithet alone, whether that epithet was a local or functional one: not Apollo Iatros, "Doctor," but just Iatros; not Zeus Meilichios, "of Propitiation," but Meilichios; or (to take a local example) not Zeus Pelinnaios, "of (Mount) Pelinna" (on Chios), but Pelinnaios;[73] such abbreviations occur in third-person references to gods as well as direct addresses to them. Several instances of such epithets without theonyms

71. The decisive criticisms of Usener's use of Lithuanian and Roman evidence are cited by A. Momigliano, in *Hermann Usener filologo della religione,* 18.

72. See J. Scheid and J. Svenbro, "Les Götternamen de Hermann Usener: Une grande théogonie," in *Nommer les dieux,* 94–104.

73. Iatros (in the form Ietros): *SEG* XXX 880; XXXII 737, 801. Meilichios: n. 90 below. Pelinnaios: Graf, *Nordionische Kulte,* 37–39 (with discussion of the phenomenon).

have been detected among the archaic rock-cut inscriptions from Thera.[74] In a particular sanctuary the epithet can completely efface the divine name, so that, for instance, it is only from figurines found in the same Spartan sanctuary that the goddess addressed merely as Kyp(h)arissia, "of Cypresses (?)," in six dedications can be identified as a form of Artemis.[75] It remains controversial whether the continued appearance of a "Maleatas" alongside Apollo Maleatas, "of (Cape) Malea," attests an originally independent *Sondergott,* or a familiar abbreviation; it is often unclear what proper name if any is to be associated with a given reference or dedication to Soteira, "(female) Saviour."[76] The uncertainty was not necessarily ours alone in such cases: according to Pausanias, ordinary people of Phigalia in southwest Arcadia took the "Eurynome" of a local shrine for an epithet of Artemis, whereas antiquaries thought she was a daughter of Ocean.[77] So the preference is often for the function over the name, or the more specific and local over the more general, and in that sense an intuition of Usener is proved correct.

A fifth-century inscription from Selinus in Sicily accompanying a dedication for victory shows theonyms and what are apparently semiautonomised epithets put in exact parallel:

> Because of the following gods the Selinuntines conquer. We conquer because of Zeus and because of Fear and because of Herakles and because of Apollo and because of Poseidon and because of the Tyndaridai and because of Athena and because of Malo-

74. Hikesios: *IG* XII.3.402–4; Stoichaios: 376; Delphinios: 537a; Lykeios: 551; cf. A. Inglese, *Thera arcaica: Le iscrizioni rupestri dell' agora degli dei* (Tivoli, 2008). Cf. Delphinios on a sixth-century graffito from Berezan, *SEG* XXXII 739; *IG* IX 1² 4.862, a sacred *horos* τᾶς Ἀκρίας, apparently Hera. See now J. Clackson and J. M. Carbon, *Kernos* 29 (2016): 131, on (Zeus) Keraunos and (Ares?) Theretas in an Arcadian calendar of ca. 500 B.C.

75. *SEG* L 395–405: the identification is supported by "Agrotera Kypharissia" in *IG* V.I.977. So there is no connection with the Messenian Athena Kyparissia (*SEG* XXIII 209–10, XLIII 143–44), where the epithet is topographic (cf. Paus. 4.36.7; N. Luraghi, *The Ancient Messenians,* Cambridge 2008, 275–76). Cf., e.g., Kolainis: *IG* II² 4731, 4817, for the Artemis Kolainis of *IG* II² 4791, 5140. Herodotus speaks indifferently of Ἀθηναίη Προνηίη at Delphi with and without the theonym (1.92.1; 8.37.2; 8.39.1). Cf. pp. 30–31n112 below on Aphrodite Kas(s)alitis, Hearing Goddess.

76. Maleatas: contrast, e.g., Kruse in *RE* and C. Auffarth in *Brill's New Pauly* s.v. Maleatas. Maleatas appears without theonym, e.g., in *IG* V.1.927 and 929, and receives an offering separate from Apollo's in *IG* II² 4962. Soteira: see T. S. F. Jim, "Can Soteira Be Named?," *ZPE* 195 (2015): 63–74. In *CIG* 1798–99 (Ambracia), a dedication to Σωτῆρι, Διΐ, 'Αφροδείτᾳ, the word Σωτῆρι should apparently (O. Picard ap. P. Cabanes, *BCH* 109, 1985: 755–57) be taken as a reference to the Ambracian Apollo Soter known from Ant. Lib. *Met.* 4.4 (= Athanadas, *FGrH* 303 F 1), and probably *SEG* XXXV 665 B 45; if so, an epithet in lieu of theonym was combined with two theonyms.

77. 8.41.4–6. Pausanias himself supposes Taraxippos of Olympia, taken as a hero by most Greeks, to be an epithet of Poseidon (6.20.15–19); Paus. 3.15.7 may imply similar uncertainty about Hippost-henes vis-à-vis Poseidon. Some ancients held Priapos to be the same as Dionysus, being just an epithet; Ath. 1.54, 30A-B (cf. H. Herter, *De Priapo,* Giessen 1932, 303).

phoros (Apple-Bringer) and because of Pasikrateia (Ruler over All) and because of the other gods, but most of all because of Zeus. (*IGDS* 78, ML 38)

Zeus, Apollo, and Athena bear their simple names without epithets; Malophoros is an epithet of Demeter in Selinus's mother city, Megara (Paus. 1.44.3), but here and in another Selinuntine dedication (*IGDS* 54) it stands alone, irrelevant though the function indicated is to military success; Pasikrateia is probably an epithet standing in euphemistically for Persephone.[78] The divinised abstraction Fear rounds out this illustration of the diverse types of divine power and diverse ways of naming them in early Greece.

An opposite but complementary phenomenon to the epithet that acquires quasi autonomy is the divine name that partially surrenders its autonomy by being juxtaposed to that of another god. Enyalios, Pai(aw)on, and Eleuthia/Eileithyia, which in the second millennium in Linear B stood on their own, in the first millennium sometimes still appear as names of independent gods, but sometimes serve as epithets or alternative names for, respectively, Ares, Apollo, and Artemis.[79] Similarly, it is generally believed that Artemis Ortheia/Orthosia, Artemis Diktynna, and Athena Alea emerged through the permanent coalescence of local figures Ortheia/Orthosia, Diktynna, and Alea (who are often so named, without any other theonym) with Artemis and Athena; in their places of origin these goddesses retain their independence, but when they travel, or are viewed from outside, tend to be assimilated to the Panhellenic figure.[80] We noted above combinations of

78. Commentators compare *Hymn Hom. Dem.* 365, where Hades promises her "rule over everything." She may be distinct from the Pasikrata of northern Greece (n. 69 above). A late dedication is still made θεᾷ Μαλοφόρῳ, *IGRB* I. 370bis, perhaps from the Megarian colony Mesambria.

79. Cf. Rougemont, "Noms des dieux en linéaire B," 332–35 (who also mentions Enesidaon as a possible precursor of similar epithets of Poseidon, and e-ri-nu/Demeter Erinys); for a possible link between Linear B di-ri-mi-jo and Apollo's epithet Δρύμας, see Hornblower, "Cult Epithets in the *Alexandra*," 107. On Pai(aw)on, see I. Rutherford, *Pindar's Paeans* (Oxford, 2001), 11 and 385 n. 12. Rutherford notes that in poetry the name is "usually a substitute for (Apollo's) name rather than a true epithet," but in cult it is already an epithet of Apollo in *IG* I³ 383.163–64. It can be applied to other gods (Rutherford, ibid.), though not apparently as a fixed cult title. In Homer, Enyalios too seems to serve as an alternative name for Ares (see *Il.* 20.69 with 21.391–92) rather than an epithet (though see *Il.* 17.210–11); even in cult the combination Ares Enyalios is rare and late (*IG* IV.717; *IG* II² 1072.5; Enyalios Ares in *IG* V.2.343.44–45 [*DGE* 665 C 16–17] is not a clear instance, given Zeus Ares just above); rather, Enyalios either stands alongside Ares (so in the Athenian ephebic oath, RO 88.17, where Enyo too appears; similarly *IG* XII.5.913.12–13, a Delphic response) or on his own (for early instances, see *LSS* 85; *SEG* LVI 470), probably in that case being seen as the same god differently named (Nilsson, *Geschichte*, 517–19: cf. Aen. Tact. 24.2).

80. "Ortheia/Orthosia" (in fact always written with an initial digamma, and in very varying forms) is always freestanding in Spartan inscriptions until the first century A.D. (J. A. Davison, *From Archilochus to Pindar*, London 1968, 169–72), as in Xen. *Resp. Lac.* 2.9; outside Sparta (for refs., see B. Kowalzig, *Brill's New Pauly* s.v. Orth[e]ia) it is an epithet of Artemis (except in Pind. *Ol.* 3.30), already in Hdt. 4.87 and *IG* I³ 1083. Dictynna is freestanding in Crete (e.g., *IC* II.XI.3.30) but outside Crete usually treated

deified abstractions with goddesses such as Athena Nike and Aphrodite Peitho, and the reduction of possible *Sondergötter* to epithets; other gods too can come together to give forms such as Artemis Hekate or Zeus Ares.[81]

The phenomenon is complex and can probably not be given a single explanation. The precondition for an assimilation such as Artemis Eileithyia was a functional overlap: Eileithyia's sole function, care for childbirth, coincided with one of the functions in Artemis's more varied portfolio. So Artemis Eileithyia was Artemis seen in relation to birth. A similar overlap between gods of different gender created comparable but grammatically distinct double names: Athena Areia, "of Ares," was Athena in her martial aspect; Athena Hephaistia was the technological Athena. A Greek who spoke of Athena Hephaistia did not believe that Hephaistos was merely an aspect of Athena, and by analogy a Greek who prayed to Artemis Eileithyia did not necessarily disbelieve in the separate existence of Eileithyia. The structure of the cultic double name invited expressions such as Artemis Eileithyia, but did not require a decision about the ontological status of the two elements. It is a mistake to agonise over such questions, when the point that we should retain from Usener is the ebb and flow of the relation between divine name and epithet. Different are the puzzling cases where a theonym is combined with what in other contexts is attested as the name of a hero or heroine: Poseidon Erechtheus, Zeus Agamemnon, Zeus Eubouleus, Artemis Iphigeneia, Artemis Andromeda.[82] We have already noted the epithet that acquires quasi autonomy, and it is arguable that Poseidon Erechtheus, say, and Artemis Iphigeneia preexisted the heroes Erechtheus and Iphigeneia; the process here would be one of division, not amalgamation.[83] Different again is the spasmodic emergence in the

---

as an alternative name or epithet of Artemis (though still independent in *IAmyzon* 14.3, joint cult with Zeus Kretagenes, and Callim. *Hymn* 3.198): see *Nommer les dieux,* 223 n. 34; *IG* II^2 4688; *IG* IX.1.5; cf. 4. On Alea, see M. Jost, in *Nommer les dieux,* 393. Britomartis and Aphaia resisted reduction to epithets in cult, though in Hellenistic narrative (Ant. Lib. *Met.* 40.4; cf. Paus. 2.30.3) Aphaia became a name for Britomartis: for independent Aphaia, see *IG* IV^2 2.1024, 1038.1 (cf. Polinskaya, *Local History,* 177–96); for Britomartis, A. Chaniotis, *Die Verträge zwischen kretischen Poleis in hellenistischer Zeit* (Stuttgart, 1996), 71 n. 387, Bruneau, *Cultes de Délos,* 203.

81. On all this, see Parker, "Artemis Ilithye." On gods "inhabiting" and "determining" other gods in Egypt, see D. Kurth, "Götter determinieren Götter," *Studien zur Altägyptischen Kultur* 5 (1977): 175–81; J. Baines, in *One God or Many?* 31–36; on juxtapositions in Phoenician, with comparative material from other ancient polytheisms, P. Xella, "'Divinités doubles' dans le monde phénico-punique," *Semitica* 39 (1990) [*Mélanges M. Sznycer* II]: 167–75.

82. For the first four (and some others), see in brief Parker, "Artemis Ilithye," 223–24; for Artemis Andromeda, *ILindos* 220. At Megara there is Dionysus Dasyllios (Paus. 1.43.5), in the Megarian colony Kallatis an apparently freestanding Dasyllios (*SEG* XLV 911 A 4): see the refs. in *SEG,* loc cit.; and Chiekova, *Pont gauche,* 87–88.

83. But in regard to Iphi-, Linear B i-pe-me-de-ja (Iphimedeia: Rougemont, "Noms des dieux en linéaire B," 338) might suggest a yet more complex process whereby two theonyms coalesced (cf. Hes. fr. 23a.15–26), and one then split off as a heroine; Burkert, *Greek Religion,* 184, takes Poseidon Erectheus as by origin a combination of theonyms.

Hellenistic period of expressions such as Zeus Amphiaraos or Zeus Trophonios, where the point is not (as with Artemis Eileithyia) to focus Zeus via Trophonios, but rather to elevate Trophonios by association with the supreme god. Such expressions had an important future: the elative use of the name of Zeus in combination with innumerable other theonyms became very common in the Greek East in the Roman imperial period.[84] What made them possible was the elasticity of the cultic double name; that elasticity, in which the logical/theological relation of the two elements was always opaque, was crucial to practical polytheism.

Usener drew attention, through the concept of the *Sondergott,* to the way in which help was sought from divine powers with very specialised tasks. The advantage of abandoning his schema of historical progress is once again that we can see how uncertainly the powers in question continued to be conceptualised. At Olympia the flies that clustered around sacrificial animals were dispelled by Zeus Apomuios, "Away with Flies," and Strabo lists a string of gods bearing "pesticide" epithets, such as Apollo Pornopion, "of Locusts." But at Aliphera, southeast of Olympia, the difficulty with flies was met by a preliminary sacrifice to a hero Mu(i) agros, "Fly-Catcher," while Aelian claims that at Cape Leucas on Leucas sacrifice was made to the flies themselves.[85] We encounter at Cyzicus in Bithynia Zeus Chalazios Sozon, "of Hail, Protector"; at Amaseia in Pontus the exotic variant Aither Alexichalazos, "Hail-Averting Ether"; but Kleonai in the Peloponnese was notorious for its special "hail wardens," who sacrificed small animals or even a little of their own blood directly to the destructive force.[86] As late as the Roman period one could not bring a "thank-offering from the quartan (*tetartaios*)" only to "Asclepius Saviour," but do the same to Tetartaios itself.[87] In different places Sosineos, "Save Ship," and Sosipolis, "Save City," were epithets and independent agents.[88] At Lebadeia in Boeotia the figure commonly known as (Zeus) Meilichios was named, in roughly the same period (3rd–2nd c.), Daimon M(e)ilichios (three times), Meilichios (once), Zeus Milichios (once); at Thespiai in Boeotia Zeus Milichios

84. See p. 94 below.

85. Olympia: Paus. 5.14.1; but Pliny, *NH* 10.28.75 speaks of a god perhaps called Myiacores; Strabo 13.1.64, 613 (but the interpretation of Apollo's epithet Smintheus as "Mouser" was judged "unseemly" by Aristarchus, who preferred to see it as toponymic, surely rightly: see Σ Hom. *Il.* 1.39, with Erbse's note ad loc.); Aliphera: Paus. 8.26.7; Aelian, *NA* 11.8.

86. Cyzicus: Cook, *Zeus,* 3:880–81, with 875–81, much further information about hail aversion; Amaseia: Cook, 879; *SEG* XLVI 1608–9; Cleonai: Sen. *Nat. quaest.* 4b.6.2–7.2; Clem. Al. *Strom.* 6.3, p. 446.11–15 Stählin. But in a passage of Proclus referring to the Theban *Daphnephoria* where Apollo's epithet varies in the MSS, Galaxios is preferable to Chalazios (Schachter, *Cults of Boiotia,* 1:48).

87. *IStrat.* 1122; *Haute terres* no. 30; cf. the altar of Tetartaios, *IG* XII.6.536, and the dedication to θεὸς πυρετός, *SEG* XXXVII 1503 (Kastabala in Cilicia).

88. Sosineos: *SEG* XXXIII 147.50 (hero); *CIRB* 30 (epithet of Poseidon, Roman period; cf. νηοσσόος for Artemis and Apollo; Ap. Rhod. *Argon.* 1.570, 2.927). Sosipolis: *IGB* 5. 5103 (hero); *LSA* 32.48 (epithet of Zeus); Paus. 6.20.2 (a δαίμων ἐπιχώριος).

was associated with a feminine Miliche (much later in Crete a Hera Melichia appears). (Similar ambiguities attach to Epidotes, "Extra Giver": the word occurs as an epithet of Zeus, but there is also a collective of Epidotai.[89]) It has often been supposed that Meilichios was in origin an independent god of the underworld, a *Sondergott* only secondarily attached to Zeus.[90] But the attachment, if attachment it was, had already occurred in some places by the fifth century, and the appearance of Daimon M(e)ilichios in Lebadeia centuries later cannot be dismissed as a mere survival; it must attest continuing or new uncertainty about the status of the figure addressed. The uncertainty did not matter, however; what mattered was to secure the assistance of the power in question, not the theological status of that power.

Function also prevailed over individual identity when appeals and dedications were made to anonymous clusters of gods: "the gods of aversion" (*apotropaioi* or *alexikakoi*), "saviour gods" (*sōtēres*), "gods who give heed" (*epēkooi*), "gods of boundaries" (*enorioi*).[91] These or closely similar titles were born individually by several gods, and it may be that in a particular case the worshipper who dedicated thus anonymously would have been able to name the specific gods he or she had in mind. But the frequency of such expressions suggests that nonspecificity may have had its attractions. The very numerous dedications to, for instance, *theos/thea epēkoos*, "god/goddess who gives heed," are different, in that, as with simple dedications to "god," the identity of the god will have been obvious from the place

89. Epithet: *SEG* XI 1002 (6th c.); *SEG* LIX 422 (both Messenia); *IG* V.2.270; and Paus. 8.9.2 (Mantinea), probably the "phiala of Epidotas" in an inventory of the Argive Heraeum, *IG* IV.526; collective Paus. 2.27.6; cf. *IG* IV^2 1.108.158, and possibly *SGDI* 342 near Pagasai. Paus. 3.17.9 speaks of Epidotes as a daimon in Sparta who averts divine wrath over Pausanias's murder (though for Hesych. s.v. Epidotas it is a Spartan title of Zeus); there is also (Paus. 2.10.2) Hypnos Epidotes at Sicyon. For an Ἐπιδώτειον at Epidaurus, see W. Peek, *Neue Inschriften aus Epidauros* (Berlin, 1972), no. 23.

90. For all the attestations, see M. H. Jameson, D. R. Jordan, and R. D. Kotansky, *A* lex sacra *from Selinous, GRBM* 11 (Durham, N.C., 1993), 81–91 (for Lebadeia and Thespiai 84). At Selinous Meilichios is commoner than Zeus Meilichios (five instances to one), but the proportion is reversed elsewhere, and even at Selinous the Zeus Meilichios is on received datings as early (6th c.) as the first Meilichios without the theonym. Unnamed Meilichioi: Paus. 10.38.8; *IG* IX.2.1329. Underworld god: J. E. Harrison, *Prolegomena to the Study of Greek Religion*, 2nd ed. (Cambridge, 1908), 17–20, taken up, e.g., by G. Murray, *Five Stages of Greek Religion*, Thinker's Library (London, 1935), 14–15.

91. *Theoi sōtēres:* e.g., the inscription by the architect of the Pharos at Alexandria; Lucian, *Hist. conscr.* 62; *SEG* XLVIII 1331, Halicarnassus; *Gephyra* 10 (2013): 133 (Pisidian Conana); Ronchi, *Lexicon theonymon*, 1068. *Apotropaioi:* e.g., Aesch. *Pers.* 203; Xen. *Symp.* 4.33; Hippoc. *Vict.* 4.89; Pl. *Leg.* 854B. *Theoi apotropaioi kai alexikakoi: IByzantion* 12 (2nd/3rd c. A.D.; epigraphically references to individual *apotropaioi* are commoner). *Theoi epekooi:* Weinreich, *Ausgewählte Schriften*, 1:150–51. *Theoi enorioi:* Robert, *Noms indigènes*, 358 (Apameia/Apollonia frontier in Pisidia). *Themelioi* ("of foundations," anti-earthquake gods): *BÉ* 1968, no. 465. *Meilichioi:* previous note. The *theoi prokuklioi* of *IErythrai* 201 d 18, 23, have not been securely explained (Graf, *Nordionische Kulte*, 365–66); the epithet may be topographic, not functional. On groups, see too Polinskaya, *Local History*, 80–81.

where the offering was made. But still the emphasis is placed on the function rather than on the name. A pair of dedications from Patara in Lycia is interesting. One altar is dedicated (*TAM* II.403) to "Theos Soter (Saving God) Hedraios Asphales and Poseidon Hedraios and Helios Apollo." The main preoccupation of the dedicator becomes obvious from Asphales, "Safe," and the repetition of Hedraios, "with a Firm Base": it is protection from earthquakes. The other (*TAM* II.404) is for Zeus Soter Hedraios: the same god as on the other altar, but with Zeus here replacing Theos. Was the dedicator of the first altar thinking of this Zeus Soter Hedraios? Or did he feel the identity of the greatest protector against earthquakes to be unknowable and leave it deliberately vague? The example chosen comes from Anatolia in the imperial period, but the drift toward vagueness is an important fact of Greek religious psychology at all periods.[92]

We can conclude this section with the god Pylon, "Gateway," who emerges in eastern Pontos and Cappadocia in (to our observation) the second century A.D., worshipped by the *beneficiarii consularis;* these were a kind of police officer often, as it seems, having their headquarters, for obvious reasons, near city gates. Another dedication from eastern Pontos by a *beneficiarius* was made to a Zeus Propylaios, "of the Gateway," no doubt envisaged as very similar to Pylon himself.[93] The late emergent *Sondergott* Pylon, with his very specialised group of worshippers, shows how the impulses located by Usener at the origins of polytheism persisted in its latest stages.

THE CREATION OF EPITHETS

The epithets confront us in the thousands,[94] but in the main we have no precise information as to how they were created; they are just there. We can see rather better how they were diffused, because we know that oracles often advised a city to introduce a god under this or that title, and we can even identify titles that they favoured in such prescriptions, such as Poseidon Asphaleios, "of Security," Zeus Hypatos, "Highest," Apollo Prostaterios, "Protector," Artemis Orthosia, "Who Sets

92. Cf. Mitchell, "Theos Hypsistos," 102, on the choice between Theos Hypsistos and Zeus Hypsistos as conscious, not arbitrary; Chaniotis, "Megatheism," 129–31, on *SEG* L 1222 (Iuliopolis in Bithynia, 2nd/3rd c.), an enthusiastic dedication to an unnamed "best greatest listening saviour god . . . lord of the kosmos"; W. Wischmeyer, *ZAC* 9 (2005): 156, on "anonymisation." For *theos* Olbios and Zeus Olbios in the same cult(s) in Hellespontine Mysia, see L. Robert, *Collection Froehner,* vol. 1, *Inscriptions grecques* (Paris, 1936), 58–61, with *Hellenica* II, 152–53 (and in brief *BÉ* 1972, no. 368); note too, e.g., "the god who lightens and thunders," *TAM* V.1.585; the "goddess good to meet" (εὐάντητος), *TAM* V.2.1185.

93. See T. M. Mitford, "The God Pylon in Eastern Pontus," *Byzantion* 36 (1966): 471–90; one Cappadocian example (Kırşehir) is published, and another (Nigde) mentioned, by D. French, *EpigAnat* 28 (1997): 118.

94. For a useful overview of those of Thessaly, see Mili, *Ancient Thessaly,* 304–24.

Straight (?)."[95] Only very occasionally can we observe the creation of an epithet. The most reliable case must be one for which we have a king's own word: Eumenes II of Pergamum explains in a letter to the Coans concerning his festival Nikephoria that he has "named her [Athena] Nikephoros (Victory-Bringer), considering this title ([προσω]νυμία) the fairest and most appropriate."[96] Herodotus tells us that when in 480 the Greeks received news of the providential storm that had wrecked much of the Persian fleet, they "prayed and made libations to Poseidon Soter (Saviour)" and have "used the title (ἐπωνυμία) of Poseidon Soter ever since from then to this day" (7.192.2). Such a naming would both commemorate a victory and express a hope for continued saving in the future, that is, for a function; such a blend of the retrospective and the prospective is found in some titles in Catholicism, such as Madonna della Salute, "of Health" (relating to help in a time of plague), and Madonna della Pace, "of Peace" (relating to a particular peace). This was not in fact a case of the creation of an epithet, but of its first application to a particular god; we hear similarly of occasions when Asclepius was acclaimed (for the first time in that place?) as Soter in Sparta, Herakles by the same title in Kassope in Epirus, Artemis Leukophryene as Nikephoros, "Victory-Bringer," in Magnesia on the Maeander.[97] Later the Mantineans founded a temple of Aphrodite Symmachia, "of Alliance," which may be an authentic new coinage, "as a memorial of their fighting alongside the Romans at Actium" (Paus. 8.9.6).

Most other claims are anecdotal and unreliable. According to a story, Sophocles founded a sanctuary of Herakles μηνυτής, "informer," when the hero revealed to him in a dream the identity of a thief who had stolen a golden crown from his shrine (or from the Acropolis). The epithet is so singular that some such occasion seems necessary to explain it, but perhaps not one involving so well-known a figure as Sophocles. We can perhaps believe Plutarch when he tells us that Themisto-

95. See, e.g., the responses H 19, 28, 29, 39, 50 and 74 in J. Fontenrose, *The Delphic Oracle* (Berkeley, 1978), with *LSA* 87 = *IKaunos* 31 (identified as an oracle by E. Lupu, *Greek Sacred Law: a Collection of New Documents,* Leiden 2004, 35 n. 162). The Delphic oracle *SGO* 02/02/01 (Tralleis) lists five titles under which Poseidon should be invoked as protector against earthquakes. Late in the third century A.D., the prophet of Apollo at Didyma, Damianos, having secured the god's permission to establish an altar of Soteira Kore, then asks the god to prescribe her εὔφημος καὶ ὑμνικὴ προσαγόρευσις: the god recommends μείλιχος (*IDidyma* 504.19–31).

96. K. Rigsby, *Asylia* (Berkeley 1996), no. 176.5–6; cf. 177.17–19. The epithet has a later history: the goddess becomes "Polias and Nikephoros," apparently via an intermediate stage, coinciding, it has been argued, with the end of the monarchy and a perception that she no longer brought victory (H. Müller, *Chiron* 33, 2003: 443–44), in which she was Polias alone.

97. Isyllos of Epidauros 82 (Powell, *Coll. Alex.*, p. 135); *SEG* XXXVI 555.6; *IMagnesia* 100 b 41 (householders of the city are urged to set up altars bearing this title on the occasion of the inauguration of the city's Parthenon—the Athena Nikephoros of Pergamum doubtless brought the epithet into view, though its popularity on Hellenistic Delos among worshippers of Egyptian gods [*ID* 1230, 1304] may relate to local circumstances: see Roussel on *ID* 1304).

cles caused annoyance by naming Artemis Aristoboule, "of Best Counsels," in disguised celebration of his own excellent strategy (another commemorative/ functional blend if so);[98] we should surely not believe that Aphrodite Kallipygos, "Fair Buttocked," in Syracuse owed her name to two young women grateful for the social advancement that their own fair buttocks had brought them (Ath. 12.80, 554C-E). Royal authority doubtless counted for much: we have seen Eumenes' creation of the title Nikephoros for Athena, and it was under the same impulse that the cult of Dionysus under the name Kathegemon, "Leader," radiated out so widely from Pergamum.[99]

Many titles must have been products of particular circumstances that we can only guess at. We can understood why Apollo was called Korax, "Raven," on an early graffito dedication from Cyrene, though the form of the double name—simple juxtaposition of a god's name and an animal's—is unusual: the epithet obviously evokes the myth that in the form of a raven he guided Battos and the first settlers when they arrived in Libya.[100] But the circumstances that caused the mythical detail to become an epithet (if it was regularly so used) escape us. At Eretria on Euboea ca. 340 a decision was taken to hold a musical competition at the Artemisia τεῖ Μεταξὺ καὶ τεῖ Φυλάκει, probably to be translated "for the One in Between and the Guardian," most unusual titles of, we assume, Artemis; they commemorated, it has been argued, her role as intermediary and conciliator in a recent civil war.[101] There was scope for creativity and individual choice. In the mid-second century, during a revolt in Syria against Demetrios II Nikator, the badly defeated army of Demetrios's general Sarpedon was saved when the enemy, pursuing them along the shore, were engulfed by a great wave. Sarpedon's men sacrificed a great mass of fishes also deposited by the wave to Poseidon Tropaios, "of Route." Poseidon receives that title, normal in the cult of Zeus, here only; it was probably a nonce invention by Sarpedon's troops to celebrate the paradoxical routing of an army on land by the god of the sea (*FGrH* 87 F 29). But the process can usually not be reconstructed. As a result, we do not know with any precision—to take a handful of examples at

98. Informer: *Vit. Soph.* (= Sophocles T 1 Radt) 12; and Cic. *Div.* 1.54 (T 167a Radt). Of Best Counsels: Plut. *Them.* 22. 2; cf. Parker, *Athenian Religion,* 155.

99. Ohlemutz, *Kulte in Pergamon,* 90–116. Alexander supposedly proclaimed that the Apollo of Tyre suspected of treachery in his favour by the Tyrians should be dubbed Philalexandros: Diod. Sic. 17.46.6 (cf. 17.41.7–8). It is not clear how seriously this was meant, or whether it had any effect.

100. *SEG* XLIV 1541, no. 1; cf. Callim. *Hymn* 2.65–68. For such a juxtaposition, cf. Artemis πῶλος in Paros and Thasos (*IG* XII Suppl. p. 106, no. 202; p. 160, nos. 382–83).

101. D. Knoepfler, in M. H. Hansen, ed., *The Polis as an Urban Centre and as a Political Community* (Copenhagen, 1997), 376–77, on RO 73.6. Before Knoepfler the expression τεῖ Μεταξὺ καὶ τεῖ Φυλάκει had been understood as a designation of place or time. Speaking of the epithets "of Hera and King and of Concord" given to Zeus in RO 85.6–7, V. Pirenne-Delforge and G. Pironti, *ZPE* 191 (2014): 30, write: "Les officiels de la cité ont construit une configuration panthéonique de circonstance." Cf. *Keltische Götternamen.*

random from Pausanias—why Herakles is Parastates, "Comrade in Arms," in Elis (5.8.1); Dionysus Aisymnetes, "Magistrate," in Patrai (7.20.1); Artemis Hymnia, "of Hymns (?)," throughout Arcadia (8.5.11); Aphrodite Melaina, "Black," near Mantinea (8.6.5); Dionysus Polites, "Citizen," at Heraia (8.26.1); Aphrodite Machanitis, "Deviser," in Megalopolis (8.31.6); Apollo Epikourios, "Helper," at Bassai (8.41.7); Ares Aphneios, "Wealthy," in the territory of Tegea (8.44.7–8); Dionysus Aigobolos, "Goat-Shooter," at Potniai in Boeotia (9.8.1–2); or Hermes Kriophoros and Promachos, "Ram-Carrier" and "Frontline Fighter," at Tanagra (9.22.1–2). Still obscurer is the origin of the many epithets the meaning of which is unknown.

AN OUTLINE HISTORY OF THE DOUBLE NAME

A form of the cultic double name is common in Linear B, though not quite of the type familiar in later Greece: most commonly it consists of Potnia, "Lady, Mistress," preceded or followed by specifications of various kinds: Potnia of Atana (or Athena Potnia), Potnia of Asia, Potnia of Horses (a remarkable preoccurrence of the familiar epithet), Corn Potnia or Potnia of Corn, and several Potnia plus toponym combinations, most notably Potnia of the Labyrinth; there is also Diktaian Zeus, of Mount Dikte.[102] Cult epithets appear in Greek literature and inscriptions from the earliest times, if not in great numbers; a good early instance is Alcaeus's account of how the Lesbians named the gods of their great central sanctuary "Zeus Antiaos ('of Suppliants'?) and you, Aeolian glorious goddess, origin of all, and third this κεμήλιον [not understood] Dionysus Omestes ('Raw Flesh Eater')" (fr. 129.5–9 Voigt). In classical texts the system is unfolded more clearly to our view, but not changed. Change, or at least a change in the language of dedicatory inscriptions, occurred in the Hellenistic period with the growing importance of acclamatory/supplicatory epithets applicable to several gods, such as "Saviour" or "Who Heeds Prayers." A tendency also arose to see the ordinary cult epithet less as an identifier than as a glorifier: where a single epithet suffices to identify, a heaping up of glorifiers heaps up glory; these two developments together could lead to the multiplication of epithets, occasionally to an extreme degree.[103] All this will be discussed in a later chapter. With the assimilation of mortals to gods in ruler cult, monarchs and their families too acquired variants of the cultic double name;[104] henceforth there

102. See Rougemont, "Noms des dieux en linéaire B," 344–60. The interpretation of u-po-jo Potnia is disputed (p. 354). Rougemont concludes (360): "The functioning of the epithets is identical to that known in the classical period."

103. Scholars and educators also became interested in epithets in the period: see *OpAth* 28 (2003): 174, and the list of epithets of Zeus inscribed (on a gymnasium wall?) in Miletus perhaps in the second century A.D. (*SEG* XLV 1612 = *IMilet* VI.3.1395).

104. See, e.g., the striking list of Alexandrian street names based on titles of Arsinoe II in Fraser, *Ptolemaic Alexandria,* 1:35.

was a certain interaction between royal titulature and divine. But these changes in the language of dedication are not reflected in Pausanias: in him, there is no trace of the acclamatory use of the epithet: it continues to identify, and is extended to identify, not merely individual sanctuaries but also individual cult images within them. Nor did all dedications adopt the new style: the intense and moving devotion of many Phrygian peasants in the second/third centuries A.D.[105] was addressed to simple combinations of Zeus and one epithet. The old system met a need, and proved remarkably durable.

NAMING MEN AND NAMING GODS

Onomastic science has pursued the psychology of naming into unexpected areas: one can inform oneself about how people name pleasure boats, pharmacies, racehorses . . .[106] But it is not easy to find theoretical reflection on the special issues involved in the naming of gods (a different matter from the naming of God, on which there is an abundant devotional literature going back at least to *On Divine Names* of Dionysius the Pseudo-Areopagite). In some respects naming of gods follows principles found in human naming or other aspects of ordinary language use: titles can replace names in addressing both human and divine superiors; euphemisms veil feared conditions as they veil feared gods. But we noted earlier that divine names tend to be either more opaque (e.g., Asclepius) or more transparently related to a function (e.g., Hygieia, "Health"), than those of mortals. And it is hard to think of analogues in secular usage for the cultic double name. For the topographic epithet one might compare "the Rokesby Venus" or the "Marks and Spencers in Summertown," but in these cases the noun refers to a class via an unstated term (images of Venus; branches of a chain store), whereas Artemis of Ephesus is not reducible to "the sanctuary of Artemis at Ephesus," though grounded in that. For the functional epithet I can think of no analogue. The various ways in which a theonym can be replaced by an epithet (theonym omitted completely; theonym replaced by *theos/thea;* singular named god replaced by anonymous group) move us into a sphere in which the object of address is not a person but a power, a danger to be averted, a benefit to be secured. Also unthinkable in human terms are conflations of related beings such as Artemis Eileithyia, however one chooses to parse the precise relation of the two names in that expression.

"We know nothing about the gods, neither about them in themselves nor about the names by which they call one another": this famous remark of Plato's Socrates

105. Those studied in *Phrygian Votive Steles:* the simplicity is stressed by Chiai, "Medien religiöser Kommunikation," 71.

106. *Namenforschung,* 2: sec. 18.

(*Crat.* 400D) explains some of the singularities of the phenomenon.[107] There is a problem in naming that which one cannot see, or know. (In a very different religious context, a quondam archbishop of Canterbury writes of the "challenge" of referring to "what is not an object among others or a definable substance that can be "isolated" and examined."[108]) Uncertainties about where one god ends and another begins (Artemis Eileithyia), about who is a god and what is a mere epithet, about whether a particular minor power is an individual or a group,[109] are one product of this fundamental unknowability; so too are vague and anonymised forms of expression that stress the good to be secured rather than the name. The cultic double name is an attempt, by contrast, to stop the gods slipping away like water through the hands. It roots them in the places visibly consecrated to them and in the urgent needs which worshippers bring to them, and which they are believed to have satisfied in the past. But it also fragments the god into pieces, so that the question arises, for us at least, whether Greeks prayed to one Zeus or several hundreds.[110]

Socrates continues, in the passage just quoted, "just as in prayers it is conventional for us to pray them as 'whoever and from wherever they like to be named,' so too we should address them, since we know nothing else" (*Crat.* 400E). From such formulae (an attested but by no means universal feature of Greek prayers) it has often been inferred that the danger of addressing a god by the wrong name was a source of anxiety for the Greeks, a danger they guarded against by such catchall formulae.[111] Little in the epigraphic evidence supports that view. What is striking is the variety of ways in which gods are named, not just in different places but within the same sanctuary. Artemis at Gonnoi in Thessaly is addressed as Artemis Eileithyia repeatedly, but also quite often as Artemis Genetaira, "of Begetting," and occasionally as Artemis (Eu)lochia, "of (Easy) Birth," or Artemis Euonymos, "of Auspicious Name"; all these dedications are likely to come from the same precinct.[112] There are many comparable examples; fixed and regulated forms of address

107. Cf. T. Harrison, *Divinity and History: The Religion of Herodotos* (Oxford, 2000), 191 nn. 31 and 33; 258 n. 29.

108. R. Williams, *The Edge of Words: God and the Habits of Language* (London, 2014), as cited in the *Times Literary Supplement,* Feb. 6, 2015, 28. The *Cratylus* position was often echoed in antiquity, e.g., *Deorum vera nomina nemo novit,* a view mooted in Serv. ad *Aen.* 4. 577; it does not matter what name is given to the supreme being, Celsus ap. Origen *C. Cels.* 1.24, 5.41 (a position rejected by Origen himself, *C. Cels.* 1.24).

109. On figures such as Pan occasionally attested as plurals, see F. Sokolowski, *Lois sacrées des cites grecques, Supplément* (Paris, 1962), 95, with references; for Eileithyiai, *BÉ* 2015, no. 574.

110. Cf. n. 64 above.

111. For exposition and critique of this view, see S. Pulleyn, *Prayer in Greek Religion* (Oxford, 1997), 96–115.

112. See B. Helly, *Gonnoi* (Amsterdam, 1973), 1:31 and 148, on inscriptions 161–96 of his vol. 2; the most questionable provenance is perhaps that of Euonymos. The sanctuaries of Artemis at Lousoi (nn. 61 and 63 above), Artemis at Blagana (M. Hatzopoulos, *BCH* 111, 1987: 397–401), Zeus at Baitokaike

evidently did not exist, and it looks as if worshippers sometimes improvised and innovated. Names and epithets were imperfect human instruments for gaining contact with the divine; fortunately they worked, and there was no need to seek an unattainable precision.

ENDNOTE: THE ALTARS AT OLYMPIA

One of the most remarkable illustrations of the varieties of Greek divine naming in a single context is Pausanias's list (5.14.4–15.10) of the sixty-five altars within the sacred precinct at Olympia on which the Eleans made offerings each month (5.14.4, 15.10), listed, as he stresses (5.14.4, 10), in the order in which the offerings were made.

It is the following: Hestia; Zeus Olympios; <Zeus Laoitas and Poseidon Laoitas>;[113] Artemis; Athena Leitis; Ergane; Alpheios (*the river*) and Artemis; Alpheios; Hephaistos or Zeus Areios; Herakles Parastates; Epimedes; Idas (or Akesidas); Paionaios; Iasos (the last four all heroes); Zeus Herkeios; Zeus Keraunios; Zeus Olympios; "unknown gods"; Zeus Katharsios; Nike (Victory); Zeus Chthonios; "all the gods"; Hera Olympia; Apollo and Hermes; Homonoia (Concord); Athena; Mother of the Gods; Hermes Enagonios; Kairos (Right Moment); Kouretes or Herakles; Ge (Earth); Themis (Divine Order); Zeus Kataibates; Dionysus and Charites (Graces); Muses; Nymphs; "all the gods"; Aphrodite; Horai (Seasons); Nymphai Kallistephanoi; Artemis Agoraia; Despoinai (Mistresses); Zeus Agoraios; Apollo Pythios; Dionysus; Moiragetas; Moirai (Fates); Hermes; Zeus Hypsistos (bis); Poseidon Hippia; Hera Hippia; Dioscuri; Ares Hippios; Athena Hippia;[114] Tyche Agathe (Good Fortune), Pan and Aphrodite; Nymphai Akmenai; Artemis; River Kladeos; Artemis; Apollo; Artemis Kokkoka; Apollo Thermios; Pan; Artemis Agrotera; Pan.

The number of gods named without epithets may surprise, but we are dealing here with altars, not separate sanctuaries; there was therefore less "bureaucratic" pressure for distinguishing epithets. Among the epithets, almost none are toponymic, except Olympios; the reason again is that these are altars within a confined space, not sanctuaries dispersed through a territory. Moiragetas was inscribed without an accompanying theonym, but Pausanias explains that anyone "who

(p. 123 below), Apollo Lairbenos in Phrygia (see the references in *SEG* LX 1440–43), Aphrodite Kas(s) alitis Hearing Goddess at Anazarbos, where various selections and combinations of those four names, most commonly just "Hearing Goddess," are found (*IAnazarbos* 29–38), and spectacularly of Asclepius at Salda and Zulmudra in Thrace (*IGB* II, index p. 247; and III.1, index p. 295), provide further examples. On personal choices by persons of some culture, see, e.g., L. Robert, *Hellenica* IX, 61–63; K. Rigsby, "A Greek Dedication at Sidon," *Tyche* 22 (2007): 143–50. Cf. n. 101 above.

113. Supplied from 5.24.1.

114. The sudden rush of Hippios epithets occurs because we are entering the hippodrome; Hermes Enagonios and Kairos are distinctively sporting too.

understands the affairs of men" will recognise here an epithet of Zeus; Ergane too presumably lacked a theonym, since Pausanias gives none, but here he perhaps finds it unnecessary to explain that Athena is meant. (The doubts about the ownership of two altars—Hephaistos or Zeus Areios, Kouretes or Herakles—seem to show that they were uninscribed or had lost their inscription; they received offerings all the same.) A large number of the epithets will have borne their meaning on their face for any Greek: Ergane, "Worker"; Areios, "of Ares"; Parastates, "Comrade in Arms"; Herkeios, "of the Courtyard"; Keraunios, "of the Thunderbolt"; Katharsios, "of Purification"; Chthonios, "of the Earth"; Enagonios, "of Competitions"; Kataibates, "Who Comes Down" (as the thunderbolt); Kallistephanoi, "with Fair Garlands"; Agoraios, "of the Marketplace"; Moiragetas, "Leader of the Fates"; Hypsistos, "Highest"; Hippios/a, "of Horses"; Agrotera, "Huntress." A few are more recherché. Leitis, "of the Spoil," is a good Homeric epithet for Athena (*Il.* 10.460), but Laoitas would suggest no more than a vague link with *laos,* "people"; and it would have taken a very good scholar to connect the Nymphai Akmenoi with a word occurring once in Homer where it apparently means "flourishing" (*Od.* 23.191). Pausanias speculates that Thermios is equivalent to Attic Thesmios, "of Laws"—the word was therefore not transparent for him—but was unable to learn any explanation of Artemis's epithet Kokkoka (5.15.7). We are reminded how wrong it is to suppose that epithets needed to convey meaning in order to remain in use.

2

Interpretatio

The previous chapter has given an outline of the "inherited conglomerate" of Greek ways of naming and addressing the divine. The next will consider the fortunes of those linguistic conventions when applied to gods encountered outside Greece. But first must be considered the practice that provided the indispensable bridge between cultures. Let us approach it via something magnificent, the resonant opening to Alexander Pope's translation of Homer's *Iliad.*

> Achilles' wrath, to Greece the direful spring
> Of Woes unnumber'd, heavenly goddess, sing!
> That wrath which hurled to Pluto's gloomy reign
> The Souls of mighty chiefs untimely slain;
> Whose Limbs unbury'd on the naked Shore
> Devouring Dogs and hungry Vultures tore.
> Since Great Achilles and Atrides strove,
> Such was the Sov'reign Doom, and such the will of Jove.

I quote it for the substitution that Pope makes in the last line whereby the Greek "plan" or "will" of "*Zeus*" becomes the "will of *Jove*." No translator today would dream of replacing the Greek god by the Roman, but for Pope nothing could have been more natural, indeed inevitable. Pope and his contemporaries inherited the tradition going back to the Romans themselves of substituting Roman divine names for their supposed Greek equivalents, a tradition that has long died out in scholarship but lives on at lower levels: most of us were probably told at an early age that Minerva was the Roman Athena, that Neptune equaled Poseidon, and so on. I recall those early experiences because, when told of such an equivalence as a child, one did not question its basis, one accepted it; and it is very important that many ancients too were very probably exposed to such identifications early in life or in other situations which did not encourage reflection on how they were justified. They were just part of common speech, of the world one was born into.

In one of the rare passages where an ancient author makes the process explicit, Tacitus says of an ancient sacred grove in east Germany that "a priest who wears female clothing presides, but they call the gods in a Roman rendering (*interpretatione Romana*) Castor and Pollux: that is their efficacy, but their name is Alci [or Alcis]."[1] In consequence of that well-known passage the process of identification has come to be known as *interpretatio;* the term occurs in this application here only, but serves usefully to name a practice that is pervasive. Such identification was applied by the Romans, as before them by the Greeks, to the gods of almost all the nations that they encountered. One can ask which of the two English derivatives from the verb *interpretari* more nearly catches Tacitus's sense: is he referring to the work of an interpreter, someone who merely translates from one language to another (but how does one "translate" the name of a god?), or is the process rather one of interpretation, the attempt to bring out the meaning of an expression in different words? A glance at a Latin dictionary shows that the question cannot be answered, since both *interpretari* and *interpretatio* are regularly used with both forces.[2] This ambiguity that inheres in the Latin word inheres, as we shall see, in the whole process.

Georg Wissowa, in a classic study of *interpretatio* within the Roman empire, confined himself to the Latin West, on the grounds that in terms of both sources and historical realities the situation in the Greek East was quite different.[3] So even within the Roman world no single framework of analysis was possible; even within the Roman West substantial differences are visible between what happens in Gaul and Germany on the one side, and in Africa on the other.[4] If one extends one's view to *interpretatio* throughout the ancient world, the relevant differences multiply greatly. An even more important difference than the regional, one might argue, is that between native speakers of Greek and Latin, giving Greek and Roman names

1. Tac. *Germ.* 43.4: *Apud Naharvalos antiquae religionis lucus ostenditur. Praesidet sacerdos muliebri ornatu, sed deos interpretatione Romana Castora Pollucemque memorant . . . ea vis numini, nomen Alcis. Nulla simulacra, nullum peregrinae superstitionis vestigium; ut fratres tamen, ut iuvenes venerantur.*

2. For the latter view, see Bettini, *Elogio del politeismo,* 66–67, stressing the element of uncertainty so often present.

3. Wissowa, "Interpretatio Romana," which built on the work of his pupil Richter, *De interpretatione.* For recent bibliography, see *Interpretatio Graeca/Romana/Indigena,* 13 n. 1.

4. In Europe juxtaposition of the different theonyms is the norm, while in Africa the Latin name often appears alone: Wissowa, "Interpretatio Romana," 18–19; more generally, *interpretatio* is commoner in Africa: Cadotte, *Romanisation des dieux,* 9–10. Different Roman gods prevail in the two places: Richter, *De interpretatione,* 41, 51–4.

to the gods of foreign nations, and the indigenous worshippers of those gods, adopting or not adopting the imported designations; and there is a subdivision within the former group between historians and ethnographers writing from a distance, and travelers and settlers meeting the foreign gods at first hand. But there was a historical continuity in the practice of *interpretatio* within that world which it is important to recognise; this was a central feature of a partially shared religious culture. The learned *interpretationes* to be found in the pages of a Dionysius of Halicarnassus are not a completely distinct phenomenon from the Phrygian villager's veneration for gods he calls Zeus and Apollo; arcane identifications do occasionally enter actual cult practice. In what follows I shall take examples from any quarter that seem to shed light on the central issues, while trying to acknowledge essential differences.

SOME HISTORY

I begin with a brief historical survey. The phenomenon is most familiar in the Greco-Roman context, but it has much earlier origins. The Egyptologist Jan Assmann has stressed the willingness of peoples of the ancient Near East to identify their gods with one another's; it forms a plank in his controversial contrast between the supposed tolerance of ancient polytheist religions and the intolerance of Mosaic monotheism.[5] The Ugaritic scholar M. S. Smith has recently surveyed the phenomenon in a monograph strikingly entitled *God in Translation*. An unfortunately fragmentary text from Ras Shamra in northern Syria lists Sumero-Akkadian, Hurrian, and Ugaritic gods in three columns, evidently identifying them on the basis of functional similarities; there are similar monolingual lists in different languages which are virtual translations of one another. An Akkadian document known from its opening line as An:anu ša ameli identifies Sumerian with Akkadian deities; it reduces their number by treating numerous Sumerian deities as subaspects of a single Akkadian god.[6] These texts and their many congeners are learned works, products of scribal schools and needs, but An:anu ša ameli

5. J. Assmann, *Die Mosaische Unterscheidung: Oder der Preis des Monotheismus* (Munich, 2003); cf. Assmann, *Moses the Egyptian*, 44–54, "Counter-Religion and Religious Translatability in the Ancient World." The contrast goes back at least to David Hume (*The Natural History of Religion*, ed. A. Wayne Colver, Oxford 1976, 58–62), though he grounded pagan "tolerance" not in *interpretatio* but in the willingness to accept foreign gods; Voltaire in *Traité sur la tolérance* (1763) spoke of ancient religions as "comme des noeuds qui les unissaient tous ensemble" (cf. C. Bonnet, *CRAI*, 2012, 503–4); Gibbon too credited the Romans with a "Universal spirit of toleration" (opening pages of vol. 1, chap. 2, of *Decline and Fall*). For J. G. Droysen, by contrast, universalism in religion emerged only in the Hellenistic period (Moyer, *Limits of Hellenism*, 145–46).

6. See Smith, *God in Translation*, 45–48 (Ras Shamra), 42 (An:anu ša ameli), though on the latter note the reservation of Allen, *Splintered Divine*, 98 n. 7 (citing Rubio).

is a reminder that identifications between Sumerian and Akkadian deities could scarcely be avoided, so thorough and durable was the interpenetration of their religious cultures.

Also relevant is the practice of transferring logograms for deities, no differently than for ordinary objects, from one language to another, so that (for instance) a Sumerian, an Akkadian, and a Hittite god of vaguely similar function would all be represented in writing by the same symbol. These identifications existed only on the tablet, not in utterance, since when read out the name of the relevant local deity would be the one pronounced. But the scribal practice could scarcely have existed without the presumption that deities could be assimilated, a presumption that it must in turn have reinforced. In a prayer, gods could be identified across national barriers, as when the Hittite queen Puduhepa reminds the sun-goddess of Arinna that though that is her name in the Hatti country, "in the land which thou madest the cedar land thou bearest the name Hebat."[7] In diplomatic contexts identification was not the norm: on the contrary, in treaties, the gods of both parties were listed separately, and in correspondence one often finds a distinction between one's own gods and the gods of one's "brother" (a foreign king).[8] An Amarna letter shows the Hurrian goddess Šauška being sent on loan to Egypt, and thus becoming, in the language of diplomatic politesse, also the god of the Egyptian king.[9] Here therefore the gods of different nations were treated as separate, though of equal worth. But crossovers could occur even in contexts such as these. In the Egyptian version of the treaty between Ramses II and the Hittite king Hattusilis I the divine witnesses even on the Hittite side are largely Egyptianized: thus the Sun-goddess of Arinna who appears so often in Hittite texts becomes "the Re of the town of Arinna," accompanied inter alios by a Seth of Hatti and a Seth of the town of Arinna. One must, however, add the essential qualification that such identifications had no effect on ritual practice: gods so identified retained their separate names and cults.[10]

What part the Greeks of the second millennium B.C. may have played in this general east Mediterranean willingness to assimilate deities is not demonstrable on the basis of direct evidence. It is not proven, though highly plausible, that the

7. *ANET,* 393; cf. an address to Ninurta in the Standard Babylonian version of the Epic of Anzu, tablet 3.131–32 (trans. B. R. Foster, *Before the Muses,* 3rd ed., Bethesda 2005, 576): "In Elam they gave you the name Hurabtil, in Susa they speak of you as Inshushinak . . ."

8. So, e.g., Tušratta of Mitanni writing to Nimmureya of Egypt, *Amarna Letters* 20.23–27.

9. Tušratta to Nimmureya, *Amarna Letters* 23.31–32: "Is Šauška for me alone my god(dess), for my brother not his goddess?"

10. Egyptian version: *ANET,* 201. Conversely, Smith, *God in Translation,* 52, notes that in the same text the Hittite "thousand god" scheme is applied also to the gods of Egypt (*ANET,* 200–201). Essential qualification: so, for instance, Puduhepa's assimilation of the sun-goddess of Arinna and Hebat, cited in the text, was not reflected in cult: Allen, *Splintered Divine,* 74.

"great king" of the Ahhiyawa who belonged to the "Great Powers' club" of the fourteenth century B.C., and thus was inevitably drawn into high-level diplomatic exchanges with brother kings, was a Greek.[11] But, whether he was or not, Mycenaean overseas contacts were extensive and encounters between Mycenaean and foreign gods inescapable; it is likely that the process of Greek *interpretatio* of ancient Near Eastern gods (and ancient Near Eastern *interpretatio* of Greek gods?) was already in full swing centuries before it first becomes visible. The first explicit instance of *interpretatio* in our literary sources is perhaps a fragment of Hecataeus early in the fifth century which boldly speaks of a temple of Leto and an island of Apollo, Greek gods both, at Bouta in Egypt;[12] *interpretatio* is then pervasive in the pages of Herodotus, and is conducted as if nothing could be more natural. But already in the *Iliad* Greeks and Trojans are shown as worshipping the same, Greek gods: is this simply the same epic disregard for realism that effaces linguistic difference between the two sides, or is it *interpretatio*?[13]

Archaeologically, Greek offerings in non-Greek sanctuaries and vice versa are commonplace at least from the eighth century, and plausible cases have been identified of transculturally appropriate dedications, such as an Egyptian mirror showing Mut (the "Egyptian Hera") found in the Heraion of Samos.[14] An explicit document of the sixth century is a dedication made by one Melanthios in Egypt on a statue base of thoroughly Egyptian aspect to "Zeus Thebaieus," Zeus of (Egyptian) Thebes, a common Greek designation for the Egyptian god Amun-Re.[15] Ptolemaic Egypt was dotted with towns such as Hermopolis and Panopolis named from Egyptian gods translated into Greek. Herodotus already speaks of two examples of

11. See, e.g., J. Latacz, *Troy and Homer,* trans. W. Windle and R. Ireland (Oxford, 2004), 121–28.

12. *FGrH* 1 F 305, quoted verbatim in Steph. Byz.

13. This paragraph depends largely on W. Burkert, "Herodot als Historiker fremder Religionen," in *Hérodote et les peuples non-grecs,* Entretiens Hardt 35 (Geneva, 1990), 1–39. Burkert suggests that the Aphrodite of *Hymn. Hom. Aphr.* is an interpretation of Kybele/Kubaba of Mount Ida. For *interpretatio* in other cultures at this date (ca. 720), see, above all, the great Luvian-Phoenician bilingual from Karatepe, Hawkins, *Hieroglyphic Luwian Inscriptions,* 1.1:48–58, with such equivalences as Tarhunzas/Baal, Runzas/Rešeph of the goats.

14. See S. Ebbinghaus, "Begegnungen mit Ägypten und Vorderasien im archaischen Heraheligtum von Samos," in A. Naso, ed., *Stranieri e non cittadini nei santuari greci* (Udine, 2006), 187–229 (with much other relevant material), at 197. J. E. Skinner, *The Invention of Greek Ethnography* (Oxford/New York, 2012), strongly emphasises contacts and mutual interest between Greeks and non-Greeks in the archaic period.

15. Jeffery, *LSAG*[2] 358, no. 49 (supposedly from Memphis, 550–525); for the base, see Vlassopoulos, *Greeks and Barbarians,* 159, fig. 17. Cf. for the god Hdt. 1.182.2, 2.42.1, 2.54.1, 4.181.2; another dedication to him is *SB* 1.2463 (Naukratis). (Herodotus, however, always calls the god of Siwa Ammon, though associating his cult with that of Zeus Thebaieus: 2.55, 4.181.2.) Vlassopoulos compares G. Roeder, *Aegyptische Bronzefiguren* (Berlin, 1956), 208, no. 2458 (Saqqara, ca. 400), as read in *SEG* XXVII 1107, a bronze statuette of Osiris dedicated by a Greek to Selene, on the basis apparently of a gender-crossing identification of Selene with Osiris in his lunar aspect.

the kind, Heliopolis and Hephaistopolis; these names were evidently not invented by him but already existed in the speech of Egyptian Greeks. Everything suggests that archaic Greeks detected their own gods very freely all round the Mediterranean; one of the most fixed and long-standing equations, that of Herakles with Phoenician Melqart, is there by implication in Herodotus (2.44.1) and may well go back much earlier.[16]

We already find in Herodotus a by-product of *interpretatio* that has an interesting history of its own, the division of one god or hero into two or more (2.43–44). He endorses the position of those Greeks who distinguish between an ancient god named Herakles and a later-born Greek hero of the same name. The historical process leading to this conclusion must have been, first, identification of Herakles with a foreign god, followed, later, by a recognition of differences between them that required them to be distinguished. That recognition ought to have to led to a total separation between the two figures, since the foreign god will not at home have been called Herakles; we saw above that the Phoenician Herakles was Melqart. But the power of the identification was such that the Greek name stuck to the foreign figure in Greek understanding, and it became necessary to explain how the Greek hero too came to bear it; thus Herodotus, who thinks the original god Herakles was Egyptian, points out that the parents of the Greek Herakles, Amphitryon and Alcmene, were both Egyptian by descent. Wherever Greeks went they encountered a new Herakles, and the process of identification followed by division was repeated again and again: Varro makes the spectacular claim that there had been forty-three different Herakleses, "since all those who were valiant were called Herakles."[17] Dionysuses multiplied too, partly for the same reason.

16. See Lane Fox, *Travelling Heroes*, 206–7. Sophocles fr. 126 Radt from *Andromeda* speaks of "barbarians" performing human sacrifice to "Kronos," i.e., Baal-Ḥamān, according to Allen, *Splintered Divine*, 231.

17. Varro ap. Serv. ad *Aen.* 8.564; Cic. *Nat. d.* 3.42 has only six; cf., e.g., Arr. *Anab.* 2.16.1–4; Macr. *Sat.* 1.20.6, and for the Indian Herakles, Arr. *Ind.* 5.13, 8.4; Diod. Sic. 2.39.1. But even a Herakles who is not the son of Amphitryon has an Iolaos as faithful companion (Eudoxus of Cnidus ap. Ath. 9.47, 392D; Bonnet, *Enfants de Cadmos*, 299). Multiple Dionysuses: e.g., Diod. Sic. 4.1.6–7; for the supposed Egyptian account of how Greeks appropriated (ἐξιδιάζεσθαι) all the best gods and heroes: Diod. Sic. 1.23.2, 1.23.8 (cf. Tac. *Ann.* 2.60.3: *Herculi, quem indigenae [of Egypt] ortum apud se et antiquissimum perhibent et eos, qui postea pari virtute fuerint, in cognomentum eius adscitos*). But Plutarch insists against Herodotus that there is only Herakles, *De Herodoti malignitate* 14, 857E-F; and Aristides predictably (*Or.* 40.10) assumes the same, to show the scope of his fame. The euhemerizing lists of homonymous gods and heroes found in Cic. *Nat. d.* 3.42, 3.53–60, and later sources are partly products of this process of conflation and division: the fourth Venus, for instance, is the one "Syria Cyproque concepta, quae Astarte vocatur" (3.59). (On their euhemerizing tendency, see W. Michaelis, "De origine indicis deorum cognominum," PhD diss., Berlin, n.d., 59–68, who ascribes the original list to a grammarian influenced by the Peripatos and Euhemeros and who was active in Rhodes in the second century A.D.; W. Bobeth, "De indicibus deorum," PhD diss., Leipzig, 1904, 14, dates the original to the first century A.D. but says little about its theological presumptions.)

Interpretatio began early and continued.[18] When Alexander traveled far beyond the bounds of traditional Greek culture, he still, according to his historians, encountered Greek gods. Sightings of, say, Herakles and Dionysus in India may initially have been based on features of landscape and flora that evoked them, combined with the emergent myths that the gods themselves had traveled to these lands, but Megasthenes, who visited the court of Chandragupta late in the fourth century, was apparently identifying the Greek gods with their Indian counterparts.[19] In the Hellenistic period the scale of the phenomenon becomes immense, and is not merely a matter of Greek observers describing the foreign gods they encounter by Greek names (what one might call external *interpretatio*). To a remarkable extent those names are adopted by the indigenous worshippers of those gods, along with the Greek language, when (but probably only when—a crucial reservation) they are using that language (accepted *interpretatio*). Not all external *interpretationes* were accepted or persisted,[20] nor were they attempted for every god, but many were attempted and accepted. In Lycia and Caria accepted *interpretatio* is already common in the fourth century, before the conquests of Alexander. The adoption of Greek names for the gods of non-Greek peoples is a crucial part of the history of what used to be called Hellenisation. It also became common for bearers of a theophoric name to "translate" their name via *interpretatio:* a Sidonian named in his own language "servant of Tinnit," son of "servant of Shamash," becomes in the Greek text of his bilingual tombstone in the Piraeus "Artemidoros son of Heliodoros" (*KAI* 53).[21]

Just as Greek divine names overspread the East, so did Roman divine names overspread North Africa and the West. In a minority of cases these were Roman gods worshipped in Roman form by expatriates; more commonly an indigenous deity had been renamed (with more or less adjustment to the cult practice).[22] As in

18. In Pseudo-Scylax, *Periplous,* apparently of the 330s (G. Shipley, *Pseudo-Scylax's Periplous,* Bristol 2011, 19), see the "island of Ares" off the territory of the Mossynoikoi (86), Mount Karmel sacred to Zeus (104.3), the shrine of Athena Tritonis in Numidia (110.8), the altar of Poseidon at "Cape Soloesa" (112.4). For a late specimen, see the Greek/Parthian bilingual attesting the deposit of a statue of Herakles/Vahram in the temple of Apollo/Tir: P. Bernard, *JSav,* 1990, 3–68 (Greek text only in *IEstrOriente* 86).

19. On all this, see appendix C.

20. The goddess Perasia of Castabala in Cilicia is identified as Perasia Artemis by Strabo 12.2.7, C 537, but not in any inscription from the site: see Robert, in Dupont-Sommer and Robert, *Déesse de Hiérapolis Castabala,* 50. Cf. p. 93n71 on D(o)usares; p. 86 on the goddess of Komana. On unsuccessful *interpretationes* of Sandan in Tarsus, see Ehling, Pohl, and Sayar, *Kulturbegegnung in einem Brückenland,* 214. Zeus at Apamea in Syria mutated (reversion? or new identification?) in the second/third century A.D. to "greatest sacred (ἅγιος) god Belos" (*SEG* XLVI 1769; Millar, *Roman Near East,* 263).

21. See further appendix B.

22. For neo-Punic/Latin bilinguals which confirm the presence of *interpretatio Latina,* see Jongeling, *Neo-Punic Inscriptions,* 25, Labdah N 17; 48, Sabratha N 16.

the Greek East, examples come not in handfuls but in thousands, and the historical survey can end here with the phenomenon in full spate throughout the Greco-Roman world.

A MULTICULTURAL SYSTEM

Interpretatio is always potentially a two-way process. If a Greek identifies Mars as Ares, it becomes very natural for a Roman to identify Ares as Mars. A silver vase of Greek manufacture (early 4th c.) which depicts the judgment of Paris bears inscriptions in Lycian and was evidently intended for the Lycian market: Athena is here identified for the Lycian viewer as Maliya, a Lycian goddess. It has also been suggested that *Maliya wedrẽñni* of two Lycian texts from Rhodiapolis is a simple translation of the familiar Greek cultic double name Athena Polias, "of the City," which she actually bore in nearby Phaselis.[23] A Nabataean made a bilingual dedication on Cos to Aphrodite (in Greek) and Astarte (in Nabataean): since the dedication was doubtless made in a sanctuary of Aphrodite, the address to Astarte represents an *interpretatio Nabataea* (*IG* XII.4.546). The two-way traffic which we only glimpse in other cases is manifest between Greece and Rome. *Interpretatio* mediated the heavy influence of Greek cult on Roman. The Romans relied in many matters of cult on the Sibylline books, written in Greek, and so could not avoid *interpretatio* in applying their teachings to their own gods;[24] the presence of the Greek god Apollo in Rome, where he was early naturalized, made it all the more tempting to assimilate the two pantheons. A fragment of a large black-figure Attic crater depicting the return of Hephaistos has been found in what was apparently the archaic sanctuary of Vulcan in Rome; it is plausibly taken to show that, for a Roman, Hephaistos was already Vulcan (thus *interpretatio Romana*) in the sixth century.[25] On the other hand Greek writers on Roman institutions give Roman gods Greek names (thus *interpretatio Graeca*). There is asymmetry between those two forms of *interpretatio,* since the Roman adoption of elements of Greek cult practice that followed on the identification of gods has no Greek equivalent at least in the republican period. But Dionysius's systematic use of Greek names for Roman gods is exactly paralleled by Livy's use of Roman names for Greek: if one writes in Greek the gods have Greek names, if in Latin, Roman.

23. The vase: R. D. Barnett, "A Silver-Head Vase with Lycian Inscriptions," in *Mélanges Mansel* (Ankara, 1974), 893–903. Maliya wedrẽñni as Athena Polias: J. D. Hawkins ap. Barnett, 902–3, on *TAM* I.149.2–3, 12; 150.6–7; see further R. Parker, "Athena in Anatolia," *Pallas* 100 (2016): 73–90, at 73. An equivalence Malis-Athena has now been detected in a Lydian-Greek bilingual dedication: A. Payne and D. Sasseville, "Die lydische Athene—eine neue Edition von LW 40, 2016," *Historische Sprachforschung* 129 (2016): 66–82.

24. Wissowa, "Interpretatio Romana," 6–7.

25. F. Coarelli, *Foro Romano* (Rome, 1983), 1:177.

Other Mediterranean polytheistic cultures were included.[26] There survives in Polybius a list of gods by whom Hannibal swore an oath to Philip V, containing both Greek gods and Carthaginian gods translated, presumably with Carthaginian approval, into Greek. He swore "before Zeus and Hera and Apollo, before the God/Fortune of the Carthaginians and Herakles and Iolaos," as well as several further powers.[27] The position of Herakles and Iolaos here shows them to stand for Melqart and a Carthaginian equivalent of Iolaos. *Interpretatio* here is perhaps a matter of diplomatic politesse, part of the ability to communicate with a different state in that state's language. The multiplicity of cultures potentially involved emerges from the case of the great temple at Pyrgi in Etruria. Greek writers differed as to whether it honoured Leukothea (perhaps joined with Apollo) or Eileithyia; characteristically they gave the goddess a Greek name, even though the temple was in Etruria.[28] But the gold tablets in two languages found in 1964 that record its dedication (or at all events a dedication) by the Etruscan king Thefarie Velianas speak of Uni in the Etruscan version and Astarte in the Phoenico-Punic.[29] In each case the goddess receives a name indigenous to the language that is being used; this really is god in translation.

A further twist may be present: how the Greek writers got from Astarte/Uni to Leukothea or Eileithyia is at first sight puzzling, and it has been suggested that there was an intermediate Roman stage: Uni was first identified as either Juno Lucina or Mater Matuta; the two Roman goddesses will then have been identified respectively as Leukothea and Eileithyia, as we know they often were.[30] That process of *interpretatio* by stages or through intermediaries is one that is known to occur: Baal of North Africa, for instance, becomes Kronos, who then becomes

26. How Achaemenid Persia stood in relation to Greek gods is hard to say. The Achaemenid rulers are now generally held to have tolerated/supported the cults of their subject peoples (P. Briant, *Histore de l'empire Perse,* Paris 1996, 66–72, 559–71); but this does not entail that they recognised similarity/identity, and the "Zeus of Bag/radates" (p. 114) may be a Greek Zeus, not an interpreted Ahura-Mazda (Briant, 1026). Greek authors, by contrast, despite recognition that Persian theology differs radically from Greek (Hdt. 1.131), are happy to speak of Ahura-Mazda as Zeus (Hdt. 1.131.2, 7.40, 8.115.4; Xen. *Cyr.* 8.3.12); in the Lydian-Greek environment Anahita becomes Artemis Persike (p. 99).

27. Polyb. 7.9.2–3: ἐναντίον Διὸς καὶ Ἥρας καὶ Ἀπόλλωνος, ἐναντίον δαίμονος Καρχηδονίων καὶ Ἡρακλέους καὶ Ἰολάου. "Official wording, probably of a Greek translation produced in Hannibal's chancellery," says F. W. Walbank in his commentary ad loc. On the gods, see S. Lancel, *Carthage, a History,* trans. A. Nevill (Oxford, 1995), 208; and n. 69 below.

28. Arist. [*Oec.*] 2.1349b34; and Polyaenus, *Strat.* 5.2.21 (Leukothea); Ael. *VH* 1.20 (Apollo and Leukothea); Strabo 5.2.8, C 226 (Eileithyia).

29. *Corpus Inscriptionum Etruscarum* II.1. 4, nos. 6313–16; *KAI* 277. G. Colonna's many relevant studies are collected in his *Italia ante Romanum Imperium,* vol. 4 (Rome, 2005). For dedications to Aphrodite and Etruscan Turan from the same building at Gravisca, see Vlassopoulos, *Greeks and Barbarians,* 95.

30. R. Bloch, "Figures divines de Pyrgi," in Bloch, ed., *Recherches sur les religions de l' Italie antique* (Geneva, 1976), 1–9; but doubted by Colonna, *Italia ante Romanum Imperium,* 4:2267–68.

Saturn.[31] Another interestingly complicated early case is the fourth-century Lycian/Greek/Aramaic trilingual from Xanthos.[32] Since the sanctuary is in Lycia, the presumption is that the Lycian version should indicate its original owners: "the pñtrennu mother of the sanctuary here and her children and the Eliyana" (Lycian), of which the Greek "Leto and her offspring and the Nymphs" will be an *interpretatio.* (But a doubt remains: Greek Leto was famously a mother, but the same is not known of pñtrennu; it is possible that it was already through assimilation to Leto that the Lycian goddess acquired her "children.") The Aramaic by contrast borrows from the languages of the nonindigenous cultures that were influential in the region, to give "Leto (and) Artemis (and) ḤŠTRPTY and ʿḤWRNYŠ." It transcribes the names Leto and Artemis (strangely, the goddess's offspring, anonymous in Greek and Lycian, are here individualised and named) into Aramaic, and apparently adopts Persian equivalents, roughly translatable as "Lord of Power" and "the Ladies," for Apollo and Eliyana/the Nymphs.[33] The eclecticism reflects the situation of someone translating into an official language, Aramaic, which is not the vernacular of his own culture.[34]

FORMS OF *INTERPRETATIO*

Interpretatio had several forms, more than one of which might be applied to the same god.[35] The simplest and most drastic was simple substitution of one theonym

31. Wissowa, "Interpretatio Romana," 34–39, also with the example of a Dalmatian god who becomes Silvanus via Pan. For Babylonian names given to gods of Syria and Cilicia, some of whom in turn received Greek names, see H. Seyrig, *Syria* 20 (1939): 302 n. 5, with the example of Δὶ Βηλέῳ θεῷ Ἀδαδθελα (the god of Adadthela becomes Bel becomes Zeus); O. Casabonne, *La Cilicie à l'époque achéménide* (Paris, 2004), 70, 126–29, 178.

32. H. Metzger at al., *Fouilles de Xanthos*, vol. VI: *La stèle trilingue du Létôon* (Paris, 1979).

33. ḤŠTRPTY in the Aramaic is generally recognised as an Aramaic rendering of the Iranian divine title **xšaθrapati-* signifying lordship. ʿḤWRNYŠ (as Elizabeth Tucker advises me) is most plausibly taken as another Iranism, *ahurānīš,* a feminine from the masculine stem *ahura-* ("lord"), rather than as first proposed, Aramaic ʿḤWRN "others"; once in the old Avesta, waters are spoken of as *ahurānīš,* "ladies," of the Lord (Mazdā): see R. Humbach, "Die aramäischen Nymphen von Xanthos," *Die Sprache* 27 (1981): 30–32. The dominant view (cf. A. Dupont-Sommer, "L'énigme du dieu 'Satrape' et le dieu Mithra," *CRAI,* 1976, 648–60) has been that **xšaθrapati* was primarily, and we must assume for the argument recognisably, a title of Mithra, who is then identified with Apollo; for a different view, see M. Schwartz, "Apollo and Khshathrapati, the Median Nergal, at Xanthos," *Bulletin of the Asia Institute* 19 (2010): 145–50. The Phoenician god Shadrapha (in Greek Σατράπης) is distinct: Bonnet, *Enfants de Cadmos,* 194–95.

34. Elsewhere too we find divine names simply transcribed into Aramaic: A. Abou-Assaf, P. Bordreuil, and A. R. Millard, *La statue de Tell Fekherye et son inscription bilingue assyro-araméenne* (Paris, 1982), probably of the ninth century A.D.; *KAI* 267 (*TADAE* 4, D.20.3), a bilingual tombstone for an Egyptian from Saqqara of 482 B.C.; and unfailingly in monolingual Aramaic texts from Egypt (e.g., *CIS* II.123, 128 [*TADAE* 4, D.20.1, D.22.17], and often).

35. See further pp. 92–95.

for another, as often seen in Herodotus: Ptah becomes Hephaistos. Whereas Herodotus confidently identifies gods between cultures on a one-to-one basis, later authors, as we shall see, sometimes introduce doubt not about the process but about its application: is goddess A of a foreign culture the equivalent of X or Y or Z of our own? Lucian even treats the "Assyrian Hera" as betraying traces of many different goddesses;[36] but this is an abnormal position.

Another regular possibility was juxtaposition, to give Zeus Ammon; in such cases, the god of the language in which the text is written normally comes first,[37] and the added name operates rather like an epithet in a typical cultic double name. This modality was extremely common in Roman Gaul and Germany,[38] the Greco-Roman Near East and Egypt, and rather less so in Africa, Thrace, and Anatolia; the earliest instance seems to be Herodotus's description of the god of Babylon as Zeus Belos. (He here diverges from his usual practice of simple substitution of one name for the other.[39]) When two divine names are juxtaposed in this way, it is not self-evident how the connection between the two was understood, and the possible relations were in fact various. The second name might restrict or define the scope of the first, as in cases internal to Greek such as Artemis Eileithyia. Or one can allow in theory (though it is hard to prove empirically) that the two names might combine as a new entity with the powers of both.[40] Such may, for instance, have been the intention of Antiochus I of Commagene's ambitious though unsuccessful attempt to synthesize Greek cult with Persian in the first century B.C., through composite figures such as Zeus Oromasdes, Apollo Mithras, and so on.[41] But the possibility that is relevant here is simple identification or equivalence, as in Herodotus's "translations" and in a variant that occurs in dedications from Egypt where Egyptian and Greek theonyms are linked by the "also known as" (ὁ καί) formula

36. See p. 47 below.

37. But for exceptions in Roman provincial inscriptions, see Richter, *De interpretatione,* 25.

38. Cf. Richter, *De interpretatione,* 31, on proportions of different types of *interpretatio.* But the status of many supposed Celtic divine names as true theonyms or as epithets (and, if the latter, as originally Celtic or as translations of Greek or Roman epithets) is problematic: see, e.g., P. Y. Lambert, "Le statut du théonyme gaulois," in A. Hofeneder and P. de Bernardo Stempel, eds., *Théonymie celtique, cultes, interpretatio* (Vienna, 2013), 113–24; cf. Lambert, in *Keltische Götternamen,* 69–76; the translation theory is advocated by P. de Bernardo Stempel in several studies (see, e.g., the bibliography to her "Method in the Analysis of Romano-Celtic Theonymic Materials," in *Celtic Religion,* 19–27, or *Keltische Götternamen,* 35–36). On the problem of discovering theonyms in the non-Indo-European "Iberic" language, see J. Velaza, "Salaeco: Un teónimo ibérico," *ZPE* 194 (2015): 101–7.

39. Hdt. 1.181.2; 3.158.2 (but the statue is of plain Zeus, 1.183.1).

40. Bonnet, *Enfants de Cadmos,* 507–8, says of *ID* 2132 (a dedication to "Isis Soteira Astarte Aphrodite Euploia Epekoos" and a similar male composite): "This ad hoc assemblage embodies a cumulative and interactive logic rather than one of fusion."

41. Cf. C. Crowther and M. Facella, "New Evidence for the Ruler Cult of Antiochus of Commagene," *AMS* 49 (2003): 41–80; for its failure, 65–68.

familiar from humans bearing two names;[42] in the Latin West the two names can be joined by an "or" (*Vertumnus sive Pisuntus*),[43] while at Dura-Europos on the Euphrates we meet "the goddess Artemis called Azzanathkona."

The commonest form of all is the combination of a divine name with a geographic epithet: the epithet can be local, as with the Thebaian Zeus mentioned earlier and innumerable further instances, or less frequently national, as with "Persian Artemis." Or the Greek theonym might be combined with a functional epithet, whether one imported from old Greece or newly coined. Some combinations of a divine name and an ordinary descriptive adjective appear first or exclusively outside the homeland of the god in question and seem to represent *interpretationes* of an indigenous god: Zeus Megistos, "Greatest Zeus," for instance, first appears at Iasos in Caria, where he apparently represents trqδ, the Carian descendant of the old Anatolian storm god Tarhunt. On Delos, the Syrian goddess Atargatis becomes consistently for her Athenian worshippers Aphrodite Hagne, "Reverend Aphrodite"; on the same island, a single instance of Herakles Halios, "of the Sea," may be a Greek rendering of Melqart.[44]

One might suppose a priori that the implications of these different modalities were different. Simple substitution of the Greek name for the indigenous equivalent effaces the latter completely; an expression such as Persian Artemis, by contrast, signals explicitly that this is the goddess in a non-Greek form; combination with a local toponym falls somewhere between these two limits. But it would be hard to prove empirically that the more thorough the assimilation of a god in name to a Greco-Roman standard, the more thorough also was the assimilation in cult practice. The cult of the god styled simply "Saturn" in Roman Africa remained distinctively African.

42. *OGIS* 111.3–6, 130.7–11. The order is [Greek deity] also known as [Egyptian deity] in *OGIS* 111, the reverse in *OGIS* 130, but in both, strangely, Ammon has come to count as Greek (as also in double names such as Ἀμμωνία ἡ καὶ Σεμνῖνις; Nock, *Essays,* 429 n. 95). Note too "the festival of Prometheus, whom they [the Egyptians] call Iphtimis," *PHib.* 27.85–86. The simple Zeus Ammon type of juxtaposition later becomes quite common both within Egypt (see, e.g., Mueller, *Megas Theos,* nos. 215, 219, 230; but cf. 228 with καί) and for Egyptian gods outside Egypt (see the index to *RICIS* s.vv. Harpokrates Apollo, Isis Aphrodite, etc.). Conflated names (Hermanoubis) occasionally occur. On the "also known as" formula and its variants, see R. Calderini, *Aegyptus* 21 (1941): 221–60.

43. P. de Bernardo Stempel and M. Hainzmann, "*Sive* in the Theonymic Formulae as a Means for Introducing Explications and Identifications," in *Celtic Religion,* 28–39 (who show, however, that this is only one, and not the commonest, use of the *sive* formula). Dura: *Dura* V, 142, no. 453.

44. Zeus Megistos/Tarhunt: cf. p. 107. Aphrodite Hagne: cf. pp. 163–64. Herakles: G. Siebert, *BCH* 112 (1988): 765, publishing a thymiaterion found in the "Maison des sceaux" on Delos, inscribed with a dedication to Zeus Pasios, Poseidon Asphaleios, Apollo Prostates, Herakles ἅλιος; cf. Bonnet, *Melqart,* 482, index s.v. "navigation, dieu de la"; Bonnet, "Héraclès en orient," 177 n. 44.

AVOIDANCE OF *INTERPRETATIO*

Greek and Latin authors typically use the divine names of their own language even when speaking of foreign gods;[45] in particular, as noted earlier, Greeks writing of Rome and vice versa constantly apply the standard Greco-Latin equations such as Artemis-Diana. Less often authors allow the foreign name to enter their text in a bilingual version such as "Enyo, whom they call Ma" (Strabo 12.2.3, C 535) or "Zeus whom the Babylonians call Belos" (Diod. Sic. 2.8.7). But though *interpretatio* was pervasive, even in literary texts it was not universal. Occasionally a foreign name appears on its own, though often within a phrase that acknowledges its unfamiliarity: a "temple of Anaitis and her altar-sharers Omanos and Anadates, Persian *daimones*" (Strabo 11.8.4, C 512); "a temple most famous among those tribes which they called Tanfana's."[46]

As for dedications in Greek and Latin, they reveal to us numerous uninterpreted foreign gods, though the frequency varies greatly from region to region.[47] Even in the Lycian trilingual from Xanthos quoted above, the Lycian (?) god Arββazuma is not interpreted but enters the Greek text in a puzzlingly imprecise transliteration, as Arkesimas. Many gods appear both partially interpreted and uninterpreted in dedications from the same region and even the same site, as, for instance, Mars Belatucadrus alongside plain Belatucadrus; in Britain the single indigenous name is much more common than the Romano-British doublet.[48] The god Elqônerâ, "god who possesses the earth," is interpreted as Poseidon in one bilingual Palmyrene dedication but elsewhere transcribed as Κόνναρος or Connarus. A single dedication may be addressed both to interpreted and to uninterpreted gods. Three merchants from Iamneia in Judaea made a dedication on Delos to "Herakles and Hauronas, the gods who hold Iamneia."[49] Evidently an established

45. J. H. Mordtmann, *Zeitschrift der deutschen morgenländischen Gesellschaft* 32 (1878): 552, long ago spoke of barbarian divine names being avoided with "ängstliche Scheu."

46. Tac. *Ann.* 1.51.2; cf. 4.73.7: *lucus quem Baduhennae vocant*. A rare epigraphic example of such an apologetic formula is *Dura* V, 112, no. 416, the god called (λεγομένος) Aphlad of the village Anath. Tacitus also names the primeval German god Tuisto and his son Mannus: *Germ.* 2.3; cf. *Hist.* 2.78, Carmelus: *ita vocant montem deumque*. Some other untranslated gods: Polyb. 10.27.12, temple said to be of Aine (i.e., Anaitis) in Media; Strabo 11.14.14, C 531, temple of Baris on road to Ecbatana—accepted by Radt, though unique; 16.1.4, C 738, temple of Anaia (?) near Arbela. Dio Cass. 36.48.1 speaks of a territory Anaitis as "consecrated to a certain god of the same name." From over a hundred Celtic theonyms known from inscriptions, ten appear in literary texts: A. Hofeneder, in *Keltische Götternamen*, 123–54.

47. They are rare in Africa (and for different reasons in the Balkans) but numerous in the Iberian Peninsula: J. B. Rives, *Religion in the Roman Empire* (Oxford, 2007), 71, 76, 80–81. For Anatolia, see appendix D.

48. See Zoll, "Double-Named Deities."

49. *ID* 2308: Ἡρακλῆ καὶ Αὐρώνᾳ, θεοῖς Ἰάμνειαν κατέχουσιν, Ζηνόδωρος, Πάτρων, Διόδοτος Ἰαμνῖται. Elqônerâ: see *IGLS* XVII.1.318, with J. B. Yon's note ad loc. (At Baalbek, Konnaros is named on a relief ὁ καὶ Βριαρής, "also called mighty," a new formation evoking the Giant Briareos: J. P. Rey-Coquais,

interpretatio existed for one of the gods whom they wished to honour but not the other: the issue is not one of preference for one or the other form of naming, rather of the expressive resources that were available at that date to the three merchants (who doubtless will not have seen it as their job to attempt speculative identifications). At Gerasa in Jordan we meet "Pakeidas and Hera" (*IGerasa* 17). Not long after Melanthios had made his offering, mentioned above, to Amun-Re under the name Thebaian Zeus, another Greek in Egypt dedicated "an image of Isis";[50] the Egyptian deity might, or might not, undergo *interpretatio*. Much later, in a Palmyrene bilingual dedication, Ḥertâ becomes Hera probably by resemblance of sound, Nanaî becomes Artemis in accord with a standard equivalence, but Rašaf remains Rasaphes.[51] In Greece itself, an extraordinary new text that reveals an imported cult in second-century B.C. Thessaly gives us four hitherto unknown Semitic gods (Mogga, Lilla/aias, Alaia, Adar[a]), one explicit product of *interpretatio* ("Pan whom the Syrians call [. . .] PLEN"), and one figure with a Greek name, Moira, whose role is so unusual that she too may be a Semitic figure in Greek dress.[52]

THE BASIS FOR IDENTIFICATIONS

On what basis were identifications made?[53] Familiarity should not disguise the difficulties that the enterprise faced, the bold disregard for apparent counterevidence that it often imposed. Lucan writes: "There [in Libya] stands oracular Juppiter, as they say, but not hurling a thunderbolt or like ours, but Hammon with twisted horns."[54] Hammon may be Juppiter, then, but does not look much like him! One

"Connaros le puissant," *Syria* 55, 1978: 361–70: a hint of *interpretatio* reappears.) Conversely, in the Phoenician part of a bilingual dedication at Arados by a Phoenician to Hermes and Herakles of the gymnasium, Herakles becomes Melqart, but Hermes is transliterated 'RM (*IGLS* VII.4001). A neo-Punic text written in Greek letters (*KAI* 176) combines Kronos with Tinnit face of Baal; Hatran bilinguals from Dura-Europos may either transcribe or interpret (T. Kaizer, in *Religious Identities in the Levant*, 24 nn. 27 and 29).

50. *SEG* XXVII 1115; the contrast is noted by Vlassopoulos, *Greeks and Barbarians*, 160. *SEG* XXVII 1116 is another dedication by a Greek in Egypt of similar date to "Panepi" (?), another uninterpreted figure; cf. *SEG* XX 644 and XLIV 1484 ("c. 400"), dedications to Osiris and Isis by the couple Alexiades and Tabo.

51. *IGLS* XVII.1.177.

52. See Jean-Claude Decourt and A. Tziaphalias, "Un règlement religieux de la région de Larissa: Cultes grecs et 'orientaux,'" *Kernos* 28 (2015): 13–51, with, on Moira, R. Parker and S. Scullion, *Kernos* 29 (2016): 217–18.

53. Cf. Smith, *God in Translation*, 46.

54. Lucan, *Pharsalia* 9.512–14: *stat sortiger illic / Iuppiter, ut memorant, sed non aut fulmina vibrans / aut similis nostro, sed tortis cornibus Hammon.* Ammon was perhaps worshipped as Zeus κεραός, "horned Zeus," in the Syrian Decapolis: Schwabl, "Nachträge," 1461.

has only to juxtapose in one's mind the image of Athena with that of the hippopotamus goddess Thueris with whom she was sometimes identified to appreciate the chasm over which the practitioner of *interpretatio* often had to leap.[55] Herodotus repeatedly makes that leap, evidently following established practice, without ever pausing to offer an explanation of the rationale.[56] But, as far as we can judge, different criteria were applied, here and elsewhere, to different cases and sometimes to the same. As criteria governing different instances of Greco-Egyptian identifications in Herodotus there have been suggested prominence within the pantheon, function, roles in myth, partial theriomorphism, ithyphallicism, family relation to male deities, and similarity of sound: among these, similarity of function would account for the most cases. Tacitus is explicit that the German Alci were not identified with Castor and Pollux on the basis of similarities in appearance but of the "power of the deity": there were no representations of the Alci, but they were honoured as brothers and young men.[57] Caesar writes of the Gauls (*BGall.* 6.17.1–2):

> Among the gods they chiefly worship Mercury. There are very many images of him, they call him the inventor of all skills, they call him guide over roads and routes, they think he has the greatest power in the acquisition of wealth and in trade. After him they worship Apollo and Mars and Minerva. About them they have much the same views as other peoples: that Apollo drives away diseases, Minerva conveys the elements of works and crafts, Juppiter holds sway over the heavens, Mars controls battles.[58]

The criterion is again here one of function: he ascribes to the Gallic Mercury all the familiar spheres of activity of the Roman, and assigns a single power to each of the other four gods. Such was probably the commonest basis for identification, especially where iconography was absent or unusable. The difficulties of iconographic *interpretatio* are brought out (perhaps with malicious intent) in Lucian's account of the image of the Dea Syria at Hierapolis: whereas Zeus, he notes, is unambiguously Zeus, Hera's form is complex: "Taken as a whole she is certainly Hera, but she has touches of Athena and Aphrodite and Selene and Rhea and Artemis and Nemesis

55. Cf. Wissowa, "Interpretatio Romana," 24–25, on the aspect of Gallic gods. Thueris/Athena: *POxy.* 579 (by a partial supplement). On Thueris, see Bernand, *Fayoum,* vol. 1: note on no. 2.

56. See K.S. Kolta, "Die Gleichsetzung ägyptischer und griechischer Götter bei Herodot" (PhD diss., Tübingen, 1968); and now Lieven, "Translating Gods"; both these careful studies show how the cases range from obvious to difficult, and to us obscure, identifications. On Perseus/Min, see A.B. Lloyd, *JHS* 89 (1969): 82.

57. *Germ.* 43.4, quoted in n. 1 above.

58. *Deorum maxime Mercurium colunt. Huius sunt simulacra plurima, hunc omnium inventorem artium ferunt, hunc viarum atque itinerum ducem, hunc ad quaestus pecuniae mercaturasque habere vim maximam arbitrantur. Post hunc Apollinem et Martem et Iovem et Minervam. De his eandem fere quam reliquae gentes habent opinionem: Apollinem morbos depellere, Minervam operum atque artificiorum initia tradere, Iovem imperium caeleste tenere, Martem bella regere.*

and the Moirai"; he goes on to specify those iconographic touches (*Syr. D.* 32). Lucian also has a discussion of the bizarre way in which the Celts depict Herakles, "whom they call Ogmios."[59] But even Lucian's play with its paradoxes implies the possibility of iconographic *interpretatio*, and it is hard to doubt that, say, the possession of shared attributes could bring gods together: Arrian says of "the Phasian goddess" "to look at, she is Rhea." It may by contrast have been similarity of sound that at Palmyra caused Ḥertâ to become Hera (as noted above), Arsu to become Ares.[60]

The range of possibilities emerges from Tacitus's comments on Serapis:

> Many infer (*coniectant*) the god to be Aesculapius, since he heals the sick, some Osiris, the most venerable deity of those peoples, very many Juppiter, as controller of everything, most of all Dis pater on the basis of emblems visible upon him or by complicated interpretation (*insignibus, quae in ipse manifesta sunt, aut per ambages*).[61]

Here we have *interpretationes* in terms of a specific power ("since he heals the sick"), of standing within a particular community ("the most venerable deity of those peoples") or general mythology ("controller of everything"), and of iconography ("on the basis of emblems visible upon him") all juxtaposed, with a hint too of allegory ("by complicated interpretation"). Lucian illustrates the difficulty of iconographic *interpretatio*, Tacitus that of conflicting criteria. Functional identification was in itself highly approximate and subjective, since the powers of, say, a Roman and a Gallic god would never overlap precisely, and even the underlying presumption that the Gallic pantheon consisted of functionally distinct powers like the Roman may have been mistaken. A Roman would very seldom have a detailed knowledge of what a Gallic god could do (nor a Gaul of a Roman) any-

59. Lucian 5.1. On the problem of Lucian's Ogmios, see *Neue Pauly* s.v. Ogmios; W. Spickermann, "Lukian und die Götter der Fremden," in J. Rüpke, ed., *Antike Religionsgeschichte in räumlicher Perspektive* (Tübingen, 2007), 147–51, at 149–50 (very sceptical). In *Syr. D.* 35 Lucian similarly discusses the unique Syrian representation of Apollo as dressed and bearded.

60. Arr. *Peripl. M. Eux.* 9.1; but Zosimus 1.32.3 speaks of what is probably the same goddess as "Phasian Artemis." Ares-Arsu: so in the Palmyrene bilinguals, *IGLS* XVII.I.150 and probably 127; Ḥertâ -Hera: *IGLS* XVII.I.177. Cf. G. W. Bowersock, in E. Gabba, ed., *Tria Corda* (Como, 1983), 45, for Ares at Areopolis (Rabbathmoba) south of Amman as deriving from the eponymous deity of the town, Ar (and then entering local mythology as father of Lycurgus [below, nn. 70–71]: Nonnos, *Dion.* 19.149; cf. 188–222).

61. *Hist.* 4.84. Just before (4.83) he has used the proximity of a statue "quam plerique Proserpinam vocent" as explanation for the identification as Juppiter Dis (note the variation in naming). The association with Plouton, on the basis of (sometimes allegorised) attributes, is common: Plut. *De Is. et Os.* 28, 362a; Porphyry, *De imaginibus* fr. 359 Smith (p. 424.59) ap. Euseb. *Praep. evang.* 3.11.28; J. E Stambaugh, *Sarapis under the Early Ptolemies* (Leiden, 1972), 27; on grounds not stated, Archemachus and Herakleides Pontikos ap. Plut. *De Is. et Os.* 27, 361E; "some people" in Diod. Sic. 1.25.2; Artem. 5.93.

way.[62] Though some identifications were easy and universal (though doubtless still simplistic), others were variable and contested.

Given the importance in cult practice of correct naming of gods, one may feel that the problem of achieving accurate *interpretatio* ought to have been a source of great anxiety to the ancients.[63] But in instances of *interpretatio* imposed by an observer there was nothing practical at issue: the Syrian goddess remained the Syrian goddess, Sarapis remained Sarapis, for their worshippers, whatever guesses outsiders might choose to make about their real identity. Similarly, when Plutarch tells how (*Sull.* 9.7) Sulla had a vision of "the goddess whom the Romans have learnt from the Cappadocians to honour, whether she is Semele or Athena or Enyo" (Strabo had more decisively called her Enyo; 12.2.3, C 535), Sulla had experienced the goddess, regardless of her name.[64] Precision was only needed where an *interpretatio* was accepted and employed by a worshipper. But even in these circumstances unease is not visible. The many uncertainties and imprecisions did not bring the whole enterprise into question.

Herodotus translates fifteen Egyptian gods into fifteen different Greek gods and heroes, and the range of standard Greco-Roman equivalences was extensive too (though Greek writers on Roman antiquities were often defeated by minor figures; Janus too was apparently too singular ever to find a partner).[65] But the palette varied from region to region, and in some parts was much more restricted: Mercury, Mars, and Hercules dominate in Gaul and Germany, Zeus and Apollo in Anatolia; it is widely held that in the Near East the name of Zeus lost specificity to the extent that an expression such as Zeus Marnas signifies "supreme god Marnas" rather than any strong association with the great Greek god.[66] The relative standardisation of Greco-Roman equivalences did not apply in other contexts: an indigenous

62. Wissowa, "Interpretatio Romana," 23–27; cf. Touléc, "Images de Silvanus" (below, n. 148), 37 n. 1.

63. As argued by C. Ando, *The Matter of the Gods* (Berkeley, 2008), 43–58.

64. Similar is Plut. *Artaxerxes* 3.2, where Anahita is described as "a warlike goddess, whom one might conjecture to be Athena (ἥν Ἀθηνᾶν ἄν τις εἰκάσειεν)."

65. Ovid, *Fast.*1.89–90: *quem tamen esse deum te dicam, Iane biformis? nam tibi par nullum Graecia numen habet;* cf. Plut. *Num.* 19.10: ὁ γὰρ Ἰανὸς ἐν τοῖς πάνυ παλαιοῖς εἴτε δαίμων εἴτε βασιλεὺς; Herodian 1.16.2, where Janus is described as "the god of Italy"; in Dio Cass. 1 fr. 6.7 (p. 14 Boissevain) he is an ancient hero who entertained Kronos. Minor figures left without Greek equivalents in Plutarch include Terminus, *Quaest. Rom.* 15; Lares, ibid., 51; Rumina, ibid., 57; Moneta, *Rom.* 20d; Bona Dea, *Quaest. Rom.* 20 (τῇ γυναικείᾳ θεῷ, ἣν Ἀγαθὴν καλοῦσιν). Dion. Hal. *Ant. Rom.* 2.50.3 ends a list of gods (given in Greek form): καὶ ἄλλοις θεοῖς, ὧν χαλεπὸν ἐξειπεῖν Ἑλλάδι γλώττῃ τὰ ὀνόματα. The Greek version of the *Res Gestae* of Augustus, however, finds a rendering for everything, even such untranslatables as Ianus Quirinus (13), Lares and Penates (19 and app. 2; the rendering of Penates varies). The normal practice with deified abstractions was simply to translate, though M. Aurelius made such a dedication which baffled Dio Cass. (72.34.3), who speaks of a temple to benefaction (εὐεργεσία) called by a very strange name (Wissowa suggested the Latin was *indulgentia*).

66. Cf. Parker, "Zeus Plus," 330.

deity could become a different Greco-Roman god in different places; the same Greco-Roman god could cover many different indigenous figures.[67] Some Greco-Roman divine names were used in some regions but not in others as equivalents of indigenous gods: Asclepius, it is claimed, is always himself, an imported Greek god, when he is worshipped in Anatolia,[68] but in the Phoenician world is a cover for the indigenous healer Eshmun.

Lesser Greek gods and heroes make rare and surprising appearances. Herakles' charioteer Iolaus accompanies him in the list of Carthaginian oath gods translated into Greek preserved in Polybius: he must stand for a Punic deity associated with Melqart, Ṣid, or Eshmun.[69] Leukothea is quite widely honoured, in company occasionally with her son Melikertes, roughly in the area of modern Lebanon: the explanation lies, it is plausibly argued, partly in perceived similarities between Ino/Leukothea's leap in the sea with her son and episodes in the myths of local deities such as Atargatis/Derketo, partly in the eagerness of Tyre and Sidon to assert their genealogical connection with Thebes via Ino/Leukothea's father Kadmos.[70] Most surprising of all at first sight are several dedications from the Roman province of Arabia to a Lycurgus. But Lycurgus the enemy of Dionysus had already

67. For many examples of both possibilities, see Cadotte, *Romanisation des dieux,* 385; Richter, *De interpretatione,* 41–42, 51. For Cocidius's different partners, see Zoll, "Double-Named Deities," 37; W. Spickermann, *Germanische Superior* (Tübingen, 2003), 8, speaks of "the dynamic and undirected process of *interpretatio*." For the multiple Egyptian interpretations of Herakles, see Fraser, *Ptolemaic Alexandria,* 1:208, of Hera: Bernand, *Inscriptions d' Alexandrie,* note on no. 32; for Aphrodite covering various Semitic deities, see Aliquot, *Vie religieuse au Liban,* 150. K. Rigsby, "A Greek Dedication at Sidon," *Tyche* 22 (2007): 143–50, argues that Eshmun at Sidon was first interpreted as Apollo, only later as Asclepius. Smith, *God in Translation,* 254, points out that Philo of Byblos in different places makes Chousor both Hephaestus and Zeus Meilichios. F. Solmsen, *Isis among the Greeks and Romans* (Cambridge, Mass., 1979), 20: "If we try to visualize the environment where Greeks and Egyptians mingled, we may easily understand how some people considered Io and Isis as one and the same figure, while at other times the same men and women, or some other group, looked at Isis as Demeter married to Dionysus-Osiris." There were anomalies even in Greco-Roman equivalences (Wissowa,"Interpretatio Romana," 8): writers and sculptors might find different Greek equivalents for the same Roman figures; Greek writers seeking a name for Consus called him Poseidon Seisichthon, "Earthshaker" (Dion. Hal. *Ant. Rom.* 2.31.2–3); Consus, however, was not a subaspect of Poseidon's normal Greek counterpart Neptune.

68. L. Robert, in Gagniers, *Laodicée,* 291–92. An exotic product of *interpretatio* by contrast is Ἀσκλήπιος λεοντοῦχος of Askalon; Marinus, *Vita Procli* 19.

69. See Bonnet, *Melqart,* 179–80; on Ṣid, see Bonnet, in *Religio Phoenicia,* 217–22; and A. Roobaert, ibid., 336–38, on Ṣid in Sardinia (an island not coincidentally associated with Iolaos; Diod. Sic. 4.29.1, 4–6: this connection makes me favour Ṣid over Eshmun for the oath).

70. See Aliquot, *Vie religieuse au Liban,* 164–71, with full details of the cult attestations. It is a delicate question whether the dedication to Mater Matuta at Deir el-Qalaa by a Roman lady honours the traditional patroness of Roman matrons, or rather Leukothea with whom Mater Matuta was sometimes identified, or a combination: T. Kaizer, *Syria* 82 (2005): 199–206; Aliquot, 167. Bonnet, *Enfants de Cadmos,* 350, thinks the cult may in different places represent an importation and an *interpretatio.*

been made a king of Arabia, no longer Thrace, by Antimachus of Colophon, and cities such as Damascus took up the theme; it might have resonated particularly during the war in the 80s B.C. between Antiochus XII "called Dionysus" and the "king of the Arabs." Nonnos in a late telling of the myth has Lycurgus eventually honoured among the Arabs with libations of blood;[71] and it is an intriguing suggestion (certainly, no alternative theory is available) that he is none other than a tribal god of the region, Shai ʿal Qaum, of whom it is said that he "doesn't drink wine."[72] The myth came, as it seems, to reflect the cultural difference between wine-drinking cities and abstemious nomads, and the name from the myth even entered cult. Here then myth underlies the identification, and the gap between learned and "on the ground" *interpretatio* vanishes.

What might seem an obvious difficulty for *interpretatio* is the embedding of an individual god within a network of family relationships and myth. One can easily identify god X with god Y, but ought this not to mean that X's relatives ought to be identified also with Y's, and their myths combined? That would have been an impossibly complex task, and the normal solution seems to have been simply to ignore the problem. When Herodotus tries to tell an Egyptian myth while substituting Greek theonyms (2.156.4–6), the genealogical consequences are in Greek terms, as he admits, unique.[73] Partial accommodations may sometimes have been sought. Pausanias tells of a conversation with a Sidonian who agreed that the father of Asclepius (long identified with Eshmun) was Apollo but denied that his mother was a mortal woman; Pausanias then appears to claim that certain Greek traditions also allow this possibility.[74] But it is rare that we can see such intercultural discussion in action.

Diverging *interpretationes* of the same god also raise an issue. Hermes according to many authors was Thoth, but there is also a composite figure Hermes Anoubis or Anoubis Hermes or even Hermanoubis. The rationale is easy to see: Hermes as god

71. Dedications: *IGLS* XV.2.311, with note. Antimachus: ap. Diod. Sic. 3.65.7–8. Antiochus's war: Joseph. *BJ* 1.99–102; *AJ* 13.387–91. Nonnos: *Dion.* 21.158–61. Damascus: Dam. Isid. ap. Phot. *Bibl.* cod. 242.200 (fr. 136b in P. Athanassiadi, *Damaskios: The Philosophical History,* Athens 1999); Steph. Byz. s.v. Damaskos; cf. Aliqout, *Vie religieuse au Liban,* 191–94.

72. See M. Gawlikowski, in *ANRW* 18.4.2669; E.A. Knauf, "Dushara and Shai ʿal-Qaum," *Aram* 2 (1990): 176–83; Healey, *Religion of the Nabataeans,* 146; J. Retsö, *The Arabs in Antiquity* (London, 2003), 610–14; Aliquot, *Vie religieuse au Liban,* 187–94; doubts in Sourdel, *Cultes du Hauran,* 83–84. Arab teetotalism is mentioned (too generally) by Diod. Sic. 19.94. "Who doesn't drink wine": Healey, *Aramaic Inscriptions,* no. 42 (for the god, cf. 16).

73. Similar bizarreries occur occasionally in Greco-Egyptian texts, such as Asklepios/Imouthes son of Hephaestus and Hermes son of Horos in *POxy.* XI 1381 (Totti, *Texte der Isis- und Sarapis-Religion,* no. 15) 228–31 (for the former, cf. *PPar.* 19.6); on the issue, cf. n. 129 below. Bonnet, "Héraclès en orient," 177, stresses that the genealogies of Greek Herakles and Herakles of Tyre were always kept separate.

74. Paus. 7.23.7–8; for claims about the parentage of a "Phoenician Asclepius," see Philo of Byblos, *FGrH* 790 F 2 (ap. Euseb. *Praep. evang.* 1.10.25 and 38); Dam. Isid. ap. Phot. *Bibl.* cod. 242.212 and 302 (fr. 142 B Athanassiadi).

of writing and communication is Thoth, as "guider of souls" Anoubis. The second name in a combination such as "Isis Aphrodite" may work in the same way, picking out one aspect within Isis's multiple personality while allowing that there are other aspects too. Lucian's division of the Syrian goddess among eight goddesses or clusters of goddesses exaggerates the same issue on the level of iconography.[75] The underlying logic, if spelt out, would no longer be that the gods of two cultures have coextensive powers, but rather that different cultures divide the powers among gods in different ways, so that a god of culture A may overlap with several gods of culture B, or may need to be broken down into several gods of culture B, and vice versa. But the underlying logic never was spelt out. As was noted above, the simple juxtaposition of two theonyms did not commit the speaker to any defined view of the relation between the two powers. Implicit models of the relation that saw it as one to one and as one to many coexisted without conflict, the former still dominant.

RATIONALE

I come now to the three central questions about *interpretatio: why, who,* and *with what effect*. I begin with *why*. It is extraordinary how little discussion of the rationale for *interpretatio* is found. Explanations are, as we have seen, occasionally found at the narrower level of why god X should be identified with god Y (shared powers, shared appearance, or whatever), as also surveys of a range of possible identifications.[76] But the question of whether and how the gods of different peoples relate to one another at all is not raised in these contexts. It can be suggested that the names of particular gods in particular places have changed over time or been corrupted;[77] but the suggestion is ad hoc, not related to a general theory. The academic Cotta in

75. The element Osiris in Phoenician theonyms can become either Dionysus or Sarapis in Greek equivalents: Bonnet, *Enfants de Cadmos*, 290.

76. In addition to Plut. *Sull.* 9.7 and Tac. *Hist.* 4.84, cited above, see, e.g., Plut. *Num.* 12.2: ἐξαιρέτως δὲ τὴν προσαγορευομένην Λιβίτιναν, ἐπίσκοπον τῶν περὶ τοὺς θνήσκοντας ὁσίων θεὸν οὖσαν, εἴτε Περσεφόνην εἴτε μᾶλλον, ὡς οἱ λογιώτατοι Ῥωμαίων ὑπολαμβάνουσιν, Ἀφροδίτην; *Crass.* 17.9–10: τὰ χρήματα τῆς ἐν Ἱεραπόλει θεοῦ . . . ἣν οἱ μὲν Ἀφροδίτην, οἱ δ' Ἥραν, οἱ δὲ τὴν ἀρχὰς καὶ σπέρματα πᾶσιν ἐξ ὑγρῶν παρασχοῦσαν αἰτίαν καὶ φύσιν νομίζουσι καὶ τὴν πάντων εἰς ἀνθρώπους ἀρχὴν ἀγαθῶν καταδείξασαν; Dion. Hal. *Ant. Rom.* 2.31.2–3 (different views of Consus); Diod. Sic. 1.25.1–2: Καθόλου δὲ πολλή τίς ἐστι διαφωνία περὶ τούτων τῶν θεῶν. τὴν αὐτὴν γὰρ οἱ μὲν Ἶσιν, οἱ δὲ Δήμητραν, οἱ δὲ Θεσμοφόρον, οἱ δὲ Σελήνην, οἱ δὲ Ἥραν, οἱ δὲ πάσαις ταῖς προσηγορίαις ὀνομάζουσι. τὸν δὲ Ὄσιριν οἱ μὲν Σάραπιν, οἱ δὲ Διόνυσον, οἱ δὲ Πλούτωνα, οἱ δὲ Ἄμμωνα, τινὲς δὲ Δία, πολλοὶ δὲ Πᾶνα τὸν αὐτὸν νενομίκασι. λέγουσι δέ τινες Σάραπιν εἶναι τὸν παρὰ τοῖς Ἕλλησι Πλούτωνα ὀνομαζόμενον.

77. Steph. Byz. s.v. Γάζα: ἔνθεν καὶ τὸ τοῦ Κρηταίου Διὸς παρ' αὐτοῖς εἶναι, ὃν καὶ καθ' ἡμᾶς ἐκάλουν Μαρνᾶν; Hyginus, *Fabulae* 197, *Venerem, quae postea dea Syria appellata est;* Agathias, *Hist.* 2.24, claiming that Persians before Zoroaster worshipped the familiar Greek gods except that their names were not properly preserved; they called Zeus "Belos," and so on. Agathias cites Berossos (*FGrH* 680 F 12) and others for the identifications; but the point about getting the names wrong may be Agathias's own.

Cicero's *On the Nature of the Gods* notes that the names of gods, unlike those of humans, vary from place to place,[78] but there does not follow a discussion of the proposition that the same god underlies the different names; had such a debate existed specifically in relation to *interpretatio,* it would certainly have left echoes in our sources. The great mathematician Eudoxus of Cnidus claimed to be puzzled as to why Isis, but not her supposed equivalent Demeter, ruled over love, and why Dionysus, unlike Osiris, did not make the Nile rise or rule the dead. He pointed therefore to the lack of fit between the powers of deities who were commonly assimilated.[79] This was potentially a sharp challenge to the convention of *interpretatio,* but not one that we see taken up. More often, where a fit was imperfect, the issue was fudged by saying that it was unclear whether god A was really X, Y, or Z.

Habit and familiarity were undoubtedly powerful preservatives of the convention of *interpretatio;* every Greek and Roman was born into a world in which the translatability of gods' names was taken for granted. But the convention could not have survived, and found constant new applications, if it was based on implicit assumptions that in a Greco-Roman context were absurd. Since the ancients do not explain their practice, we must attempt an explanation for them. Two rationales seem to be possible. According to one, the gods of different peoples would be distinct, but functionally similar; in the words of the most eloquent advocate of this understanding, Elias Bickerman,

> The statement of Herodotus that Pan is called Mendes in Egyptian is on a par with such statements by modern writers as that the Japanese bonze signifies the clergy, or that the Hebrew mishpaha denotes a clan, or that Arabian jinn are totems. What is apparently an identification is rather an explanation or interpretation of a foreign phenomenon.[80]

We can call this the similarity or equivalence model. The other rationale would be, very simply, that gods are the same throughout the world, but names for them differ: the identity model.

A first approach to choosing between the two might be through the language with which *interpretationes* are introduced. In literary texts, the commonest method, as we have noted, is simply to use the divine name belonging to the writer's own language without explanation. Stylistic purists, it is argued, classed foreign theonyms along with foreign words as things to be shunned by the fastidious writer;[81] a conception of the divine must, however, underlie that stylistic choice.

78. *Nat. d.* 1.84, *quot hominum linguae tot nomina deorum;* cf. Pliny, *HN* 2.15, *nomina alia aliis gentibus et numina in iisdem innumerabilia invenimus.* I am not clear what the younger Pliny means when, in *Ep.* 8.24.3, he advises the administrator of Achaea "reverere conditores deos et nomina deorum."

79. Eudoxus fr. 298 Lasserre ap. Plut. *De Is. et Os.* 64, 377A; cf. Diodorus's recognition of only approximate fit in 1.13.5, mentioned below.

80. Bikerman, "Anonymous Gods," 188.

81. Wissowa, "Interpretatio Romana," 4–5, 11.

Simple substitution of the Greek name for the foreign is Herodotus's practice in many passages; but sometimes he adds a note on the lines of "In Egyptian Apollo is Horos, Demeter Isis, Artemis Boubastis" (2.156.5), or retains a native name such as Osiris and offers a translation into Greek, or an explanation: "Osiris is Dionysus in Greek,"[82] "the Assyrians call Aphrodite Mylitta" (1.199.3); once he gives the supposed Assyrian, Arab, and Persian names for Aphrodite (1.131.3). Similar "bilinguals" occur from time to time in later historians; Caesar was unusual in succeeding in keeping his text completely pure of foreign names.

A rare variant is Herodian's reference to a "local god [at Aquileia], whom they call Belen and reverence greatly and claim to be Apollo," where the identification is ascribed but not endorsed; in the subsequent narrative, however, Herodian speaks of this god simply as Apollo, and Roman soldiers equate him with Apollo in the account of the same incident in the *Historia Augusta*.[83] Unusual too is Strabo's statement that Cleon, a bandit by origin, among other undeserved successes became "priest of Zeus Abrettenos, a Mysian god" (12.8.9, C 574). This seems to deny the universality of Zeus Abrettenos and reduce him to a Mysian level: perhaps Strabo's distaste for Cleon has spilt over onto the god. These two passages are outliers and contrast with the norm, which is simple substitution of one name for the other, or mention of both. That norm appears to take the interchangeability of the names in question for granted; it seems therefore to favour the identity model.

But other passages recognise that a process of translation (ἑρμηνεύω and its compounds; μεταφράζω) is required which might be problematic; such language is found particularly in Greek writers, Dionysius of Halicarnassus and Plutarch, who discuss lesser figures of the Roman divine world for whom agreed *interpretationes* were lacking. Dionysius writes:

> The Romans call these gods Penates. Among those who translate the name into Greek some call them Patrooi ["Ancestral"], some Genethlioi ["of Descent"], some Ktesioi ["of Property"], others Muchioi ["of Recesses"], others Herkeioi ["of the Courtyard"]. Each seems to be choosing the title in accord with one of their characteristics, and they all seem to mean roughly the same.[84]

82. 2.144.2; cf. 2.42.5, 2.46.3, 2.59.2, and especially the six Greek-Scythian translations in 4.59.2 (where note his approval for Papaios as a rendering of Zeus).

83. Herodian 8.3.8: τοῦ ἐπιχωρίου θεοῦ . . . Βέλεν δὲ καλοῦσι τοῦτον, σέβουσί τε ὑπερφυῶς, Ἀπόλλωνα εἶναι ἐθέλοντες; SHA *Max.* 22.1–2: adverse oracles of *deus Belenus* are reported, *unde etiam postea Maximiniani milites iactasse dicuntur Apollinem contra se pugnasse [debere]*. Dedications are made both to Belenus and to Apollo Belenus, e.g., *ILS* 4866–74.

84. Dion. Hal. *Ant. Rom.* 1.67.3: τοὺς δὲ θεοὺς τούτους Ῥωμαῖοι μὲν Πενάτας καλοῦσιν· οἱ δ' ἐξερμηνεύοντες εἰς τὴν Ἑλλάδα γλῶσσαν τοὔνομα οἱ μὲν Πατρῴους ἀποφαίνουσιν, οἱ δὲ Γενεθλίους, εἰσὶ δ' οἳ Κτησίους, ἄλλοι δὲ Μυχίους, οἱ δὲ Ἑρκείους. ἔοικε δὲ τούτων ἕκαστος κατά τινος τῶν συμβεβηκότων αὐτοῖς ποιεῖσθαι τὴν ἐπίκλησιν κινδυνεύουσί τε πάντες ἀμωσγέπως τὸ αὐτὸ λέγειν.

The uncertainty here is not quite the same as to whether, say, Cappadocian Ma is Semele or Athena or Enyo. There the assumption is that Enyo is one of the three, though we cannot be sure which; in the Dionysius passage the suggestion is that all the suggested equivalences are partial and imperfect. Elsewhere Dionysius speaks of "a particularly venerable shrine jointly honoured by the Sabines and the Latins of a goddess called Pheroneia; some translators into Greek call her 'Flower-Bringer,' some 'Garland-Lover,' some 'Persephone.'"[85] The first two suggestions are epithets, the third a familiar divine name: the challenge for the translator seems to be not to identify the known Greek power of whom Pheroneia is the Roman form, but to find a way of suggesting in Greek, if necessary by invention, the nature of a distinctive Italian deity.[86] A casual phrase in Diodorus points up the problem of exact "translation": Osiris translates as Dionysus, he writes (1.13.5), whereas the "nearest equivalent" (ἔγγιστά πως) of Isis is Demeter. Here then we find support for the "equivalence without identity" model.

The approach via the language used in the context of *interpretatio* proves inconclusive. A broader approach involves a larger, indeed almost the largest possible, theological question: in ancient thought, did the same gods rule the whole world, or did different nations have different if comparable gods?

Explicit statements that the gods of different peoples are separate are late to emerge in theoretical or reflective contexts; they depend on the Stoic-influenced conception of a supreme and universal god under whom the others serve as provincial governors. The point could then be brutally simplified in Christian polemic: "Each province and city has its own god."[87] But in practical religion, and not the comic fantasy of Aristophanes in *Birds* alone, usages implying that view abound at

85. 3.32.1: ἱερόν ἐστι κοινῇ τιμώμενον ὑπὸ Σαβίνων τε καὶ Λατίνων ἅγιον ἐν τοῖς πάνυ θεᾶς Φερωνείας ὀνομαζομένης, ἣν οἱ μεταφράζοντες εἰς τὴν Ἑλλάδα γλῶσσαν οἱ μὲν Ἀνθοφόρον, οἱ δὲ Φιλοστέφανον, οἱ δὲ Φερσεφόνην καλοῦσιν· Translation language also, e.g., in 2.31.2–3 (with a contested identification), 2.70.4 (where Dionysius is clear that the identification of Salioi with Kouretes is correct); Diod. Sic. 1.25.7.

86. One might compare the combination of transliteration and translation to which Philo of Byblos has recourse in translating old Phoenician theonyms into Greek: cf. A. I. Baumgarten, *The Phoenician History of Philo of Byblos* (Leiden, 1981), 144–46, 152, and passim.

87. "Each province and city has its own god: Syria has Astarte, Arabia Dusares, Noricum Belenus, Africa Caelestis, Mauretania its own princes": Tert. *Apol.* 24.7; and other texts cited by MacMullen, *Paganism in the Roman Empire*, 185 n. 31; in Min. Fel. *Oct.* 6.1 the concept of *dei municipes* is put in the mouth of the pagan spokesman by the Christian author. On the pagan side, see Artem. 1.8; Marcus Aurelius to Fronto (p. 47 Naber; p. 50 in the Loeb Fronto, vol. 1): *omnes omnium populorum praesides deos;* Celsus ap. Origen, *C. Cels.* 5.25–26 (cf. A. Fürst. in *One God*, 91–92); Iambl. *Myst.* 5.25; Julian, *Against the Galilaeans* 143A (καθ' ἕκαστον ἔθνος ἐθνάρχης τις θεὸς ἐπιτροπεύων); Symmachus, *Relat.* 3.8: *ut animae nascentibus, ita populis fatales genii dividuntur;* and the astrological variant in Manilius 4.710 (*alias aliud terras sibi vindicat astrum*). Assmann, *Moses the Egyptian*, 45, contrasts "ethnocentric" tribal religions with the universalising ancient polytheisms; but the two things coexist.

all dates. Different parties to treaties swear by their own gods: such is the rule in the treaties of the ancient Near East (despite the many instances of translation in other contexts noted above), and such continued to be the practice in the Greco-Roman world; one swore the "local oath" (ἐπιχώριος ὅρκος),[88] or by a mixture of gods from both sides (as in Hannibal's oath to Philip V quoted above). The gods of both parties are as a collective functionally equivalent, since both serve as adequate sanctions against violation, but remain distinct. There are "gods of the Greeks" and of other nations,[89] and "foreign gods," expressions which do not readily invite reinterpretation as meaning "universal gods worshipped according to the particular cultural understandings of the people concerned"; such was certainly not the understanding of the Kaunians, of whom Herodotus reports (admittedly he finds their behaviour eccentric): "All the Kaunians of military age donned their armour and followed beating the air with their spears as far as the boundaries of Kalynda, and said that they were expelling the foreign gods" (1.172.2).

Herodotus himself once speaks of Apis as "the god of the Egyptians" (3.64.3). Alexander supposedly swore an oath by "Zeus of the Greeks and Ammon of the Libyans," despite often treating them as identical (Arr. *Indica* 35.8). Gods felt particular love for particular cities, which, however, they would abandon in the event of sack; in Etrusco-Roman usage they could be lured from one to another by the ritual of *evocatio.*[90] Then there are the innumerable further subdivisions among gods introduced by the "god plus toponym" combination, which at the limit imply that not just each people and each city but almost each village has individual gods. And though one can in principle understand expressions of "Zeus plus village name" type as "Zeus as active in a particular village," we find in historians and geographers the concept of the "local" god or power: Herodotus speaks in this way of Kybebe at Sardis and Pleistoros in Thrace (5.102.1; 9.119.1); according to Dionysius, Faunus, Sangkos, and Janus were "local powers" (*epichōrioi daimones*) of the

88. ἐπιχώριος ὅρκος, Thuc. 5.47.8; cf. A. J. Bayliss, in A. H. Sommerstein and Bayliss, *Oath and State in Ancient Greece* (Berlin, 2013), 160–67; Polyb. 3.25.6 (different oaths of Carthaginians and Romans). Dio Chrysostom, urging Nicomedia to be friendly with Nicaea, says: "I pray to your gods and to theirs" (38.9; but in 38.22, cf. 46, they worship the same gods). Nero tells the Greeks that he is grateful for the favour "your gods" have shown him (*IG* VII.2713.21–22; cf. 36, "our gods" in the Greek response).

89. Versnel, *Coping with the Gods,* 102–10; even the measure of universalism (i.e., common to all Greece) in that expression is untypical according to Polinskaya, "Shared Sanctuaries." A hymn to Telepinu reminds him that he is honored only in Hatti : E. Laroche, *Catalogue des textes hittites* (Paris, 1971), 377; trans. in I. Singer, *Hittite Prayers* (Leiden, 1002), no. 9.

90. As described by Livy 5.21.1–3 and 22.3–7 (Veii); Furius as quoted by Sammonicus Serenus in Macr. *Sat.* 3.9.6–9 (Carthage); cf. A. Blomart, in *Reichsreligion und Provinzialreligion,* 99–101. But Apollo, stolen from Gela to Tyre, remained hostile to the Tyrians: Diod. Sic. 13.108.2–4, 17.41.7–8, 46.6: the *interpretatio* of Apollo as Reshef apparently did not occur in this case (C. Bonnet, *CRAI,* 2012, 513–14).

Romans;[91] above we met Herodian's description of Belen as a local god; Adranos is for Plutarch "a certain god (*theos tis*) honoured much in all Sicily" (*Tim.* 12.2); Strabo tells (17.2.3, C 822) how the people of Meroe in Egypt worship Herakles, Pan, Isis, and another "barbarian god"; such instances could no doubt be multiplied. The perspective is that of an outsider, who sees a particular god as known or operative within a confined region. "In spite of *interpretatio,* foreign gods were foreign gods, for Cicero and Lucian alike," writes A. D. Nock.[92] Their entry into a state might therefore be resisted, we must add. Again, in spite of *interpretatio,* foreign gods such as Ammon and Isis entered the Greek world under their own names; the argument "We have Zeus already, what need of Ammon?" was not used.

The first explicit statement of the other view comes in the Roman senate's criticism of Fulvius Flaccus's behaviour in 173 as reported in Livy: by stripping an Italian temple to adorn a Roman one he was "exposing the Roman people to religious danger, building temples from the destruction of temples as if there were not the same immortal gods everywhere, but some gods ought to be honoured and adorned with spoils taken from others."[93] As a statement of general principle this remains crucial even if there are problems in reconciling it with actual Roman behaviour in other contexts. Late though this testimony is, it has been strongly argued that this must be the assumption that underlies Herodotus's whole treatment of foreign religion. In his remarks on the naming of gods, Herodotus implies a process whereby different cultures learnt by stages to identify and name gods who had always been present, there and everywhere.[94] These gods are worshipped according to different cultural conventions in different places, but this does not mean that they are different gods: forms of worship fall within the category of customs, a familiar variable of human society that implies nothing about the gods themselves. Since these differences in convention are satisfactory to the gods, they should with some exceptions be respected; acknowledging that the gods are the same everywhere does not entail that forms of worship should be standardised.

91. Dion. Hal. *Ant. Rom.* 1.31.2, 2.49.2, 3.22.7 ("or god"); in 1.57.4 an *epichōrios daimon* appears to Latinus in a dream. The concept of ἐγχώριοι θεοί (Polinskaya, *Local History,* 36–41) is slightly different, since they were not necessarily exclusive to the place concerned.

92. Nock, *Essays,* 558.

93. Livy 42.3.9: *obstringere religione populum Romanum, ruinis templorum templa aedificantem, tamquam non iidem ubique dii immortales sint, sed spoliis aliorum alii colendi exornandique.*

94. J. Rudhardt, "Les attitudes des Grecs à l'égard des religions étrangères," *RHR* 209 (1992): 219–38. Rudhardt stresses that Herodotus's belief is not based on diffusionism. Cf. Assmann, *Moses the Egyptian,* 47, summarising what he takes to be the common conception underlying translatability in the second millennium: "Cultures, and political systems, may be different. But as long as they have a religion and worship some definite and identifiable gods, they are comparable and contactable because these gods must necessarily be the same as those worshipped by other nations but under different names. The names, iconographies and rites—in short, the cultures—differ, but the gods are the same."

The god(s) of the philosophers had almost necessarily been universal from the beginning, certainly since Xenophanes had attacked the blue-eyed, red-haired gods of the Thracians (F 16 Diels-Kranz) as shaped in their own image; if it is hard to find texts in the philosophers that say as much in so many words, that is presumably because for them the point was too obvious to need making. Plutarch is heir to the philosophical tradition when he insists that all the world has always known the power of Isis and the gods associated with her, even if some regions have only recently learnt their names; for gods are not

> different among different peoples nor barbarian and Greek nor southern and northern, but just as sun and moon and heaven and earth and sea are common to all but are differently named by different people, so too, for the one reason which orders all this and the one providence that supervises it and the subordinate powers set over everything, different forms of worship and of address have arisen among different peoples according to their customs.[95]

A dedication probably of the first century A.D. from the Asclepieum at Pergamum is addressed, uniquely, θεοῖς τοῖς πανταχοῦ, "to the gods everywhere."[96] These genuinely universal claims are to be distinguished from myths and theories that treated the gods of one country as borrowings or transformations of those of another: a Greek myth told how the Greek gods fled from Typhoeus to Egypt and there assumed animal disguise, that is, became Egyptian gods; for another myth, Io on reaching Egypt became Isis and her son Epaphus was the sacred bull Apis;[97] it was the conviction of Dionysius of Halicarnassus that the Romans were Greeks by origin and their language a dialect of Greek, their gods too therefore Greek; there were also theories of diffusion from the most ancient culture, Egypt.[98] These

95. *De Is. et Os.* 67, 377D-378A: οὐχ ἑτέρους παρ' ἑτέροις οὐδὲ βαρβάρους καὶ Ἕλληνας οὐδὲ νοτίους καὶ βορείους· ἀλλ' ὥσπερ ἥλιος καὶ σελήνη καὶ οὐρανὸς καὶ γῆ καὶ θάλασσα κοινὰ πᾶσιν, ὀνομάζεται δ' ἄλλως ὑπ' ἄλλων, οὕτως ἑνὸς λόγου τοῦ ταῦτα κοσμοῦντος καὶ μιᾶς προνοίας ἐπιτροπευούσης καὶ δυνάμεων ὑπουργῶν ἐπὶ πάντα τεταγμένων ἕτεραι παρ' ἑτέροις κατὰ νόμους γεγόνασι τιμαὶ καὶ προσηγορίαι; cf. Artem. 1.8, p. 17.5–8 Pack. In late antiquity, arguing for tolerance, Themistios claimed (Or. 5.70a) that "the originator of everything" (ὁ τοῦ παντὸς ἀρχηγέτης) desired different regions to have different social systems (πολιτεύεσθαι); the implied reference is to different forms of worship for the same gods.

96. Habicht, *Asklepieion,* no. 133.

97. Flight from Typhoeus: e.g., Ant. Lib. *Met.* 28, and already apparently Pind. fr. 91 S/M ap. Porph. *Abst.* 3.16; for its aetiological character, Lucian, *De sacrificiis* 14. Io renamed Isis: Apollod. *Bibl.* 2.1.3; and J. G. Frazer's note (Hdt. 2.41.2 merely notes the iconographic resemblance); the equation is pervasive in Latin poetry, and the theme is mocked in Lucian 78.11.2. Epaphus is already identified with Apis in Aesch. *Supp.* 41–48; Hdt. 2.153; Egyptians reject the equation on chronological grounds according to Ael. *NA* 11.10. Cf. Hippolytus's reappearance in Italy, renamed as Virbius (Verg. *Aen.* 7.765–77), or Medea's as Angitia (Serv. ad Virg. *Aen.* 7.750; for related traditions, see Ogden, *Drakon,* 207–8).

98. Most clearly expressed for us in Diod. Sic. book 1: see, e.g., 1.23.8. Dionysius: the theory is pervasive in *Ant. Rom.*; on the gods, see, e.g., 7.72.13–14.

myths and theories are (except perhaps the last) more specific, relating only to connections between particular cultures; but they certainly did not conflict with the impulse to identify gods more generally.

One can ask which of the models better fits cases where *interpretatio* is not attempted, or is qualified in some way, or where doubt exists about an identification. If the gods of different peoples are separate, but comparable (the similarity or equivalence model), these things would occur when no satisfactory fit with the god of a different culture was possible: Vulcan is like enough to Hephaistos to receive his name, but there is no equivalent of Janus.[99] The possibility of introducing a foreign god would obviously pose no difficulty to this model. On the other view, problems of translation would arise because, though the gods are always the same, apprehensions of them differ: some gods have not yet been discovered by some peoples; every people's understanding of them is different and, very probably, in some measure imperfect. A supposedly foreign god might be identical with one of one's own without that identity being obvious to mortal knowledge. The powers of a foreign god might always have been available, but unrecognised, in one's own country;[100] the introduction of a foreign god would represent not in fact its introduction but the introduction of its cult, the activation of a neglected potential. There would indeed be no point in importing a foreign god if that god were truly local, powerful only within its place of origin.

Expressions such as Artemis Persike and Zeus Ammon might seem to imply the similarity model: Anaitis is the Persian equivalent of Artemis, not the same goddess. But she could be understood as "that aspect of Artemis which is emphasized in Persia," or as "Artemis as worshipped according to ancestral Persian tradition." Zeus Ammon similarly might be "that aspect of Zeus which is emphasized under the name Ammon," or more simply, "Zeus also known as Ammon." As for "divine name plus local adjective" and all the other "divine name plus qualifier" combinations, their proliferation in *interpretationes* is exactly comparable to their proliferation in domestic contexts: Zeus Thebaieus (i.e., Amun–Re) at a formal level is no more distinct from all the other forms of Zeus than is say Zeus Hymettios, Zeus as worshipped on Mount Hymettos in Attica. If Zeus Hymettios is both different from Zeus Parnesios, "of Parnes" (atop another Attic mountain) and also the same, at a linguistic level the same ambiguity applies to the relation between Zeus Olympios and Zeus Ammon or Zeus Belos.

99. See for Gaul C. Letta, *Rivista Storica Italiana* 96 (1984): 1005–7.

100. As Arnobius 2.73 argues against the Romans, using examples from their own acceptance of "foreign" gods; for progressive learning of gods' names, see, e.g., Hdt. 2.146.2. Diod. Sic. notes in Arabia "three altars of gods unknown to the Greeks but greatly honoured by the natives" (3.45.2). The willingness of the Hittites to incorporate foreign gods en masse is cited as exceptional, an oddity (Allen, *Splintered Divine*, 71–93, esp. 72), but it is exceptional only in its extremism.

What of rejection of foreign gods and abuse of the religious practices of other peoples? Neither phenomenon is very common, but both, it must be acknowledged, occur. But the objection seems essentially to have been to rites deemed barbarous and forms of worship deemed socially disruptive, or to offensive conceptions about the nature of the gods (the standard objection to the supposed Egyptian worship of animals). The gods so worshipped and conceived in such misguided ways might nonetheless have existed.[101] So the issue with a foreign god who is resisted is no different from that discussed above with one who is accepted. Notoriously, many foreign gods were resisted by some sectors of society just because they were enthusiastically accepted by others.

So the modalities of *interpretatio* provide no decisive criterion, though the identity model perhaps has to work harder to account for them. Authors who introduce *interpretationes* never justify their practice, as we noted. But where Caesar deploys them the implicit model is that of identity. About the Gauls, as we have seen, he says: "Among the gods they chiefly worship Mercury. . . . After him they worship Apollo and Mars and Minerva. About them they have much the same views as other peoples" (*BGall.* 6.17.1–2); and of the Germans: "They count among the gods only those they can see and those by whose resources they are visibly benefited, the sun and Vulcan and the moon; as for the rest, they have not even heard of them."[102] Different peoples may hold different opinions about the gods, though in fact there is considerable convergence, and primitive peoples may not even have heard of the gods whose power is less readily perceptible; but expressions such as "Among gods, they worship Mercury most of all" and "They have not even heard of the other gods" suggest that everywhere the same gods exist, waiting for mortals to gain a sense of them. The view that I detect in Caesar is the same as that ascribed explicitly to Roman senators by Livy, and suggests a consensus of the educated elite at Rome. Very similar expressions are already found in Herodotus. Of the Massagetai he writes that "among the gods, they honour Helios alone," and of the Arabs, "among the gods, they believe only in Dionysus and Ouranie."[103] Bickerman explained formulations such as "Pan is Mendes in Egyptian" by arguing: "What is apparently an identification is rather an

101. Cf. J. P. Davies, *Rome's Religious History* (Cambridge, 2004), 80–81: "The designation of a god as foreign or non-Roman has no bearing on its existence or power. . . . Where Livy differentiates people in religious terms, it is predominantly with respect not to deities but to rites." Offensive conceptions: note, e.g., Augustus's respect for Sarapis but contempt for Apis: Dio Cass. 51.16.4–5.

102. Caesar, *BGall.* 6.21.2: *Deorum numero eos solos ducunt quos cernunt et quorum aperte opibus iuvantur, solem et Vulcanum et lunam, reliquos ne fama quidem acceperunt.*

103. Hdt. 1.216.4, 3.8.3, cited by Rudhardt, *RHR* 209 (1992): 228. "The acceptance of such identifications depended not on any Kulturpolitik but on assumptions which were axiomatic to Greeks and Romans and which were accepted as such by the new peoples who came within the orbit of their intellectual world": Nock, *Essays*, 752. Cf. Max. Tyr. 11.5a: the one subject on which all people agree is that there is one top god and his offspring, who corule with him.

explanation or interpretation of a foreign phenomenon." But that suggestion does not account for the universalising assumptions that underlie these expressions in Herodotus and Caesar. And "apparent identifications" between gods of two peoples occur throughout antiquity:[104] is it right to explain away this form of speech into which ancients slipped so easily as being not what it seems?

The identity model could also be applied by historical agents. According to the Greek sources, when Alexander asked permission to enter Tyre in Phoenicia to sacrifice to Herakles, the presupposition, not disputed by the Tyrians despite their refusal to admit him, was that the Tyrian Herakles (whom they called Melqart) and the ancestor of the Argeads were the same. In the many regions of the archaic Mediterranean where Greeks and Phoenicians rubbed shoulders, the same equation between Herakles and Melqart must have provided a cultural "middle ground," even if at too early a date to leave unambiguous traces in our sources.[105] In Alexander's dealings with Ammon, that god and Zeus are in the main treated as interchangeable (though we have recently learnt that it was as Ammon that he dedicated to him in Egypt[106]). Antiochus I, it has been argued, in the cylinder from Borsippa near Babylon exploits the identification of Apollo with Nabû to give his piety a Janus-faced aspect. Some believe that attention paid by the Romans to gods with Roman names during the Hannibalic war period was designed to propitiate gods friendly to Carthage, thus Tanit under the guise of Juno, Melqart as Herakles, Eshmun as Apollo, Ba'al Hammon as Saturn, Astarte as Venus of Eryx. If so, the identity model was powerfully influential at Rome in that time of crisis; but the *interpretationes* in this case are owed to modern scholarship.[107]

The assumption of identity conflicts drastically, as we have seen, with the many contexts and locutions which treated the gods of different nations as distinct, and also, if less drastically, with language sometimes used in reference to the problem of "translating" the names of lesser deities. Possibly some authors worked with a largely unconscious two-tier model whereby great gods were universal, but there also existed "local powers," who would be resistant to *interpretatio;* one may

104. I could readily list fifty.

105. For a strong cumulative case, see I. Malkin, *A Small Greek World* (Oxford, 2011), chap. 4; he notes, e.g., the "shrines honoured by Greeks" in Phoenician Motya (Diod. Sic. 14.53.2). On Alexander and the Herakles of Tyre, see Bonnet, *Enfants de Cadmos,* 66–80, 94–97, esp. 95 n. 186 (the postvictory games in his honour).

106. *SEG* LIX 1764. At Cyrene he was always Ammon, and probably worshipped in a separate temple from Zeus: see F. A. Mohamed et al., in L. Gasperini and S. M. Marengo, eds., *Cirene e la cirenaica nell'antichità* (Rome, 2007), 20–22; *SEG* LVII 2004.

107. See, above all, R. E. A. Palmer, *Rome and Carthage at Peace* (Stuttgart, 1997), 53–72; for Venus Erycina also, I. Stark, in *Reichsreligion und Provinzialreligion,* 24–33; cf. L. Rawlings, in Rawlings and H. Bowden, *Herakles and Hercules* (Swansea, 2005), 162–63. Antiochus and Nabû: P. Kosmin, "Seeing Double in Seleucid Babylonia," in A. Moreno and R. Thomas, eds., *Patterns of the Past: Epitēdeumata in the Greek Tradition* (Oxford, 2015), 173–98.

wonder how such "local gods," who certainly appear in texts, were understood—real gods of restricted scope, gods mistakenly believed in (but were there false gods in ancient polytheism?) by certain peoples?[108] But these complications and contradictions can be added to the long list of those inherent in ancient practical theology: there are many different Zeuses and there is one Zeus, there are many different gods, and there is "the god" or the divine, the gods are present everywhere but you have to visit their temples to contact them all the same: this is one more to add to the list, the gods of different peoples are different and the same. We can appeal here to the distinction between theoretical and practical theology drawn by the cognitive science of religion: while believers on their best behaviour deny that God is (for instance) anthropomorphic, restricted in space, and vengeful, in practice they treat him as having just those qualities.[109] *Interpretatio* was lived, not discussed; there was neither any possibility of imposing a common understanding nor any need to do so. So in our case, in theory the gods of all nations are perhaps the same. But in practice an individual worshipper was still free to distinguish Isis from Demeter, to see Zeus Thebaieus or Mercurius Visucius or whoever as a distinct entity especially powerful over a particular region. There might at one level be no need for Isis, if she was merely Demeter under another name. But the ritual in her cult, the experience of participation, and the hopes attached to it were so different from Demeter's as to give Isis a separate aura. Where the cultic double name was used, it provided a perfect way of fudging the issue: Zeus made Zeus Thebaieus universal and international, Thebaieus restored local particularity: Zeus Thebaieus was a glocalized god.[110]

One crucial qualification is needed with respect to what has been said thus far. Evidence in favour of the identity model comes from Greek and Roman authors, not from any of the Gauls and Germans and Phrygians and Syrians and Nabatae-

108. Some texts which raise such questions: Plut. *Amat.* 13, 756C: Eros isn't a foreign god derived from barbarian superstition like Attis and Adonis; Lucian 24.27: "Pan and the Korybantes and Attis and Sabazios, these immigrant and questionable gods" (τοὺς μετοίκους τούτους καὶ ἀμφιβόλους θεούς: foreign gods also 21.8); Strabo 13.1.12, C 587: Priapus was unknown to Hesiod but revealed as a god (ἀπεδείχθη) by later generations; Dio Cass. 40.47.3, 53 B.C.: the senate voted to destroy privately established shrines of Sarapis and Isis: "for for a long time they did not accept (οὐκ . . . ἐνόμισαν) these gods, and even when it came to their being honoured publicly, they located them outside the pomerium." No false gods in polytheism: J. Scheid, *HSCP* 97 (1995): 18.

109. See, e.g., J. L. Barrett, "Theological Correctness: Cognitive Constraint and the Study of Religion," *Method and Theory in the Study of Religion* 11 (1999): 325–39; T. Tremlin, *Minds and Gods* (Oxford, 2006), 169–96.

110. For another fudge, see the argument in Lucian 11.5 that gods are universal but also have local preferences: χαίρουσι καὶ θεοὶ πατρίσι καὶ πάντα μέν, ὡς εἰκός, ἐφορῶσι τὰ τῶν ἀνθρώπων, αὑτῶν ἡγούμενοι κτήματα πᾶσαν γῆν καὶ θάλασσαν, ἐφ' ἧς δὲ ἕκαστος αὐτῶν ἐγένετο, προτιμᾷ τῶν ἄλλων ἁπασῶν πόλεων. Cf. Polinskaya, *Local History*, 514: gods are "neither essentially local or panhellenic, but can be either in a given context."

ans and Egyptians and all the others who in some measure accepted the practice of *interpretatio* when using Latin or Greek. The homogenizing effect of the acceptance of Greco-Roman *interpretatio* is an important fact of cultural history, but we cannot assume that the recipients shared the assumptions of philosophers, Herodotus, and the educated elite of Rome. One might think here of the simplifications noted above whereby in some regions a small selection of Greek or Roman theonyms was used to cover a wide range of local deities, or "Zeus" and "Jupiter" became signifiers for "greatest god." This broad brush application of *interpretatio* differs from the much more thoroughly worked-out equations between Greek and Roman gods, even if the difference is in part due to the character of the pantheons to which it was applied.

Can we escape from the Greco-Roman perspective? We have some evidence for attitudes ascribed by Greek authors to non-Greeks. When Herodotus speaks of "Osiris, whom they [the Egyptians] say is Dionysus," he is crediting the Egyptians with the same identity model that governs his own writing. Lucian presents a Gaul who cheerfully accepts that the god known to him as Ogmios, and depicted in ways alien to the Greco-Roman Herakles, is Herakles nonetheless, but argues that it is the conception of Herakles revealed in the Gallic iconography and not the Greco-Roman that is correct.[111] Herodotus and Lucian take it for granted that their interlocutors (real or imagined) saw things as they did. But the unreliability of such evidence is obvious. The closest we can come to an authentically external perspective may be that of two learned men originating in what once had been Phoenicia: Philo of Byblos, who in the first century A.D. wrote a cosmogony and prehistory of the world which mingled euhemerism with Phoenician legends; and the anonymous Sidonian who discussed the genealogy of Asclepius with Pausanias in the second.

Both assert Phoenician traditions against Greek, the former extensively and aggressively. But when the Sidonian seeks to correct the Greek genealogy of Asclepius from the Phoenician, he is assuming that a single god common to both cultures is at issue. Philo too is full of equations between supposed gods (for him as a euhemerist in fact dead men), such as "Kronos, whom the Phoenicians call El," and regularly speaks of Phoenician gods by Greek names. It is true that Philo cannot carry through this policy consistently, since he cannot find existing Greek equivalents for all the Phoenician deities he refers to and is often reduced to transliteration or ad hoc translation.[112] But his approach is diffusionist (Greeks appropriated and distorted Phoenician traditions), and he cannot be supposing that

111. Herodotus: see, e.g., 2.42.2: Ὀσίριος, τὸν δὴ Διόνυσον εἶναι λέγουσι; 2.123.1: ἀχηγετεύειν δὲ τῶν κάτω Αἰγύπτιοι λέγουσι Δήμητρα καὶ Διόνυσον. Lucian: Hercules (cf. n. 59 above).

112. See n. 86 above. An instance of translation is *FGrH* 790 F 2 ap. Euseb. *Praep. evang.* 1.10.16, Δαγών, ὅς ἐστι Σίτων: this sounds like an equation between existing theonyms, but Dagon is a common noun meaning "corn" as well as a theonym, and Siton is an ad hoc translation.

different nations had distinct sets of supposed gods. To this extent the assumptions ascribed by Lucian to his perhaps imaginary Gaul are also those of actual (neo-) Phoenicians: both the Gaul and the Phoenicians accept the identity model, but insist that their own local understandings of the shared god are correct in contrast to the Greco-Roman tradition. And this no doubt was what mattered to them most. It was natural for the Sidonian and Philo, themselves products of the Greco-Roman culture which they partially resisted, to take over the practice of *interpretatio* along with the Greek language in which they conversed or wrote. It was also easy to understand *interpretatio* as a claim of identity between gods, simply because an opposite view was never formulated. And it was positively desirable to accept the identity model if it made it possible not merely to assert the validity of local traditions—the crucial issue—but also to use them to correct (in thought and word at least) those of the imperial culture.

WHO DOES IT?

The second great question after the *why* of *interpretatio* is *who*. Who made the identifications, who propagated them, who accepted and used them? The agents easiest to identify are unfortunately also the least interesting. Scholars were evidently responsible for some identifications found in some contexts.[113] But in historical terms what is interesting is *interpretatio* as a form of cultural interaction, not as a product of antiquarian study. To take the Roman case first, Tacitus in the famous passage about the German gods Alci quoted above (43.4) writes that "they call them in an *interpretatio* Romana Castor and Pollux." But who are the unspecified subject of Tacitus's verb, these "they"? *Interpretatio Romana* in the West, it used to be supposed, was initially the work of such Romans as first came into contact with Gauls and Germans and their gods, thus merchants and soldiers and their like; natives then in various measure took over the imported names along with the imported language.[114] More recently, scholarship on the Roman West has transferred responsibility to the local elites, if not necessarily for the first identifications, at least for the form in which they persisted in cult practice.[115] We have

113. See Wissowa, "Interpretatio Romana," 7, on "learned ingenuity" ("gelehrte Tüftelei").

114. So Wissowa, "Interpretatio Romana," 23–28. J. G. Anderson in his commentary on the *Germania* passage makes such people subject of Tacitus's verb, whereas Nock, *Essays,* 752, has no doubt that they are the Naharvali. J. Webster, "Translation and Subjection: *Interpretatio* and the Celtic Gods," in J. D. Hill and C. G. Cumberbatch, eds., *Different Iron Ages* (Oxford, 1995), 175–83, still ascribes responsibility to the colonial masters.

115. C. Letta, *Rivista Storica Italiana* 96 (1984): 1002–5; T. Derks, "The Perception of the Roman Pantheon by a Native Elite: The Example of Votive Inscriptions from Lower Germany," in N. Roymans and F. Theuws, eds., *Images of the Past* (Amsterdam, 1991), 235–65, esp. 254–55 (in 237 n. 13 he cites other scholars who have taken that view); Cadotte, *Romanisation des dieux,* 11–12, 419.

been reminded that if a community gained colonial or municipial status, the first *decuriones* were required to choose its gods; here leading natives might have sought the advice of public slaves well versed in Roman religious antiquities.[116] An issue of perspective arises here, since a Roman surveying the gods of Gaul would not necessarily find the same equivalences as a Gaul surveying those of Rome. Are we dealing with an invader's attempt to interpret an alien environment in terms of familiar divine powers, or with a conquered people seeking among the conquerors' powerful gods those most adapted to its own habits and needs?[117]

It is doubtless impossible to say from which side a particular *interpretatio* first came. But two things that matter more are clear. First, the process was not imposed top down from Rome, or the *interpretationes* in different regions would not have varied as they do, native gods would not have been left untranslated, and Romans of high status would not have been happy (occasionally) to make dedications to untranslated native gods.[118] It was simply not a concern of central policy at Rome to regulate the gods worshipped in the provinces.[119] Second, the majority of dedications to translated (e.g., Mars) or part translated (e.g., Mars Lenus) indigenous gods were made by their original worshippers: the system did not primarily serve the needs of Romans abroad seeking a god to honour under a familiar divine name, but of provincials accommodating their old gods to new circumstances. It was by local decision that figures such as Lenus Mars, Mars Mullo, Mars Camulus, and many like them became the main civic gods of important groupings in the Three Gauls.[120]

If one attempts to nuance these broad propositions further and differentiate between urban and rural, rich and poor, more and less Romanized Gallo-Romans (or Spanish Romans or African Romans . . .), in their respective commitment to (a) wholly translated, (b) partly translated, and (c) untranslated gods, many

116. J. Scheid, in M. Dondin-Payre and M.T. Raepsaet-Charlier, eds., *Cités, municipes, colonies* (Paris, 1999), 418. But for the later role of subaltern elements, see R. Haeussler, "Religion and Individualisation in Southern Gaul," in *Keltische Götternamen*, 185–211.

117. See Wissowa's striking pages on the Roman perceptions and expectations underlying *interpretatio*, "Interpretatio Romana," 23–28, whence his conclusion, 28, that *interpretatio* teaches us more about Rome than about the realities of pre-Roman cult. Derks (n. 115 above), arguing the opposite, must distinguish the kind of *interpretatio* already found in Caesar (and presumably of Roman origin) from that attested in cult, though Wissowa stressed their continuity. Woolf's comment, *Becoming Roman*, 217, is sensible: "It will never be possible to reconstruct all the preliminary exchanges, conversations, requests, hints and negotiations that preceded the establishment of these cults in the form in which we first see them. . . . But the active participation of both sides in remodelling Gallic religion is evident."

118. Wissowa,"Interpretatio Romana," 23; T. Derks, *Gods, Temples, and Ritual Practices* (Amsterdam, 1998), 100–101.

119. See C. Ando, "Die Riten der Anderen," in *Interpretatio Graeca/Romana/Indigena*, 31–50.

120. Van Andringa, *Religion en Gaule romaine*, 141–49.

complications arise.[121] I shall not pursue them here, but turn back instead to the Greek East. The same problem over first *interpretationes* arises here as in the West, though the contacts between Greece and Near Eastern cultures were so extensive and varied from so early that it becomes still more intractable: some *interpretationes* might—who knows?—go back to the Bronze Age. What cannot be mistaken is the extraordinary extent to which, in Anatolia, Greek theonyms swept the board. Whereas many non-Roman gods' names are found in Latin inscriptions of the West, in the Greek inscriptions of Anatolia non-Greek names are a rarity, whether as freestanding items or even as the second element in a double name. From Caria one can quote, for instance, Sinuri, the obscure Kanebos, and Zeus Osogo; from Lycia, Eleuthera and Kakasbos and a cluster of occasionally attested figures; from Cilicia, Perasia; in Lydia, Men and Mother defy assimilation (but they had already entered Greek cult under those names), Anaitis is only sometimes linked with Artemis, while a taste for anonymity in the form of *theos/thea* plus epithet combinations can be observed; and the list can be extended within these regions and across others.[122]

But against all this must be set the innumerable instances of Zeus, Apollo, Artemis, Herakles, and (if in lesser numbers) the rest. Here too then the essential point is that the Greek names were adopted by Anatolian peoples.[123] A precise chronology cannot be established—we cannot, for instance, know whether when Herodotus speaks of the Carians as worshipping "Zeus Stratios" early in the fifth century this is his *interpretatio*, or theirs (5.119.2)—but probably in the main, Greek names came in along with the use of Greek language. That at all events is what is suggested by the Xanthos trilingual of 337 B.C., where, as we saw, Lycian theonyms in the Lycian text become "Leto and her children and the Nymphs" in Greek. Various Greek gods had already appeared in the poems composed for Lycian dynasts late in the fifth or early in the fourth century.[124] (In Lydia, some Greek divine names occur in Lydian language texts even earlier, but they presumably refer to gods

121. It has been noted that in southern Gaul dedications to plain Mars are characteristic of persons of high status and of towns; poorer and rural Gauls dedicate to Mars plus a local divine name or epithet, or simply to an uninterpreted indigenous god (H. Lavagne, "Les dieux de la Gaulle narbonnais: "Romanité" et Romanisation," *JSav*, 1979, 155–97, esp. 164–65; for further nuances on the region, see R. Carré, "Cultes et ideologie religieuse en Gaule méridionale," *Memorias de Historia Antigua* 5, 1981, 131–42). But the same pattern is not observable for Mercury (Lavagne, 175–76, and on indigenous gods in towns, 193–94), nor as noted in the text in the Three Gauls, where Mars Mullus–type figures are of central importance (on Gallia Belgica, see too E. Wightman, *ANRW* 18.1.583). Woolf writes, *Becoming Roman*, 228: "The Gallo-Roman élite were among the most active syncretizers." For "little people" dedicating to Diana (*frugifera*) in Spain, see G. Alföldy, "Das Diana-Heiligtum von Segobirga," *ZPE* 58 (1985): 139–59, at 157.

122. For details, see appendix D and chapter 3.

123. D. Piras, in *Hellenistic Karia*, 233, speaks of "self-hellenisation undertaken with a view to political compensation and social advantage."

124. *CEG* 178 (ML 93), 888.

perceived by the Lydians as Greek, not to Lydian gods who had already received Greek names.[125]) At all events, crucially, in western Anatolia the process antedated Alexander's conquests. Gauls worshipped Roman gods, it is sometimes argued, because these had bested their own in battle. That argument will not be transferable to the Greek East, and thereby is brought into question. Zeus and Apollo invaded Anatolia before Alexander did.

In Syria, it has been argued, *interpretatio* came initially from the Greek side only; only later, perhaps not before the Roman period, did native Syrians adopt the Greek theonyms.[126] Sometimes in the Middle East one can draw geographical ("not beyond X") or chronological ("not before") limits to the intrusion of a Greek theonym.[127] In Egypt, Greek travelers and settlers were certainly interpreting Egyptian gods by the sixth century, as we have seen, and had very likely already been doing so earlier. In the Ptolemaic period, traces have been sought of a centrally promoted *interpretatio.*[128] A cluster of formal dedications from roughly the same region of Upper Egypt from roughly the third quarter of the second century B.C. are of interest here. Two from the same site and with the same chief dedicator (though a few years apart) are dedicated to the royal family and to pairs of Greek and Egyptian gods; the pairs are linked by the "also known as" (ὁ καί) formula which is regularly applied to mortals who bear two names. The earlier contains four pairs in which the Greek god precedes ("Ammon also known as Chnoubis, and Hera also known as Satis, and Hestia also known as Anoukis, and Dionysus also known as Petempamentis"); the later lists the same four pairs in the same sequence but reverses the order within the pair to put the Egyptian god first, and adds two more pairs. A similarly formal dedication from Elephantine lists the four same gods, again in the same order, but gives their names in one language only; the languages, however, alternate, to give Chnoubis and Hera and Anoukis and Dionysus. A very similar text from the same place adds Isis between Anoukis and Dionysus; she can perhaps be considered common to both languages. There is a self-consciousness about all this that might suggest policy, though it is one of coexistence rather than replacement.[129] Earlier, the trilingual Rosetta stone had to

125. See p. 78n6.

126. M. Sartre, *The Middle East under Rome*, trans. C. Porter et al. (Cambridge, Mass., 2005), 306.

127. "Zeus" is common in northern Hauran, but "in the South does not pass Bostra and Salkhad": Sourdel, *Les cultes du Hauran,* 21; an altar of Zeus Manaph is of the fourth century A.D., which explains the easy assimilation of Manaph to Zeus (85).

128. Otto, *Priester und Tempel,* 2:267–68, followed by Rübsam, *Götter und Kulte in Faijum,* 5–8.

129. Bernand, *De Thèbes à Syène*, nos. 302 (= *OGIS* 111), 303 (= *OGIS* 130), 242, 243 (with the last, cf. 246 [*SEG* XXXIX 1697]). Bernand, no. 188 (*OGIS* 114), from the same area and period, addressed to Aroeres and Apollo (Ἀροήρει θεῶι μεγάλωι Ἀπόλλωνι καὶ τοῖς συννάοις θεοῖς), is probably an instance of identification by juxtaposition (for Aroeres and Apollo, see Bernand, *Pan du désert,* no. 59, with ὁ καί; Plut. *De Is. et Os.* 12, 356A); for another such juxtaposition, see Bernand, *Delta,* 926:

some extent applied the "to each language its own gods" principle, though some Egyptian theonyms were allowed in the Greek text (where inconsistently both Hephaistos and his double Phtha appear). What we see in the ὁ καί dedications might look like an attempt to preserve even-handedness in monolingual texts. The self-presentation of Antony as Osiris and Dionysus, of Cleopatra as Selene and Isis, can be seen as a conscious facing in both directions.[130]

But such suggestions of policy are isolated. An astronomical calendar written in Greek and dated by its editors "301–240" includes what are apparently traditional Egyptian festivals of the Saite nome but mixes Egyptian (Osiris, Phitorois, Edu [?], Bubastis, Anubis, Isis) and Greek (Athena, Hera, Apollo) theonyms for the gods honoured, and once provides a translation ("festival of Prometheus, whom they call Iphthimis"). This is far from systematic. It shows, as does much further evidence, that a plain Greek theonym can be used even where in all appearance the cult will have followed traditional Egyptian norms.[131] The ὁ καί formula continues to appear spasmodically, as does a variant X "called" (καλούμενος) Y.[132] Simple juxtapositions of the divine names from the two cultures are commoner, to give such double names as Hermes Pautnouphis; the same pairing of names will occur both with and without the "also known as" linking, and such couplings are often uncoupled to give, for instance, dedications to both Hermes and Pautnouphis at the same site.[133] The long papyrus that recounts the healing miracles of Imouthes appears to call him that and Asclepius indifferently.[134] The same Greek theonym covers for different Egyptian gods in different places, but there remain huge num-

---

Σαράπιδι Διονύσωι,῎Ισιδι Ἀφροδίτηι, θεοῖς σωτῆρσι καὶ πολυφόροις (cf. Fraser, *Ptolemaic Alexandria,* 1:197–98: "probably mid 3rd c B.C."). A dedication by two high Ptolemaic officials of 133 B.C. (*SEG* XLIX 2251: Διὶ Σωτῆρι Σαράπιδι,῎Ισιδι Μεγάλῃ Μητρὶ Θεῶν, Ἁρποχράτει, Πανὶ Εὐόδωι, Κρόνῳ, Διονύσωι, Διοσκούροις,Ἡρακλεῖ Καλλινίκωι) is made with a mixture of juxtaposition (the first two items), probable *interpretatio* (Πανὶ Εὐόδωι, Κρόνῳ,Ἡρακλεῖ Καλλινίκωι [?]) and non-*interpretatio* (Ἁρποχράτει).

130. Otto, *Priester und Tempel,* 2:267 n. 2, on Dio Cass. 50.5.3. Rosetta stone: *OGIS* 90.

131. Otto, *Priester und Tempel,* 1:7–9. The calendar: Προμηθέως ἑορτὴ ὅν καλοῦσιν Ἰφθῖμιν (*PHib.* 27.85–86).

132. See *SEG* XL 1560 (52 A.D.), "Ammon who is also Chnoubis"; Bernand, *De Thèbes à Syène,* 171, 106–7 A.D., "Zeus called greatest Nephotes"; *PLeid.* I, p. 124.14–16, "the one called Onourei in Egyptian and Ares in Greek."

133. There are many examples of separate naming of these gods in E. G. Ruppel, *Der Tempel von Dakke* (Cairo, 1930), but Gr. 67 (p. 52) combines them, Gr. W (p. 1) links them with ὁ καί. Isis Demeter, a familiar concept to art historians, first as it seems appeared in an inscription, by supplement, in *SEG* LII 1775. *SEG* XLIX 2251 (Koptos, 133 B.C.) blends Greek, Egyptian, and juxtaposed names.

134. *POxy.* XI 1381 (Totti, *Texte der Isis- und Sarapis-Religion,* no. 15). A. Lajtar, *Deir el-Bahari in the Hellenistic and Roman Periods* (Warsaw, 2006), 46–49, takes the "Asclepius" (always so named) worshipped with Amenothes at Deir el-Bahari as Imouthes (Grecized form of Imhotep), but, as he observes, this Asclepius is called "son of Phoebus" and associated with Hygieia; if Imouthes does underlie him, then Imouthes has acquired family associations from Asclepius. Lajtar also notes that at the centre of his cult in Memphis Imhotep remains Imouthes, son of "Hephaistos" (i.e., Ptah).

bers of Egyptian gods who appear in Greek texts only under their own names. A cheerful disorder prevails, which appears to confuse or embarrass nobody.

HOW DOES IT MATTER?

What difference did the practice of *interpretatio* make? A drastic and disastrous one, for scholars interested in indigenous pantheons before the advent of the Greco-Roman gods: there is usually no way of telling how many different gods in a given region may all have been relabeled Mars or Apollo or Herakles, still less of recovering their original names and perhaps distinct[135] identities; only in Egypt can most originals (not all) be identified. It may be impossible to tell whether a given "Apollo" has succeeded a native or is in fact a wholly imported god.[136] But for the ancients the practice was much less problematic. It is a striking fact that from the whole of antiquity there survives, as we have seen, no single text that discusses the rationale for *interpretatio,* and barely any that express disquiet about it or question or problematise it in any way. Scholars sometimes seek traces of cultural resistance in, say, the decision to address a dedication to Leherennus and not to Mars Leherennus. But in such cases persons apparently of similar background and social position may go either way: the feeling may rather have been that the choice mattered little, since the same god was being addressed under whatever name.[137] One situation where a differential pattern in naming can be observed is in the cult of the Syrian goddess on Delos during the Athenian occupation of the island after 166: Syrian worshippers tend to address the goddess as Atargatis, Athenians by the name she bore when the cult was officially adopted by Athens, Hagne ("Reverend") Aphrodite.[138] But exceptions occur, and there is no reason to think that the matter was politically sensitive. We noted earlier that Philo of Byblos and Pausanias's Sidonian asserted Phoenician traditions against Greek, but apparently in regard to gods whom they accepted, via *interpretatio,* to be common to the two cultures.

One case needs to be mentioned, and set aside as exceptional. The festival Hanukkah still commemorates the overthrow of the edict supposedly sent to

135. So, e.g., Sourdel, *Cultes du Hauran,* 17.

136. This problem seems to me acute in Thrace, where arguably most indigenous gods live on under their own name preceded by Hero (cf. p. 93n70), while most gods with Greek names are indeed imports (who may, however, be assimilated to indigenous traditions iconographically): but for a small number of "Zeus Ammon"-type formations from Thrace, see p. 93n70; and there may be intrusion of Greek gods on hitherto anonymous gods, p. 87n39.

137. Van Andringa, *Religion en Gaule romaine,* 151–52, notes that at Saint Placard private dedications are addressed indifferently to Sutugius, Sutugius deus, and Mars Sutugius, even by citizens. In Ardiège there are eleven dedications to Leherennus, nine to Mars Leherennus, almost all by *peregrini.* On Roman Britain, see the tables in Zoll, "Double-Named Deities."

138. See pp. 163–64 below.

Jerusalem by Antiochus IV in 167 requiring, among other attacks on Jewish religious traditions, the god of the great temple to be worshipped under the name Zeus Olympios (2 Macc. 6.2). The circumstances leading to that governmental intervention have been much discussed. That it was a response to Jewish disloyalty (actual or suspected) to the Seleucid state is agreed. Bickerman contended that Antiochus was encouraged to it by an opportunistic appeal by a Hellenising faction of Jews, who wished to end the isolation of the Jews from the rest of the Hellenistic world (but by a change of name, not of actual cult practice). There were certainly some Jews who would have accepted the view that they worshipped the same god as the one known to Greeks as Zeus.[139] (Some Greeks thought the same, and most supposed that the Jewish god was, at bottom, one they also knew; a surprisingly popular candidate was Dionysus.) But there is no evidence that it was they who provoked Antiochus's intervention.[140] Ma has argued that the measure follows attested Seleucid patterns of behaviour toward rebellious cities, which included depriving them of religious autonomy: the temple was taken out of Jewish control and attached to a (somewhat hypothetical) new Greek city of Antiocheia at Jerusalem; the renaming was a necessary consequence, particularly given Antiochus's (supposed) standardising promotion of the cult of Zeus Olympios throughout the empire. On this view, "the "abomination of the desolation," the interdiction of the Mosaic Law, and the obligation to participate in "pagan" cult were matters of administrative history. They were not religious."[141] What is clear is

139. As is explicitly stated, e.g., in the letter of Aristeas, 16; cf. Hengel, *Judaism and Hellenism*, 1:264–67 (also documenting the opposite view); for the pagan converse, see Julian, *Ep.* 89, 454a Bidez/Cumont: the Jewish god as "the very powerful and excellent god, who controls the sensible world, and whom we ourselves revere . . . under other names." On one view of a problematic passage of Valerius Maximus (1.3.3), Jews sought to introduce the cult of Jahwe to Italy under the name of Zeus Sabazius: so Bickerman, *Studies*, 2:601–6; for the other view, see E. N. Lane, "Sabazius and the Jews in Valerius Maximus," *JRS* 69 (1979): 35–38. On Greek *interpretationes* of the Jewish god, see Hengel, *Judaism and Hellenism*, 2:175 n. 47: Plut. *Quaest. conv.* 4.6 asks, "Who is the Jewish god?" and the main speaker answers, "Dionysus."

140. See F. Millar, "The Maccabaean Revolution," *Journal of Jewish Studies* 29 (1978): 1–21 (also in his *Rome, The Greek World, and the East*, vol. 3, *The Greek World, the Jews, and the East*, Chapel Hill, N.C., 2002, 67–90), dissenting from Bickerman, *God of the Maccabees*, and Hengel, *Judaism and Hellenism*, 1:267–309.

141. J. Ma, http://marginalia.lareviewofbooks.org/re-examining-hanukkah/ (July 9, 2013), summarising his "Relire les institutions des Séleucides de Bikerman," in S. Benoist, ed., *Rome, a City and Its Empire in Perspective: The Impact of the Roman World through Fergus Millar's Research* (Leiden, 2012), 59–84; and Ma, "Notes on the Restoration of the Temple," in R. Oetjen and F. X. Ryan, eds., *Seleukeia: Studies in Seleucid History, Archaeology, and Numismatics in Honor of Getzel M. Cohen* (forthcoming). The evidence for the supposed promotion of the cult of Zeus Olympios, apart from Antiochus's financing of continuing work on the temple at Athens, is numismatic (see references in Ma, "Relire," 82 n. 71). One must ask what precisely is envisaged: New cults? (so explicitly Robert, in Dupont-Sommer and Robert, *Déesse de Hiérapolis Castabala*, 96–99: Zeus introduced as *paredros* to Perasia) Renaming? Or just coin types? If the last, the parallel with Jerusalem is much weakened.

that the attempt to impose an *interpretatio* by compulsion was unique, and formed part of a much broader attack (whether designed as such, or a by-product of standard Seleucid responses to disloyalty) on Jewish particularity. It provoked from the Samaritans an equally unique reaction. According to Josephus, wishing to distance themselves from the troubles of their neighbours, they sought permission from Antiochus to rename what they called the "anonymous" shrine of Gerizim as "of Zeus Hellenios."[142] No other petition to a ruling power on such a matter is known. *Interpretatio* was here politicised as never before or afterward.[143]

I revert to more normal situations. If one presses the strict logic of the identity model of *interpretatio,* there was no reason why the local conception of a god should be affected by a change of name, since the two names were merely different descriptions of the same god. A strong tradition of scholarship assumes that, all over Anatolia and the Near East, local gods lived on essentially unchanged even though redescribed (in the Greek texts available to us) in the form of Greek divine name plus local, usually toponymic, epithet.[144] The point, it is sometimes said, was precisely to allow the cult to persist in its traditional form beneath this superficial Greco-Roman veneer.[145] In relation to the Latin West one sometimes finds the same phenomenon—supposed predominance of the (e.g.) Gallic essence over the

142. So Joseph. *Ant. Jud.* 12.261–63 (but "Zeus Xenios" in 2 Macc. 6.2.). On the temple at Gerizim, see J. Kirkpatrick, "How to Be a Bad Samaritan," in T. Kaizer, ed., *The Variety of Local Religious Life in the Near East* (Leiden, 2008), 155–78, esp. 167–72; for two dedications from it to Zeus Olympios, see R. J. Bull, "Er-Ras, Tell (Mount Gerizim)," in M. Yonah and E. Stern, eds., *Enyclopaedia of Archaeological Excavations in the Holy Land* (London, 1978), 4:1015–22, at 1018.

143. In antiquity, that is: the permissibility of using Chinese terms to indicate the Christian God formed part of the bitter Ricci or "Chinese rites" controversy in the seventeenth century, and the relation of God to Allah is controversial today: see in brief Bettini, *Elogio del politeismo,* 60–62, 63 n. 8.

144. This belief is common in, e.g., Laumonier, *Cultes indigènes;* Paz de Hoz, *Die lydischen Kulte;* R. Lebrun, "Les permanences culturelles louvites dans la Lycie hellénistique," in *Asie mineure dans l'antiquité,* 379–88; and to a large extent even in L. Robert.

145. "The representation of non-Greek deities in Syria and Phoenicia with some Greek attributes is agreed to be a mechanism by which the character of local cults could be maintained": Kuhrt and Sherwin-White, *From Samarkhand to Sardis,* 147; Macmullen, *Paganism in the Roman Empire,* 117: "Underneath the worshipper of Mars is a worshipper really of Lenus or Segomo"; more cautiously, Otto, *Priester und Tempel,* 2:221: change of name has no necessary effect on practice. Contrast Fraser, *Ptolemaic Alexandria,* 1:190: "The adoption of Greek speech was an external expression of a personal readjustment of cultural standards and religious beliefs," though Fraser concedes variability from case to case "both in belief and practice"; Aliquot, *Vie religieuse au Liban,* 139 (endorsed by Bonnet, *Enfants de Cadmos,* 145 n. 116): "The Greek and Roman interpretatio of Balmarcod, far from being formal, brings with it also a redefinition of the activity of the god." In the claim of Kuhrt and Sherwin-White quoted above, "mechanism" is presumably a mechanism of the indigenous peoples; but Rübsam, *Götter und Kulte in Faijum,* 5–8, postulates that the Ptolemies promoted *interpretatio* to encourage Greek settlers to worship Egyptian gods.

Roman name—described, confusingly, as *interpretatio Gallica* or *Celtica*.[146] It was the supposed irrelevance, in practical terms, of a change of name that led Bickerman to support the similarity against the identity model of *interpretatio*:[147] he came to it via his argument that Jahwe could have been renamed Zeus Olympios without the cult practice being affected one jot. But the conclusion was not a necessary one: there was no reason why believing that god A was the same as god B also entailed that the rituals by which the two were honoured should be the same; ancestral tradition, doubtless pleasing to the god in question, retained its validity. Customs differ and should be respected, but metaphysical reality is the same everywhere.

There is a danger also of being blinded by particular continuities observable in a particular cult. The persistence of certain elements should not be confused with thoroughgoing continuity. Scholars who have taken a broader view of religion in, say, Roman Gaul have argued that there emerges a new religious system with a newly structured pantheon in which few preexistent elements persist completely unchanged.[148] Conditions in the East were certainly very different, not least because there Roman imperialism had been preceded by Persian and Macedonian: nonetheless, it is hard to doubt that here too there emerged a new mixed religion or rather a series of regionally varying mixed religions.[149] The question is what part *interpretatio* played in these processes.

A name is only a small part of that elusive thing, a divine personality, of which Sourvinou-Inwood writes:

146. So, e.g., A. Grenier, "La triade Capitoline en Provence," in *Studi . . . Calderini* (Milan, 1956), 1:139–42. Confusingly, because, by analogy with *interpretatio Romana, interpretatio Gallica* ought to refer to a speaker of Gallic giving a Gaulish name to a Roman god, not a Gaulish essence to a god with Roman name.

147. See p. 60 above. The same considerations lead F. Dunand to speak in relation to Herodotus's identifications of "a procedure of equivalence, intended to 'make comprehensible' to his Greek readers what the goddess is": F. Dunand, "Syncrétisme ou coexistence: Images du religieux dans l'Égypte tardive," in *Syncrétismes religieux*, 97–116, at 99. The nuance is different in Baslez, *Religions orientales à Délos*, 247: because the nature of gods is unknown, *interpretatio* can have free rein as long as the rites remain unchanged.

148. See, e.g., M. Clavel-Leveque, "Le syncrétisme gallo-romain: Structures et finalités," in F. Sartori, ed., *Praelectiones Patavinae* (Rome, 1972), 92–134; D. Touléc, "Images de Silvanus dans l'Occident romain," in C. Auvray-Assayas, ed., *Images romaines* (Paris, 1998), 37–60; Woolf, *Becoming Roman*, 206–29; W. Van Andringa, "New Combinations and New Statuses: The Indigenous Gods in the Pantheons of the Cities of Roman Gaul," in *Religious History of the Roman Empire*, 109–38; Van Andringa, *La religion en Gaule romaine*, 135–41; M. Sartre, "Les signes du changement: Réalités et faux-semblants," in *Religious Identities in the Levant*, 11–18 (an excellent nuanced summary).

149. As is well shown for Lebanon by Aliquot, *Vie religieuse au Liban*, especially chap. 6, "La voie romaine." When more than one god was identified with Zeus, the question must arise whether those gods could retain their separate identities: cf., e.g., Sourdel, *Cultes du Hauran*, 20, on the convergence of Baalshamin and Hadad in Zeus.

> The (interrelated) spheres in which a divine personality manifests itself are the following. The sphere of divine name with its subordinate sphere of epithet, that of Bildvorstellung including the attributes, the sphere of myth, the sphere of cult, involving a deity as a recipient of worship, that of theology in the sense of sets of beliefs about the functions and areas of activity of the deity, and finally, the sphere of "ideology," derivative from the previous one, primarily through the agency of literature, involving the deity as an embodiment of certain ideas and concepts.[150]

To this useful list perhaps one should add the architectural form of the sanctuary, which can powerfully shape the worshipper's relation to the deity.[151] Change then can occur or fail to occur at many different levels (to say nothing of changes in the structures within which religious activity occurs and by which it is controlled). This is one main reason why the concept of "syncretism" is such a blunt instrument; another is that it obscures the crucial question of who is doing the "blending" that most of us hear in the word (though its real origin is quite different).[152] The phenomenon at issue is the selective adaptation by one culture of its religious system to that of another culture: an adaptation self-generated, not imposed, through constant interaction with members of the other culture, who are involved in the process of assimilation.

Perhaps the only safe general formula is that *interpretatio* creates the possibility of further changes. Iconography,[153] myths, and powers at once become in principle transferable. It was said above that renaming a local god Apollo did not need to entail any change in the conception of his nature, since ex hypothesi they were both the same god. But the local conception of Apollo's nature might have been

150. C. Sourvinou-Inwood, *"Reading" Greek Culture* (Oxford, 1991), 181 n. 3 (from *JHS* 98, 1978, 101 n. 3).

151. Bickerman, *God of the Maccabees*, 1099–1106.

152. The only occurrence in antiquity is in Plutarch, *De frat. amor.* 19, 490B, where συγκρητισμός is used of the coming together of the Cretan cities for common purposes (an odd idea in itself, since they did so so seldom). When first introduced into Europe, apparently by Erasmus, the concept retained this sense of "cessation of hostilities between warring (theological) factions," but the origin seems to have been forgotten and the connection with συγκεράννυμι later assumed: see U. Berner, "Synkretismus," in H. Cancik et al.,eds., *Handbuch religionswissenschaftlicher Grundbegriffe* (Stuttgart, 2001), 5:143–52, at 144. The concept obscures other important differences too: between internal assimilations of gods within a culture, and cross-cultural assimilations; within the latter, between assimilation between individual gods, and partial blending of whole panthea: cf. Berner, "Synkretismus," 144; R. Gordon, in Brill's *New Pauly* s.v. Syncretism. For critique of the concept, see, e.g., Lightfoot, *Syrian Goddess*, 82; Healey, *Religion of the Nabataeans*, 14–15; Moyer, *Limits of Hellenism*, 151–53; and already Peterson, *ΕΙΣ ΘΕΟΣ*², 586–87 (from an unpublished essay of ca. 1927).

153. See, e.g., J. Quaegebeur, "Cultes égyptiens et grecs en Égypte," in E. van 't Dack et al., eds., *Egypt and the Hellenistic World* (Leuven, 1983), 303–24, at 308–11, on Neith coming to be represented as Athena. On coins, iconographic transfer is extremely common, even where cult practice has probably not changed: see, e.g., F. Duyrat, "Interpretatio graeca et identité sémitique," in N. Belayche and J. D. Dubois, eds., *L'oiseau et le poisson* (Paris, 2011), 329–67.

partial and inadequate. Once identification between two gods has occurred, the possibility arises for a worshipper of enriching his original understanding of a god's nature with the traits associated with its new alter ego. It is at this point that power relations and cultural prestige enter the picture, since traffic seems to be one-way: it would be hard to show that, say, conceptions of Apollo's nature were reshaped for Greece or Rome by Gallic or Anatolian conceptions of the gods identified with him,[154] whereas the Greco-Roman Apollo became familiar throughout the Mediterranean. But traffic did go the other way in the form of the numerous "foreign gods" taken up in the Greco-Roman world.

I will give just two examples out of thousands to illustrate the idea of *interpretatio* as a bridge over which ideas can pass.[155] A decree survives of the Tyrian Herakleiastai on Delos in which they resolve to "send an embassy to the Athenian people so that they may be granted a site on which to found a precinct of Herakles, he who was responsible for the greatest of blessings to mankind, and is the founding father of their homeland."[156] The founding father of the Tyrians' homeland is the Tyrian Herakles, that is, Melqart, but the figure generally credited with responsibility "for the greatest of blessings to mankind" is the Greek Herakles. Another example: a Greek (we assume) gives thanks to "Pan Euagros and Epekoos" ("of Good Hunting and Hearkening to Prayer") probably late in the third century at the sanctuary of Pan at El-Kanaïs in Upper Egypt for "saving him from the (land of the) Troglodytes." Pan is a far more prominent figure in Egypt, where he is often designated Euodos, "of Fair Journeys," than he ever was in Greece, and we assume that an indigenous figure, Min, underlies him; the way in which he helps the dedicator, sending a fair wind when his ship was "driven astray" in the Red Sea, is also unusual for Pan. But the way in which he sent the fair wind, συρίζων λ̣ιγυροῖς πνεῦ̣μασιν ἐγ δονάκ[ων], "whistling with shrill breath from his pipes," is thoroughly Greek.[157] The cultural transfer, it may be objected, has not gone very far in these two cases: there are clear tactical reasons why the Tyrians, seeking a favour

154. This is not to deny that an Anatolian "Zeus" or Gallic "Silvanus" might differ very profoundly from their Greco-Roman originals, nor that the new image might be accepted by people of Greco-Roman origin. And for two-way traffic on Delos, see Baslez, *Religions orientales à Délos*, chap. 4, "L'orientalisation des figures divines"; note, e.g., the statue of a god on the back of an animal that accompanies the dedication to Zeus Kynthios and Athena Kynthia, good Greek gods, *ID* 2428 (see *ID* ad loc.). Note too the *enkōmion* of Isis from Maroneia (Totti, *Texte der Isis- und Sarapis-Religion*, no. 19) which, though written from a Greek perspective, makes Isis responsible for the "revelation of crops," the traditional benefaction of Demeter.

155. For its influence on Balmarcod, see Aliquot, *Vie religieuse au Liban*, 139.

156. ἐξαποστεῖλαι πρεσβείαν πρὸς τὸν δῆμον τὸν Ἀθηναίων ὅπως δοθῇ αὐτοῖς τόπος ἐν ὧι κατασκευάσουσιν τέμενος τοῦ πλείστων [ἀγαθ]ῶν παραιτίου γ[ε]γονότος τοῖς ἀνθρώποις, ἀρχηγοῦ δὲ τῆς πατρίδος ὑπάρχοντος: *IDelos* 1519.12–16. On the force of *archēgos* in relation to Melqart, see Bonnet, *Enfants de Cadmos*, 487.

157. A. Bernand, *Le Paneion d' El Kanaïs* (Leiden, 1972), no. 8; Page, *FGE*, CL.

from the Athenian authorities on Delos, claim for their god the great deeds of the Greek benefactor of mankind; it is a Greek, not an Egyptian, who hears an Egyptian god piping like the Greek Pan. But the point is that the transfer can occur: the Tyrians are not weakening their appeal by a ridiculous claim; the Greek believes that the familiar god has helped him in a distant place.

The concept of *interpretatio*, it can be argued, is too blunt an instrument to capture the oscillation between assimilation and recognition of difference, the raft of different possible attitudes and understandings, that can be observed in contexts where one god receives the name of another. In different places and situations, "*interpretatio* brings together or separates, hierarchises or juxtaposes; it is adopted, rejected, imposed, played with, countered. . . . To speak of *interpretatio* as if it was a question of a single and easily defined process, a kind of norm inherent in the intercultural mechanism of polytheisms is certainly a delusion."[158] That warning is very apt. Melqart does not cease to be Melqart, does not lose all the traditions associated with him, because he is sometimes termed Herakles. The new name may enhance him; it does not efface him. "A god never siphons off another."[159] But this does not mean that one should fall back, with Bickerman, to seeing the apparent identification between two entities created by *interpretatio* as always a mere comparison (even if it may have been that for some agents in some contexts). Melqart when known as Herakles has ceased to be exclusively rooted in Phoenician tradition, has become exposed to change.

THE UNIVERSAL POLYTHEISM

Thus there is no general answer to the third central question about *interpretatio* posed above: one must proceed on a case-by-case basis.[160] I leave it therefore and revert to the contrast drawn by Jan Assmann between the easygoing attitude of polytheisms and the difficulty experienced by revelation-based monotheisms in living with alternatives. Assmann's thesis has been sufficiently influential to draw a response from the scholar who later became Pope Benedict.[161] It is important to take some of the heat out of the debate by stressing that it ought not to be about a supposed moral superiority of polytheisms, grounded in liberal tolerance. Ancient polytheisms were not tolerant, except by default; those who have searched for an

158. C. Bonnet, "Comme des noeuds qui les unissaient tous ensemble (Voltaire): Le processus d' interpretatio en Phénicie à l'époque hellénistique," *CRAI*, 2012, 503–15, at 512–13.

159. Bonnet, *Enfants de Cadmos*, 514; cf. 484, 533.

160. So Rives, *Religion in the Roman Empire*, 147.

161. J. Ratzinger, *Truth and Tolerance*, trans. H. Taylor (San Francisco, 2004), 210–31: Ratzinger observes that many ancient polytheisms were warlike, and that there could be conflicting values (symbolised by warring gods) within a single polytheism. But the question is whether religion in itself led to war between polytheisms or repression within a polytheism.

ideal of religious freedom or tolerance in mainstream ancient thought have failed to find it until, late in antiquity, pagans faced by triumphant Christianity took up arguments earlier developed by persecuted Christians.[162] In particular, polytheisms are not tolerant of monotheism; Roman persecution of Christians in the second and third centuries finds its parallel in Japanese persecution of Christian missionaries from the late sixteenth century onward,[163] and no doubt there are many other examples. But what does appear a defensible claim is that ancient polytheisms, though not committed to religious toleration, found it very easy to get on with one another. I know no instance in the ancient Near Eastern or classical worlds of intercommunal violence between polytheist groups based on religious difference.[164] The grounds for this easy coexistence are doubtless various; the absence of doctrine based on revelation is one; the loose link in polytheisms between religion and the moral code is another; but a third is surely the shared assumption, grounded in *interpretatio,* that at bottom the gods you worship are also the gods I do or might worship. Think how different, for instance, relations between Greeks and Romans would inevitably have been without the easy acceptance of the other's gods as the same as one's own. Perhaps it is a mistake to speak of ancient polytheisms in the plural at all. From an actor's perspective the world was divided between different countries and tribes and political systems, but it was not divided between different gods: there was only one ancient polytheism,[165] one set of gods ruling the entire world.

162. See P. Garnsey, "Religious Toleration in Classical Antiquity," in W. J. Sheils, ed., *Persecution and Toleration* (Oxford, 1984), 1–27; cf. G. G. Stroumsa, *The End of Sacrifice,* trans. S. Emanuel (Chicago, 2009), 105–6.

163. The issues, it is true, were far from purely theological: see J. Elisonas, "Christianity and the daimyo," in J. W. Hall, ed., *The Cambridge History of Japan* (Cambridge, 1991), 4:301–72.

164. So too Bettini, *Elogio del politeismo,* 43. But Dio Cass. 42.34.2 claims that there were internal religious conflicts within Egypt based on differences in cult; notable here is *PGiss.* 99 (cf. Fraser, *Ptolemaic Alexandria,* 1:280–81), a petition complaining that the Idumaean worshippers of Apollo (taken to be their Kos) at Hermoupolis use a foreign language and sacrifice [sheep] and goats in ways "most opposed" to local norms.

165. Rudhardt, *RHR* 1992, 230, speaks of "a single religion, even if scarcely open to definition" ("une religion unique, serait-elle mal définissable"). This position is firmly rejected by Bonnet, *Enfants de Cadmos,* 419 n. 17: "One must . . . enter within the extreme variety of practices, representations, beliefs, which shatter the pretended unity and which alone permitted men to handle the complexity of relations with the divine world." Of course; but there is also great permeability between the rites and representations and beliefs of different peoples; and the idea that Herakles is Melqart is theirs, not ours. The great differences that certainly existed were not perceived as unbridgeable.

3

Gods of Many Nations and Their Naming in Greek

Non-Greek Naming Traditions

Then were they known to men by various Names,
And various Idols through the Heathen World.

MILTON, *PARADISE LOST*, 1.374–75

This chapter is about the interaction between the Greek naming practices described in chapter 1 and the traditions of the many cultures with which Greeks came into contact, and which in varying degrees adopted the Greek language. *Interpretatio* was central to that interaction, but, as we have seen, what occurred was far from being a simple translation via *interpretatio* from one language to another. Unfortunately, this is a cultural exchange which in many regions we observe from one side only. In Egypt and where Semitic languages were spoken, a contemporary indigenous naming practice is available for us to observe and compare with the Greek which comes to join it; this is evidently the ideal case. But even in the Semitic language regions, their written use is unevenly attested; whereas forms of Aramaic (or Hebrew) are found abundantly in (for instance) Judaea/Palaestina, Dura-Europos on the Euphrates, Palmyra, and places on the margins of the steppe, Phoenician vanishes from the written record of the coastal cities around the start of the Common Era, and inscriptions in languages other than Greek and Roman are never found in northern Syria or in the so-called Dekapolis.[1] If Semitic languages continued to be spoken in these regions, we lack actual examples to compare with texts in Greek and Latin. In Anatolia the situation is much worse. Contemporary evidence in languages other than Greek (and Latin) is entirely absent, apart from a few scraps of Aramaic and the small corpora of Lydian, Lycian, and Carian inscriptions, too limited and too partially understood to yield thus far more than fragments of knowledge about divine names and naming. Of the languages of Anatolia, some died out, but others persisted

1. Millar, *Roman Near East*, 503–5, 514; cf. 233, 275. For the Dekapolis, see map 2.

in spoken form. Though those of the western and southern fringes, such as Lydian, Lycian, Carian, and Sidetic, eventually disappeared, many in inner and northeastern Anatolia lasted—or so it is commonly supposed—into the Byzantine period.[2]

This persistence would have important implications. As we saw in chapter 2, where an established *interpretatio* existed, the choice of a name for a god depended on the language being used. This is clear, for instance, from many bilinguals from many places: on Cyprus, the god who is Apollo in Greek is Resheph in Phoenician; Apollo does not appear in Phoenician texts, nor Resheph in Greek.[3] A name alien to the language being used occurs only when a god's cult has been imported, as, for instance, both cult and name of Isis were introduced throughout the Greek world. But the principle that when speaking one's native language one uses native designations for native gods may be invariable: no Phoenician-speaking Phoenician, for instance, would substitute Apollo for Resheph. When a text in Palmyrene speaks of Nemesis, that is because Nemesis is a newcomer to the Palmyrene pantheon, with no Palmyrene equivalent. The recent emergence of Abolo = Apollo in a Thracian text is explained by the context, dedications made presumably by Thracians but in a Greek sanctuary of the Greek god at Zone.[4] Tmtr3 in demotic texts from Egypt is Demeter, whose cult has been adopted, not Isis in disguise.[5] The same principle (adoption of the cult) will probably explain the occurrence of the names of Artemis, Zeus, and Demeter in Lydian texts, and Artemis in Lycian.[6] I know no certain instance from the classical world where the name of a god from language A replaces that of language B in a text written in B.

2. K. Holl, "Das Fortleben der Volkssprachen in Kleinasien in nachchristlicher Zeit," *Hermes* 43 (1908): 240–54; P. Charanis, *DOP* 13 (1959): 25–27; S. Mitchell, *Anatolia: Land, Men, and Gods in Asia Minor* (Oxford, 1993), 1:172–74; on Carian, see Hornblower, *Mausolus*, 341–44; on Pisidian, see Horsley, in *Burdur Museum*, p. 268 with n. 17; *ISelge*, pp. 18–19. On the imperfect Greek of some members of the Galatian elite in the second century A.D., see *IAnkyra* pp. 27, 30.

3. Cf. appendix F.

4. C. Brixhe, "Zôné et Samothrace: Lueurs sur la langue Thrace et nouveau chapitre de la grammaire comparée?," *CRAI*, 2006, 121–46 (material of the 6th/5th c.), at 131. Noteworthy is the Απολοδορε = Ἀπολλόδωρος of a Thracian graffito from Samothrace (Brixhe, 133). But his identification of Thracian υνεσω as adaptation of a hypothetical epithet *ὄνησος = ὀνήσιμος (133) is very insecure. Nemesis: Du Mesnil du Buisson, *Inscriptions palmyréniennes de Doura-Europos*, no. 12.

5. Quaegebeur, "Cultes égyptiens et grecs," 306: "In a Greek papyrus Demeter can indicate an indigenous deity, such as Isis, but in a demotic document Demeter must be the Greek goddess"; cf. Thompson, "Demeter in Graeco-Roman Egypt," 701. If the KRW' who has a priestess at Carthage (*CIS* I.5987; *RÉS* 2001) is indeed Kore, this will be the Greek goddess, not "a Greek varnish covering a local goddess" (as tentatively suggested by C. Bonnet, *Pallas* 70, 2006, 374).

6. R. Gusmani, *Lydisches Wörterbuch* (Heidelberg, 1964); cf. *Ergänzungsband*, 1980–82, s.v. artimu (already found ca. 500 B.C.), lev- (= Zeus), lamẽtru (Demeter); G. Neumann, *Glossar des Lykischen* (Wiesbaden, 2007), 72, s.v. ertemi; and in the Xanthos stele *TAM* I.44 c 8. In the case, e.g., of Artemis of the Ephesians, the reference is certainly to a Greek cult frequented by Lydians, not an interpreted Lydian god. A neo-Punic text, *KAI* 176 (not in Jongeling, *Neo-Punic Inscriptions*), addresses Kronos and Tinnit face of Baal, but it is written in Greek letters, which may explain the exception.

All this may be relevant to a famous incident in Acts of the Apostles. We are told how at Lystra in Lykaonia Paul healed a man lame from birth, and "the masses, seeing what Paul had done, raised a cry, speaking in Lykaonian, 'The gods have come down to us in human form.' And they called Barnabas Zeus, and Paul Hermes, because he it was who began to speak" (14.11). The incident is famous as attesting both the use of what is taken to be an indigenous language and the pagan belief in gods walking the earth in human form. A different issue has been less considered. Did the masses, speaking Lykaonian, use the Greek theonyms Zeus and Hermes, or rather whatever may have been their rough Lykaonian equivalents? If we assume that "Zeus" and "Hermes" were *interpretationes* of Lykaonian gods, they will have used Lykaonian names which only became Zeus and Hermes thanks to the author of Acts. The issue is not one of naming alone: the preservation of the indigenous name for the god would evidently have helped to preserve a sense of that god's separate identity, the link with any stories that may have been told about it under that name. Nor is the issue one concerning Lykaonia alone: it arises for all the other regions of Anatolia where, as noted above, indigenous languages are thought to have survived into the Greco-Roman period.[7] In particular, if, as has been plausibly argued, Phrygian persisted as a spoken language into the Roman period, the possibility exists that the vast proliferation of divine names and epithets attested in Greek Phrygian inscriptions bears an uncertain relation to different designations for the gods used in the vernacular; uncertain and problematic traces of indigenous Phrygian theonyms may in fact survive in neo-Phrygian inscriptions.[8] A sanctuary in Isauria seems to have restored to us the name of one Isaurian god: it contains dedications to both Zeus Astrenos and Stallos Astrenos, and we infer that, as Herodotus would have said, "Stallos is Isaurian for Zeus."[9]

7. Rives, *Religion in the Roman Empire,* 146, notes the same problem for the Roman West.

8. A Greek/neo-Phrygian bilingual inscription (O. Haas, *Die phrygischen Sprachdenkmäler,* Sofia 1966, 120, no. 48; V. Orel, *The Language of Phrygians,* New York 1997, 97 W 34: both languages are written in Greek characters) contains theonyms, because the Greek portion refers back to "aforementioned gods": A. Lubotsky, in R. Gusmani et al., eds., *Frigi e Frigio* (Rome, 1997), 115–30, identifies them as Μιτραφατα, Μας Τεμρογειος, Πουντας Βας (*alii alia*) and interprets the two latter as "God Mas of the Temrog region" (cf. the river Tembris, Tembrogios, etc.) and Pontic Bas. C. Brixhe, in J.N. Adams et al., eds., *Bilingualism in Ancient Society* (Oxford, 2002), 257, interprets Ουανακταν Ουρανιον in Haas, 126, no. 88 (Orel, 108 W 42) as "*wanax* (lord) of the sky," taking Ουανακταν as probably Phrygian but Ουρανιον as borrowed from Greek. Τιος in Haas, 126, nos. 32–34, 36, and often is generally taken as genitive of a theonym; because of the sound change most linguists (but see A. Lubotsky, in *Historische Sprachforschung* 117.2, 2004: 229–37; accepted in T. Drew-Bear, A. Lubotsky, M. Üyümez, "Three New Phrygian Inscriptions," *Cadmos* 47, 2008: 109–16) deny that it could be Zeus (despite Demosthenes of Bithynia ap. Steph. Byz. s.v. Τίος, who anyway does not say quite that).

9. A. Royer and H. Bahar, "Astra en Isaurie," *Anatolia Antiqua* 19 (2011): 149–98, p. 169, no. 20, with the comment "This inscription reveals to us the Anatolian form of the name of Zeus."

The postulated large-scale survival of indigenous languages into the Roman period has, however, been challenged.[10] Neumann has argued, on the basis of plausible parallels, that the "in Lycaonian" of Acts refers not to a separate language but to "Greek as spoken (strangely, incorrectly) by a Lycaonian," and that most of the supposed later evidence is to be similarly explained. Lycaonian divine names could still have survived in the distinctive Lycaonian Greek on that view, but the case is weakened.

Theophoric names shed a dim light on this issue. Houwink ten Cate in 1961 showed that many names deriving from ultimately Luvian gods (names such as Trokondas, Rondas, and Armapias) survived along the southern coast of Asia Minor, particularly in Lycia and rough Cilicia. A single inscription from Cilicia contains five different names based on those of such old gods.[11] But, as is well known, many pagan theophorics survived into the Christian centuries, and a few such as Denis (Dionysios) survive today. The continued use of a theophoric name does not prove continued use of an actual theonym.

The extent to which two sets of theonyms continued to coexist in worshippers' heads is uncertain, therefore. What is certain is that we know very little of non-Greek ways of referring to the gods in Anatolia. To set the material that we have in a context, the best that can be done is to look back to languages and religious systems attested much earlier in the region, Hittite in the second millennium and the Luvian of the Luvian-speaking kingdoms in the first half of the first. Linguistic continuities between Hittite-Luvian and Lydian, Lycian, and Carian are well established, and the postulate of continuity in religious terminology is a reasonable one;[12] but the chronological hiatus between the Hittite texts and the bulk of the Greek evidence is enormous, less but still great for the material from the Luvian-speaking kingdoms (which are anyway concentrated in a confined region of Anatolia, the southeast). One can also try to look more broadly at forms of divine naming throughout the polytheisms of the ancient Near East, on the assumption that they had all been in direct or mediated interaction since at least the second millennium B.C.

A detailed survey of these various comparanda would be a huge task, and one requiring a team of specialists. But some general points are clear. First, something resembling the "cultic double name" is present in virtually all the ancient polytheisms. The combination of a divine name with a place-name is too pervasive to need much comment: exempli gratia the twenty or so localised storm gods, the

10. G. Neumann, "Kleinasien," in Neumann and J. Untermann, eds., *Die Sprachen im römischen Reich der Kaiserzeit* (Bonn, 1980), 167–85, at 178–79. His doubt does not extend to Phrygian, however (174).

11. P. H. J. Houwink ten Cate, *The Luwian Population Groups of Lycia and Cilicia Aspera during the Hellenistic Period* (Leiden, 1961). Single inscription: *ICilicie* 11, with the note on p. 34.

12. The best candidate for a theonym that survives is Hittite Hebat, reemerging as Ipta (see refs. in *SEG* XLV 2330); Phrygian Matar doubtless in part lies behind Mother.

three Telepinus, three war-gods, and two Ishtars of the treaty between Mursili II of Hatti and Tuppi-Teshshup of Amurru can be called in witness.[13] Comparable expressions are still common much later in, for instance, Nabataean texts of the Roman period.[14] The slight difference that in Greek the second element is an epithet but in other languages commonly a noun (Zeus Hymettios as against Baal of Saphon) is not important; a more important variable is that of the locality referred to (mountain, cult site, place of habitation?).[15]

One also finds narrower local specifications such as "of the Garden" or "of the Vineyard," and what look like close parallels to the functional epithets familiar from Greek. An Akkadian text from Emar in northeastern Syria of the fourteenth/thirteenth century, for instance, contains an impressive list of different aspects of the god Dagan: "Dagan Lord of the Offspring, Dagan Lord of the Brickwork, Dagan of the Palace, Dagan Lord of . . ., Dagan Lord of Creation (?), Dagan Lord of the Camp, Dagan Lord of Habitations . . ."; some fifteen aspects are listed in all; a Late Babylonian text reduces a succession of deities to aspects of Marduk: Marduk of Planting, of the Abyss, of the Pickax, of Battle; in Hittite texts we find weather-gods of rain, of thunderstorm, of lightning, of growth, of the vineyard, of the meadow, Teshubs (i.e., forms of Teshub, the Hurrian storm-god) of the Camp, of Rescue, of Invocation, and so on.[16]

The qualification that these forms of expression "look like" Greek functional epithets is necessary because the verbal similarity does not prove that their role in cult practice was the same. The issue with cult epithets is always not just to list them but also to see how they were used, in what contexts, and for what purposes the god was given a limiting description. To take the fifteen Dagans of Emar, there are many references within the ritual in question to the moving of gods, that is, their images, hither and thither, and it has been argued convincingly that the fifteen Dagans are so many cult images.[17] The multiplication of divine images was characteristic of many of the ancient Near Eastern religions; a spectacular illustration comes from what is perhaps the most engaging text to survive from the eighteenth century B.C.,

13. G. Beckman, *Hittite Diplomatic Texts*[2] (Atlanta, 1999), no. 8 (p. 63). For Egypt note, e.g., Assmann, *Search for God*, 19: "An Egyptian city was always the city of a deity" and "The concept of a deity was characterised by the city he or she ruled." For local Baals, see Day, *Yahweh*, 68–69; the ninth-/eighth-century *pithoi* from Kuntillet 'Ajrud even give localised Yahwes (Day, 49), though the significance of the localisations is debatable (Allen, *Splintered Divine*, chap. 6).

14. So, e.g., Allat "who is at Salkhad," or "mistress of the place" (*CIS* II.1.182; *RÉS* 2052).

15. On this, see Allen, *Splintered Divine*, 4, 232–37; this work is a major study of issues relating to epithets in the ancient Near East. The tables in the appendix (pp. 351–412) are a storehouse of examples.

16. D.E. Fleming, *Time at Emar: The Cultic Calendar and the Rituals from the Diviner's Archive* (Winona Lake, Ind., 2000), 235–51; W.G. Lambert, in H. Goedicke and J.J.M. Roberts, eds., *Unity and Diversity: Essays in the History, Literature, and Religion of the Ancient Near East* (Baltimore, 1975), 198; Haas, *Hethitische Religion*, 295, 327, 333.

17. By Fleming, *Time at Emar*, 235–51; for a Babylonian case, see Allen, *Splintered Divine*, 39–40.

the fierce reproaches of the king of Mari on the middle Euphrates to his daughter for her extravagant spending on images, each of which is likely to have acquired a separate title.[18] "I set up this Tarhunzas of the Vineyard . . . Tarhunzas shall make this vineyard grow, and the vine shall grow, and it shall put forth leaves"[19] shows how an image of the weather-god of the vineyard might be set up actually in a vineyard. In an Akkadian denunciation of a perjurer we hear how the offender swore falsely "inside the Main Court facing Ekishnugal, facing Ningal of the Egadi, before Ninshubur . . . of the Main Court, before Alammush, before Nanna-Vanguard and Nanna-Reinforcement": the deities mentioned will be different images or symbols with different attributes.[20] The Greek cult epithet was not attached to particular cult images in this way, though it was to particular sanctuaries; and detailed study by specialists in the many languages and religions involved would doubtless reveal more differences in the ways in which the double name operated in cult practice.

Nonetheless, the similarities are such that a Greek who heard talk of, as it might be, a "weather-god of the army" would not find the form of expression quaint or bizarre. One may indeed wonder whether the similarity between the cultic double name as seen in Greece and in the ancient Near East is a product of cultural contact in the second millennium. That possibility is doubtless unverifiable, when so many other explanations are available: since Latin and other Italic dialects[21] have cultic double names which are not necessarily all owed to Greek influence, one can think of an

18. "Tell Yasmah-Addu: thus (speaks) Samsî-Addu, your father. The works concerning the gods that you have had undertaken, object of the message that you have sent me via Šu-aya, this last has told me about them in my presence. You have had six gods made. The six gods you have had made, goodness, that amounts to ten! Your servants out of fear don't talk to you, in heaven's name, about your works. I want you to think about your protection, but what's the sense in these gods you have had made? Where does your silver come from, where does your gold come from, from which you count on having these gods made? . . . Are you a baby? Why make so many gods? . . . Heavens, where will the cattle and sheep come from that you'll have to offer to these gods every time on the occasion of the monthly festival? Already you send me letter after letter, about sheep and lambs, telling me 'I don't have any!' And now you've filled the town with gods, when all the sheep that there are aren't sufficient for their sacrifices . . . The town of Mari is (already) full of gods and there's none as full as it . . . Heavens, why do you want to have gods made? Just recently, you had Bêlet-Agadé made. It's not me who's made you make gods. And now you're at it again . . ." A 3609 = FM VIII 1, translated from the French trans. of J. M. Durand, in G. del Olmo Lete, ed., *Mythologie et religion des sémites occidentaux* (Leuven, 2008), 1:329.

19. Hawkins, *Hieroglyphic Luwian Inscriptions,* 1.2:463–67, X 14 (his trans.).

20. B. R. Foster, *Before the Muses: An Anthology of Akkadian Literature*³ (Bethesda, Md., 2005), 216 (I am grateful to Stephanie Dalley for so explaining this text to me).

21. Cf. F. Poli, "Jupiter Versor et Hercule Trinnianus: Les épiclèses dans le monde Oscophone," in *Nommer les dieux,* 291–304. Greek influence is revealed by the case of Jupiter Milichius (meeílíkiieís), but, if the system described by Poli derives from the Greek, it has gone its own way. For Latin the excellent old dissertation of Carter, *De deorum romanorum cognominibus,* has not to my knowledge been replaced, though there are observations in G. Radke, *Zur Entwicklung der Gottesvorstellung und der Gottesverehrung in Rom* (Darmstadt, 1987), 173–218.

Indo-European inheritance; or this may be such a useful form of religious expression that it can emerge independently in many different cultures. Whatever its origins, it is clear that it persists and acquires new applications because it answers needs of worshippers; among new applications one can note the three hundred epithets borne in the Greek Orthodox Church by the Mother of God, which equal the three hundred names of Aphrodite mentioned by John the Lydian, but are far outdone by the six thousand names for the Virgin Mary known to one Catholic website.[22]

To revert to antiquity, another locution pervasive in the ancient Near East is the one which one might call, from its commonest exemplar, the "lady/mistress of Byblos" type: the substitution of a title for a name not merely in direct address but also in third-person references. So, for instance, "goddess of Assur, lady of Apûm, lady of Nineveh" are listed as witnesses to a treaty in Mari.[23] The substitution may be occasional, or it may be so regular that it becomes unclear whether we are still dealing with a title or in effect with a name: as is well known, in Judaism 'DNY, *adonay,* "lord," in speech though not in writing came to replace Jahwe; in Phoenicia named gods are in a minority alongside combinations of Ba'al ("Lord") and Ba'alit ("Lady") plus a local or functional specification such as Baalshamin, "Lord of Heaven."[24] Expressions of "lady of Byblos" type are known from Mycenaean, Cypriot, and Pamphylian but drop out of the mainstream of classical Greek usage.[25]

A final widespread type to note, this one wholly absent from standard Greek, is the "personalized god": the god addressed or referred to as "my God/Lord," "our God/Lord," "God of our fathers," or comparable expressions. It occurs regularly in all the Semitic languages and in Hittite.[26]

INTERPRETATIO AVOIDED: ISIS, HEMITHEA, GODS WITHOUT NAME

How then does Greek cope with the various naming practices with which it came in contact? In what follows I shall treat the question synoptically. Without a doubt

22. "Orthodoxwiki" s.v. theotokounomia; Lyd. *Mens.* 4.44 [44], p. 118.16–17 Wünsch; http://campus.udayton.edu/mary//resources/titles/HailMary.html. These last admittedly are very uncritically assembled: for better information on multiple Madonnas, see Allen, *Splintered Divine,* 59–70, who cites several monographs by M. Carroll.

23. Durand, in Olmo Lete, *Mythologie et religion,* 1:276. Mistress of Byblos: see, e.g., *KAI* 4.3–4 and note ad loc.

24. Healey, *Religion of the Nabataeans,* 86–87. On Ba'al as title or name, see Allen, *Splintered Divine,* 215–21. I retain the old transliteration Baalshamin; the final consonant varies, and one often now sees Baal-Shamem velsim.

25. See p. 5 above.

26. See, e.g., Baudissin, *Kyrios,* 2:279–80; Burkert, "'Mein Gott'?," 3–14, at 7–8; in extenso Vorländer, *Mein Gott.*

large variations existed within the "Greek East": the indigenous naming traditions that Greek encountered were different from place to place, as were the indigenous languages and religious systems; familiarity with Greek culture and gods was another variable. From an observer's perspective another key difference is that a parallel naming tradition in a language other than Greek is only sometimes available. But enough phenomena were shared to make an arrangement by regions cumbersome and repetitive. A regional arrangement would also be unstable in that there would often be a case for further fragmentation: not Anatolia but Caria, Lycia, Pisidia, and so on, not quite ad infinitum but to an unmanageable degree. I shall generalise therefore, while trying to show some necessary respect for the variety of the phenomena.

One option was "no change": the non-Greek god retained his or her own name. Indigenous divine names, rare in Greek literary texts, are not so in Greek inscriptions, though their frequency varies. Most of these gods who retained their names were confined to a restricted region, but Anatolian Men is an instance of a major figure who never received an *interpretatio Graeca,* though Roman settlers at Antioch in Pisidia occasionally called him Luna (his functional connection with the moon causing a rare change of gender in the change of languages).[27] In Syria and Thrace no individual name is as dominant as Men (and Mother, if we count her as indigenous) in Anatolia, but several are found[28]. In Egypt, Isis and Serapis receive many more Greek dedications than does any god with a Greek name, and many further Egyptian gods retain their names (sometimes or always) in Greek.

There were also foreign or supposedly foreign gods whose cult with rare exceptions[29] spread under their own names even if they had been subjected to *interpretatio* when viewed from outside: on a major scale Isis (with her Egyptian team) and Mithras, rather less conspicuously Thracian Bendis and Gallic Epona. It would have been easy to rename, say, Bendis as "Thracian Artemis," while as for Isis Herodotus had already declared (2.59.2) that "in the language of the Greeks she is Demeter,"[30] an association that remained active in many contexts; these figures

27. N. Belayche, "Luna/ΜΗΝ ΑΣΚΑΗΝΟΣ: Un dieu romain à Antioche (Pisidie)," in *Ritual Dynamics,* 327–48. The Roman settlers also tried Mensis for him. For non-Greek theonyms attested in Anatolia, see appendix D.

28. For a (now rather ancient) list of Thracian theonyms, see G. I. Kazarow, in *RE* (1936), s.v. Thrake (Religion), 475–76. Gad, usually rendered Tyche, has now appeared untranslated in the plain of Trachon: *IGLS* XV.47.

29. The disguise at Epidaurus of an imported triad as "Egyptian Hygieia, Asclepius, and Apollo" (Paus. 2.27.6) is very unusual, a product no doubt of local Epidaurian pride in Asclepius. But note Hera Ammonia at Elis and elsewhere (Paus. 5.15.11; *ILind* 77; cf. *SEG* XLVI 68, 72; *SEG* XLVII 2138), presumably an unidentified consort of Ammon at Siwa (F. Colin, *BIFAO* 97, 1997: 98).

30. Herodotus's identification of the two goddesses gives way to a perception of similarity: Thompson, "Demeter in Graeco-Roman Egypt," 706–7; but for continuing cross-fertilisation, see, e.g., J. G. Griffiths, *Plutarch's De Iside et Osiride* (Cardiff, 1970), 309, 324–25; *LIMC* s.v. Isis, 781. Mueller, *Megas*

nonetheless retained their separate identity. Ammon too, despite his early association with Zeus, came to Greece not as Zeus Ammon but simply under his own name.

But a few foreign gods entered the Greco-Roman world under Greco-Roman double names: the Zeuses Dolichenos and Heliopolitanos (better known in Latin as Iuppiter Optimus Maximus Dolichenus and Heliopolitanus), the Syrian goddess (also called Aphrodite Hagne and Syrian Aphrodite), Persian Artemis. The two former cases are plausibly explained on the view that they had no names or no names in common use in their Syrian home: in inscriptions from near Doliche (in Kommagene) itself, the god is known simply as the "Dolichenian god." Greco-Roman worshippers, however, were used to named gods and so supplied the lack.[31] But the Syrian goddess and Persian Artemis had names, respectively Atargatis and Anahita, which were in fact sometime used in Greek (Anahita in the form Anaitis) but often supplanted by partial *interpretatio* (Aphrodite Hagne, Artemis Persike) or periphrasis (Syrian goddess). The "Dolichenian god" acquires a name through travel, Atargatis loses hers. Or, to establish a more useful comparison, the name Isis proves more resistant to *interpretatio* than that of Atargatis or Anahita. The different contexts of encounter between the cultures in question must explain the difference, though in ways that are no longer clear. We know that Syrian worshippers of the Syrian goddess were more disposed to call her Atargatis, Greek worshippers Aphrodite Hagne.[32] But it remains obscure why worshippers of Isis did not divide on similar lines.

From the retained indigenous theonym, I continue with some other forms which avoid *interpretatio*. The broad principle, as we have seen, is that Greek gods, with some euphemistic exceptions, have names, whereas those of the ancient Near East often lack one, at least in ordinary usage. Many gods of the East remain

---

Theos, 374, detects an isolated early Ptolemaic instance of Isis as Demeter in Mitteis, *Chr.*, no. 33 col. II. 21; for 'Isis Demeter' see p. 68n133. Conversely, Herodotus's claim (2.50.2) that Hera and Hestia were unknown in Egypt is contradicted by evidence for Satis-Hera, Anoukis-Hestia equations (p. 67 above; and literary texts cited by Dittenberger, *OGIS* 130 n. 11; cf. Mueller, *Megas Theos*, 352–53); Dittenberger inferred that the process of *interpretatio* was extended to new deities after Herodotus. Isis remains Isis in Nabataean in Petra: Healey, *Religion of the Nabataeans*, 138.

31. Bikerman, "Anonymous Gods," 191–92 (= *Studies*, 952–54). Even within Syria Baalshamin regularly became a form of Zeus while Dousares stayed Dousares (Sourdel, *Cultes du Hauran*, 28). The former was perhaps perceived as a title, the latter a name. Baudissin, *Kyrios*, 3:72 n. 5, notes that Greeks gave *interpretationes* of Semitic more often than of Egyptian gods because the character of the latters' names as theonyms was clearer. On the "Dolichenian god," see below, p. 87. The Heliopolitan too could be addressed just as "greatest Heliopolitan god," *IGR* III 1069 (Hajjar, *Triade d' Héliopolis-Baalbek*, no. 16). Interesting is the bilingual *Syria* 62 (1985): 182–83 (Greek text only in *SEG* XXXV 1490), where "lady of Byblos" in Phoenician becomes "Astarte greatest goddess" in Greek (a theonym is supplied, but not a Greek one; cf. Bonnet, *Enfants de Cadmos*, 165–67).

32. Atargatis: see pp. 163–64. Anaitis: e.g., Paz de Hoz, *Die lydischen Kulte*, 73–76.

anonymous even when written of in Greek, at least in their homelands. Formulae such as "Arenthenian god," "god at Aoultha," and "great god of Nazala" which combine the word "god" with a local specification are common in Syria (and have equivalents in Phoenician and Aramaic texts),[33] and not unknown in Thrace and Anatolia.[34] Goddesses known merely by a toponymic adjective with or without "goddess" are extremely common in Anatolia, while there are also goddesses known interchangeably as "goddess" plus toponym and "Mother" plus toponym; "Mother" here barely specifies more than does "goddess."[35]

Different explanations will apply in different cases. Where we are dealing with a single dedication, it is always possible that the worshipper knew, but saw no need to use, a specific name, since it was obvious from the context; such cases are common throughout the Greek world. There might be a reverent reluctance to name a god even where a name was known, as in the "ancestral" or "tribal" naming (god of X, god of the X) common in the Hauran region south of Damascus.[36] A worshipper might have wished to avoid an indigenous theonym but have known no convincing Greek equivalent. This is perhaps the case with the chief goddess of Komana in Cappadocia, the warrior goddess Ma. Greek authors were uncertain which of their goddesses she was, variously identifying her with Enyo, Athena, Artemis, and even Semele; in inscriptions her Komanan worshippers avoided the issue by naming her from her function: "the victory-bringing goddess." In 151 A.D. in a metrical inscription an important Roman hesitates about whether to invoke the goddess of Kastabala in Cilicia as Semele, Artemis, Torch-Bearing Hekate,

33. θεὸς Ἀρεμθηνός, *CIL* III 6668; θεὸς ἐν Ἀουλθα, *IGLS* V.2221; θεὸς μέγας Ναζαλῶν, *IGLS* V.2697, 2700, 2702; cf. VI 2946 (apparently a Latin example), XI A/12 (a bilingual, which as printed makes the Greek a direct translation from the [supplemented] Aramaic; on the site, see Wright, *Sacred Spaces,* 118–19), XVII 1 153; θεοῖς ἐν Θεβνειν, Rey-Coquais, *Inscriptions de Tyr,* 6; θεῷ ἁγίῳ Σαραπτηνῷ, Rey-Coquais, 7 (on the last, cf. *SEG* LXI 188); cf. Rey-Coquais, *Topoi* 16 (2009): 227. An early example is the Greek form of the bilingual *SEG* XXVI 1684 (cf. *BÉ* 1977, no. 542; *SEG* XLII 1384), θεῶι τῶι ἐν Δανοις, from Tel Dan. Astarte becomes merely "goddess of Sidon" on Hellenistic coins of the city inscribed in Greek: Bonnet, *Enfants de Cadmos,* 251. For a Phoenician example, see *KAI* 19.4.

34. Thrace: e.g., *IGBR* I² 353, 358 (but cf. 357, with a theonym), II 867–69; Anatolia: Marek, *Stadt, Ära und Territorium,* 143, no. 27 (Pompeiopolis region), θεῷ Δουμουισηνῷ; *SEG* LVI 1709 (Pamphylia), "great Mamblasenian god" (but depicted like Apollo); below p. 87n38.

35. See, e.g., Paz de Hoz, *Die lydischen Kulte,* 32–33, for Nenenene, Adytene, Kornene, Oleine; 29 for Aliane, Matyene, Notene, who appear interchangeably as, e.g., "goddess Aliane" and "Mother Aliane." C. Tanrıver, "The Cult of Theos Peismatene in Mysia," *EpigAnat* 45 (2012): 93–99, publishes twelve very similar, probably late Hellenistic, reliefs all dedicated to this new toponymic goddess.

36. On this, see Sourdel, *Cultes du Hauran,* 95–96; and pp. 117–18 below. For the Hauran, see map 2. But such a god might acquire a name: a goddess of Rakhlé in Lebanon named in her first appearance "goddess of Moithos" later seems to become "goddess Leukothea [of Rakhlé]": *IGLS* XI.20, 21, 23, with Aliquot, *Vie religieuse au Liban,* 165.

Kypris, or Deo, but leaves her indigenous name Perasia unmentioned.[37] Or the god may have been absolutely, or at least in all normal usage, anonymous. As noted above, a likely instance of this last case is the god known in the West as Juppiter Dolichenus: five texts from near Doliche itself spread over a considerable chronological span name him "Dolichenian god" or something similar, while one dedicated "to the reverend god" (*theos hagios*) probably also honours him. Thus he was left unnamed at home though turned into Juppiter abroad; no indigenous name appears anywhere, perhaps because he never bore one.[38] In Thrace by contrast the kind of alternation whereby, for instance, the same god appears as *theos* Eisenos and Apollo Eisenos is frequent:[39] here the foreign theonym intrudes, but coexists with the local designation, which merely identified the power by reference to its place of operation.

A cluster of early-attested figures in Caria likewise, if in a different way, avoid *interpretatio.* We here encounter three deities referred to in Greek simply by title: Hemithea, "Demi-goddess," Parthenos, "Maiden," and, at Kaunos, Basileus Kaunios, "Kaunian King," also known as βασιλεύς ὁ θεός, "king the god." (A divine "king" is found also at Priene, outside Ephesus, in Miletus, and perhaps on Chios. Whether these figures have anything to do with the god of Kaunos or one another is obscure.[40]) These are presumably indigenous figures who have acquired Greek descriptors but have been spared full *interpretatio.* The discovery of the trilingual stele of Xanthos, in which the Kaunian king appears, has allowed him also to be uncovered in an earlier Lycian text, which appears to show that his indigenous title

37. See *IKomana* 67, 75, 76; and cf. 96 ("greatest goddess of the country"). For the various *interpretationes* of Ma, see A. Hartmann, in *RE* s.v. Ma, 88–89; App. *Mith.* 114 (560) calls her simply "the goddess at Komana." Perasia: *SGO* 4, 19/18/01; cf. chapter 2, n. 20. Robert, *OMS,* 1:418, writes: "The thing is common in all regions of Asia Minor, where some ancient local deity, god or goddess, had remained the local god, the master of a particular place, without anyone having chosen to assimilate him to one Greek god rather than another to the detriment of his distinctive character."

38. *SEG* XXXII 1386, 1388 ("reverend god"), 1391; *SEG* XLVIII 1871, LVIII 1683, and LVI 1840 (?). He may be the anonymous "ancestral god" of *AMS* 73 (2014): 354, no. 1. On the cult, see M. Blömer, in *Religious Identities in the Levant,* 129–41. He can remain anonymous abroad, e.g., *IGB* III 1590. Marinus, *Vita Procli* 32, has an extraordinary passage about the "god at Adrotta" (in Lydia) and the uncertainties of the locals about his identity and even number.

39. See the entries Aularkenos, Aulousadas, Dabatopeios (but this might be a theonym or nonlocal epithet), Kendreisenos, in the index 3.3 to Oppermann, *Thrakische Reiter.*

40. Hemithea: Diod. Sic. 5.62–63; dedications in J. M. Cook and W. H. Plommer, *The Sanctuary of Hemithea at Kastabos* (Cambridge, 1966), 58–65. Parthenos: p. 7n28 above. βασιλεὺς ὁ θεός: *IKaunos* 35 C 16, E 4; 139 III c 4; 142.7; *IG* XII.4.547 (βασιλεὺς Καύνος ὁ θεός); *SEG* XXVII 942.7, 15–16 (Xanthos trilingual, βασιλεὺς Καύνιος). Priene etc.: references and discussion in Graf, "Gods in Greek Inscriptions," 74–77. The "ladies" (*Wanassai*) of Perge and Cyprus are apparently a different case: not a response to a non-Greek deity (or at all events not a recent response), but the late survival within a branch of Greek of a much older tradition of respectful naming (p. 5n16 above).

was Kaunian king;[41] the Greek title was therefore a simple translation, and one which proved very tenacious. He is possibly identical with a sacred stone familiar from coins of Kaunos, and that aniconic representation might have discouraged the identification with Zeus which would otherwise have lain close to hand. In the case of Hemithea too, there may have been an obstacle to *interpretatio:* she is a healing goddess, but the main Greek healers were masculine, and the gender gap was one which *interpretatio* very seldom crossed. Much later there are attestations of Theos Basileus, "God King," in Lydia and Phrygia; of "King and Queen" in Ancyra in Galatia; of a "Good God" and a "Good Goddess" near Anazarbos in Cilicia; of a "Good God" in Bithynia; and of "Great God" at Termessos in Pisidia and at Kestros in Cilicia; whether these are comparable cases, or rather titles of respect substituting for a theonym known to the worshippers, is hard to say.[42] At the other end of the Greek world are the "Paides and Anna" (sometimes linked with Apollo) from near Akrai in Sicily:[43] here we have a nonspecific Greek descriptor, Paides, "Children," an indigenous theonym, Anna, and a Greek theonym, Apollo (which may or may not be an *interpretatio*). One dedication in this cult perhaps gives the names of the "children," manifestly non-Greek, Neme, Mene, Pedai.

41. C. Melchert, in J. Borchhardt and G. Dobesch, eds., *Akten des II. Internationalen Lykien-Symposions* (Vienna, 1993), 1:31–34, on *TAM* I.44 c 9; followed by E. Raimond, "Dieux-rois du sud-ouest anatolien, de Kos et de Karpathos: Le roi kaunien et arĝĝazuméen," *Res Antiquae* 1 (2004): 389–408, at 391. Whether *TAM* I.44.c 9 refers to Xanthos or Kaunos is obscure; if the former, the trilingual *SEG* XXVII 942 does not in fact introduce the cult to Xanthos for the first time, despite appearances. Sacred stone: Raimond, 394. The Kaunian god is identified as Apollo by C. Marek, *AMS* 39 (2000): 195–99, but inscriptions from Kaunos treat them as separate.

42. *Theos basileus* in Lydia: *SEG* XLIX 1624, Nisyra; *TAM* V.I.167, south of Saittai; in Phrygia: Drew-Bear and Naour, *Divinités de Phrygie,* 2035 n. 491, Synnada and "southern Abbaïtis"; cf. Θεοῖς βασιλεῦσι, *SEG* XLIII 949 (= Delemen, *Rider-Gods,* 199, no. 386), near Antioch in Pisidia. King and Queen: *IAnkyra* 202–4, "perhaps to be identified with Men and Cybele" (Mitchell, *Anatolia,* 1:191). Good god and goddess: *AMS* 53 (2004): 245–46, nos. 52 and 54. Good god: *BÉ* 2014, no. 478. Great God: *SEG* XLIX 1868, LIII 1609–10, Termessos region (cf. a rider relief in the Antalya museum, *SEG* XVII 592); *Westkilikische Inschriften,* Kestros 1: cf. Bean-Mitford, *Rough Cilicia,* 155 n. 14, 162, plausibly taking this god as identical with the Zeus Megas, "great Zeus," of *Westkilikische Inschriften,* Kestros 5a and 7a; cf. Zeus Megas, *Rough Cilicia,* nos. 172 and 174, Macar Kalesi; [Zeus] Megas Lamotes, no. 191, Adanda (*Westkilikische Inschriften,* Adanda 16); Zeus Megistos, "greatest Zeus," *Westkilikische Inschriften,* Laertes 4. Bean-Mitford, *Rough Cilicia,* 162, speak of "a native Great God, later identified with Zeus." Greatest goddess: Bean-Mitford, *Rough Cilicia,* no. 69, Sinekkalesi in eastern Pamphylia. Note too the "great gods who share a temple" of *IKibyra* 94 (= *Burdur Museum,* no. 184), and the "great gods who possess Nea Kome," Petzl, *Beichtinschriften,* no. 47. All these texts are of the imperial period. Cf. p. 143 below.

43. See G. Manganaro, in L. Gasperini, ed., *Rupes loquentes* (Rome, 1992), 455–87 (*SEG* XLII 825–36); for the possible names, see Manganaro, 468, no. 10 (*SEG* XLII 834). Sicily also offers a flourishing cult of "Mothers" centered at Engyon (Diod. Sic. 4.79.7–80). It has long been thought that in the Carthaginian dedication *KAI* 83 the imported deities (Diod. Sic. 14.77.5) Demeter and Kore were similarly anonymised as "Mother" and "Mistress of the Chamber/the Underworld," but the matter remains obscure: S. Ribichini, *RStFen* 23 (1995): 14–16.

FOREIGN GODS WITH GREEK NAMES: HOSIOS AND DIKAIOS AND OTHERS

Interpretatio is again avoided when an expression is coined which, though Greek, has never been applied to a god of old Greece. In Lycia and Pisidia there are several triads of gods known as *theoi* plus a Greek or semi-Greek adjective: θεοὶ σκληροί, θεοὶ σκιρροί, the "hard" gods; θεοὶ ἄγριοι, ἀγρεῖς, ἀγρότεροι, conventionally rendered "wild" gods but perhaps rather "of the field," who apparently do not differ from θεοὶ δίκαιοι, "just gods"; to these have now been added πρώϊμοι θεοί, who defy confident translation.[44] We know these groups partly from a few literary references, more from reliefs, usually rock-cut, that depict them. They are clearly indigenous groupings; the Greek pantheon contained many female triads but few male collectives (though the Kouretes might have been a possibility), so no *interpretatio* lay close to hand. Instead they acquired these vague adjectival descriptions. It is possible in fact that the strange epithets "hard" and "wild" (if that is the right rendering of ἄγριοι) are either translated or transliterated (with subsequent assimilation to Greek) from an indigenous language.[45] The important South Lycian goddess Eleuthera, "Free," may also be a translation[46] or an adjusted transcription; "Free" makes little sense as a name for a goddess in a Greek context. There are also groups that acquire Greek titles but do not correspond to the traditional bearers of those titles: the "Twelve Gods" of Lycia (male hunters, indistinguishable one from another); the six horse-riding "Immortal Gods" of a region on both sides of the Sultan Daği on the border between Pisidia and Phrygia.[47]

We face a linguistically similar phenomenon but with much wider distribution with the Anatolian (and marginally Thracian) gods Sozon, "Saving One,"[48] and

44. See Frei, *Götterkulte Lykiens*, 1793–94, 1828–29; R. Fleischer, "Felsheiligtümer in Termessos," *IstMitt* 58 (2008): 197–242 ("just gods"); *SEG* LVIII 1605 (πρώϊμοι θεοί); R. van Bremen, in R. Parker, ed., *Personal Names in Ancient Anatolia* (Oxford, 2013), 153, 162–63. The relevance of the Thracian dedication ἀγρίῳ θεῷ ἐπ[ηκόῳ], B. Filow, *AA* 30 (1915): 222, is not clear.

45. Cf. Frei, *Götterkulte Lykiens*, 1829: "Can one infer from all these data a divine group of Luvian origin, whose name was translated into Greek in various ways (ἄγριοι etc, σκληροί, bzw. σκιρ[ρ]]οί)?"; the recurrence of θεοὶ ἄγριοι in Cilicia (*IAnazarbos* 52.6–7) is significant in this regard. This is more persuasive than the view of D. French (in French, ed., *Studies in the History and Topography of Lycia and Pisidia*, London 1994, 55) that waterless conditions create "hard" gods.

46. This is the general view: see Frei, *Götterkulte Lykiens*, 1790–91 (with the evidence for the cult).

47. For both, see Robert, *Documents*, 428–37; for the twelve gods, also B. Freyer-Schauenberg, *Die lykischen Zwölfgötter-Reliefs* (Bonn, 1994); T. Drew-Bear and G. Labarre, "Une dédicace aux Douze Dieux lyciens et la question de leur origine," in G. Labarre, ed., *Les cultes locaux dans les mondes grec et romain* (Lyon, 2004), 81–102.

48. See Weinreich, *Ausgewählte Schriften*, 2:221–32 (from *RE* s.v. Sozon, 1248–56); Delemen, *Ride-Gods*, 39–43. For Thracian evidence, unknown to Weinreich, see *IGB* III 1811, θεῷ σώζοντι Γικεντιήνῳ εὐχαριστήριον; and cf. 1594, Φοίβῳ σώζοντι Λούκιος Κορνήλιος Ῥοῦφος κατ᾽ εὐχήν; *IThrakAig* E 459, Ἀ̣π̣όλλωνι Κερσ[ηνῷ] Σώζοντι; for other addenda, see Delemen, 39–43; and *SEG* XVIII 803

Hosios kai Dikaios, "Holy and Just," sometimes without the connective: Hosios Dikaios.[49] An almost unanimous opinion accepts that Sozon is simply the participle of the Greek word *sōzō,* "I save";[50] the participle in fact appears from time to time as an epithet applied to several gods and goddesses, a variant on the commoner Soter/Soteira.[51] Even if a non-Greek word in fact underlay it, that origin had been obliterated by assimilation to Greek and forgotten; as for Hosios kai Dikaios, no one has ever denied that this is simply a very familiar pairing of Greek adjectives, a cliché in fact, originally used to describe a person or conduct that is proper both with respect to gods—*hosios,* "pious, religiously appropriate"—and to men—*dikaios,* "just." How these figures came to function as gods in the way they do is not clear. Sozon has a close relation to Apollo: thrice Sozon appears as an epithet of Apollo, dedications are made to him in a sanctuary of Apollo, he can be depicted with typical attributes of Apollo, and has no clear iconography of his own;[52] but against this must be set more than fifteen dedications to Sozon alone. That distribution would normally suggest partial assimilation between originally distinct figures, typically an indigenous god and a newcomer; but the hypothetical indigenous figure here has a Greek name, Sozon. Possibly Sozon is a translation of an indigenous name; possibly Sozon started as an epithet which acquired autonomy.[53]

(Hadrianoi in Pisidia); *MAMA* 1.8, Laodicea Combusta; *Burdur Museum,* nos. 116 (a dedication to "Mother of Sozon"), 180.

49. See Ricl, "Hosios kai Dikaios," "Nouveaux Monuments," and "New Inscriptions." New texts: *SEG* LX 1366 (Pontos: to Hosios Dikaios Alexandros, i.e., quasi-funerary); N. Erhan and S. Eskalen, *EpigAnat* 44 (2011): 185–96 (Akçakaya, seventeen kilometers southwest of Eskişehir); n. 55 below.

50. "Unanimously agreed," Delemen, *Rider-Gods,* 40, with references; but this was written before Horsley, in *Burdur Museum,* p. 101, where a connection (J. S. Sheldon) with a "Zoroastrian divine being called Saošyant- (Avestan) or Sūdōmand- (Pahlavi), 'the saving one at the end of the world,'" is tentatively mooted. Delemen also mentions Ramsay's rejected linking with Saoazos/Sabazios.

51. Eileithyia: *IG* XII.5.1022 (wrongly included also as *IG* II^2 4793), Εἰλειθυίῃ σωζούσῃ ἐπισωζούσῃ; Moirai (*IGB* I^2 305quater), Panakia (Habicht, *Asklepieion,* no. 128), Zeus (*SEG* XXXIV 1321, 1384 [?], Lykaonia), Zeus Chalazios (*JHS* 24, 1904: 21, no. 4, Kyzikos), Zeus Kasios (*SEG* XXXVI 1582, an inscribed anchor), Aphrodite (*AA* 21, 1906: 168, an anchor), Isityche (*RICIS* 504/0216), *theoi* (*IAssos* no. 28a; *BCH* 36, 1962: 609, Thasos: perhaps the Dioskouroi [?]), "all the gods who save the city" (*IG* XII.4.542); cf. Berenike Sozousa, Zenobius 3.94; Arsinoe Sozousa, *SB* V.7630.6.

52. Epithet: Delemen, *Rider-Gods,* no. 292 (Lycaonia), and two Thracian examples in note 48 above; note too *SGO* 16/55/01 (Philomelion) which makes Sozon a Letoid. J. Nollé, *Gephyra* 6 (2009): 64–70, argues that the LUK SOZON of a coin of Themisonion is Lykeios, i.e., Apollo. Sanctuary: Apollo Sourios, G. Bean, *AnzWien* 99 (1962): nos. 7 and 8; Apollo of the Perminoundeis, *ICentral Pisidia* 104a, 104g (*Burdur Museum,* nos. 181–82); in the region of Akmonia, *SEG* XXVI 1357 to be read with 1358–60 from the same site (T. Drew-Bear, *GRBS* 17, 1976: 249–51). Iconography: Weinreich, *Ausgewählte Schriften,* 2:227 (bottom); Delemen, *Rider-Gods,* 42 nn. 176–77; *Konya Museum,* no. 16.

53. As happened to Frugifer in Africa. A god unknown in Rome with a Latin name is the Aquitanian Fagus: on these, see *RE* s.vv. Fagus and Frugifer.

As for Hosios kai Dikaios, these epithets are occasionally applied to other gods, but not on any consistent basis; there is no reason to suppose that he or they originated as a pair of epithets attached to any particular god.[54] I write "he or they" because there is syntactic variety as to whether Hosios kai Dikaios is/are one god or two. There is also iconographic variety: in Lydia "he" is a single mounted horseman, whereas in Phrygia "they" are typically a pair of figures somewhat recalling Tweedledum and Tweedledee. (An extraordinary variant from the Kibyratis region north of Lycia makes "him" a "new-born child of Herakles."[55]) The question is how this cliché originally applied to proper human conduct came to be a name for a god or gods. They are occasionally referred to as "messengers" (*angeloi*), of greater gods,[56] but the earliest dated occurrence of the pair of adjectives in a religious context is an acclamation (58 A.D.) in the neuter in honour of "Heavenly Men of Artemidorus who possesses Axiotta": "great is your *hosion*," "great your *dikaion*, great the victory, great your vengeances . . ."[57] It may be that out of these proclamations of the power of righteousness in the world emerge the personified figures, by a process that slightly recalls Max Müller's notorious theory of religion as a disease of language. In another early instance the vengeance of Hosios Dikaios had been invoked against violation of a tomb.[58] What is interesting is that these figures of inland western Anatolia are unmistakably products of religious discourse conducted in Greek. Whether they had indigenous precursors is unknown; the chronology of their emergence allows the possibility that they never existed except in Greek.

A minor Egyptian god who may have emerged and been named in a Greek environment and in Greek is the Hydreios known from four dedications on Delos: he is thought to be the embodiment of the *hydreion*, "pitcher," used to carry the sacred water of the Nile or of its supposed local derivatives.[59] A little-known

54. On this I follow Ricl, "Hosios kai Dikaios," 75; for earlier opinions see Ricl, 75 n. 14. A recently published dedication from central Mysia: [θεῷ/Δι(ε)ὶ] ὀλβίῳ καὶ σωτῆρι, [ὁσί]ῳ καὶ δικαίῳ κριτῇ (C. P. Jones, *Chiron* 44, 2014: 31–33, no. 6, $2^{nd}/3^{rd}$ c.); this may be a secondary reapplication.

55. T. Corsten and M. Ricl, *Gephyra* 9 (2012): 143–51. A new Phrygian text, *EpigAnat* 44 (2011): 185–96, is addressed to Helios Hosios Dikaios Apollo and shows three figures; it apparently, against the Phrygian norm, takes Hosios Dikaios as singular.

56. Ricl, "Nouveaux Monuments," no. 1 (probable supplement, plural); Ricl, "Hosios kai Dikaios," no. 1 (singular); "not messengers but avatars or projections": Belayche, "Angeloi," 63, with an important discussion.

57. Ricl, "New Inscriptions," no. 1 (*SEG* LIII 1344, Lydia); for similar acclamations, see Ricl, "Hosios kai Dikaios," nos. 105, 111; Ricl, "New Inscriptions," no. 26.

58. G. Petzl, *EpigAnat* 20 (1992): 143–47 (*SEG* XLI 1060; Ricl, "New Inscriptions," no. 35, from Mysia). For *hosios* and *dikaios* as divine epithets, but not in combination, in Macedonia, see *IBer* 23, with note.

59. *ID* 2087.13, 2155.2–3, 2160.1 (?) (*RICIS* 202/0323, 0342, 0344); see Baslez, *Religions orientales à Délos*, 270–71; and for the ὑδρεῖον, Clem. Al. *Strom.* 6.4.37 (p. 449.21 Staehlin). Building C in Sarapieion C, "un vaste puits situé à l'emplacement du *naos*," has been interpreted as a temple of Hydreios by H. Siard, in *Guide de Délos*^{4}, 279. Hydreios is addressed as ἐπήκοος in *ID* 2155, 2160 (?); Siard explains this from the healing properties of water, *RA*, 2009, 158–61.

companion of Thracian Bendis too bears a name, Deloptes, that looks more Greek than Thracian,[60] though nothing reveals why "Clear-Seeing" would have been an appropriate title for him.

An apparent case of an "Arab" god with a Greek name is Theandrios/Theandrites, a god particularly known from the Trachon plateau south of Damascus, if we accept that the expansion once found, Θεὸς ἀνδριός, "manly god,"[61] is the actual origin of the name and not a secondary interpretation of an Arab expression assimilated to Greek. The inevitably vague guess is that he was an austere and manly god, like Shai ʿal Qaum, "who abstains from wine," and received his Greek name because of that character.[62] He is once named as "god of Ouaseathos Theandrios," once as "god Theandrios of Rabbos,"[63] that is to say, he is the kind of god inherited within lineages: the presumption then is, in contrast perhaps to the case of Hosios kai Dikaios, that he preexisted his naming in Greek. As a final instance of Greek naming of non-Greek gods, let us note the possibility that the "Two Brothers" who share a shrine at Oxyrhynchus are not, as widely supposed, Castor and Pollux, but rather the two crocodiles Pnepheros and Soxis.[64]

VARIETIES OF *INTERPRETATIO*: ZEUS, ZEUS AMMON, ZENOPOSEIDON

I turn to cases where Greek theonyms are introduced. But the distinction between this category and the previous one is often merely formal, since the same god in the same sanctuary might be addressed with or without a Greek theonym: in Palmyrene bilinguals what in Aramaic is plain BL comes out in Greek as any one of Belos, Zeus Belos, Belos *theos, theos* Belos, *theos megistos* Zeus Belos.[65] A possibility not to be forgotten is the Greek divine name standing on its own without accompanying epithet. Such simple substitution is common in literary texts, Herodotus above all, and is not unknown in inscriptions. Sometimes the plain divine name occurs alongside dedications in which it is accompanied by an epithet;[66] but

60. I owe this point to Dan Dana. D. Detschew, *Die thrakischen Sprachreste* (Vienna, 1957), 129, lists the name, but no other word in his repertory begins Δη-. On this god, attested only in Greece, see Z. Gočeva, in *LIMC* III.I (1986), 367 s.v. Deloptes.

61. *IGLS* XI.41; for the Trachon, see *IGLS* XV.1.172 with note; from the Golan *SEG* XLVI 1999.

62. So Aliquot, *Vie religieuse au Liban*, 187–88 (cf. Aliquot, *REA* 112, 2010, 363–74), with the evidence on his cult. On the problematic Syrian god (?) Γενναῖος (Aliquot, *Vie religieuse au Liban*, 202–5) I risk no opinion.

63. See p. 117n21 below.

64. *POxy.* 254, as discussed by Quaegebeur, "Cultes égyptiens et grecs," 312–16.

65. *IGLS* XVII.1 p. 424, index s.v. Bel.

66. So, e.g., *ICentral Pisidia* 100–101, amid many deds. to Apollo of the Perminoundeis vel. sim. For plain Zeus in the Hauran, see Sourdel, *Cultes du Hauran*, 22 n. 2.

there are also a few cults in which an epithet is never attested. Thus, for instance, in the cult at Amyzon in Caria, Apollo and Artemis are always just that, Apollo and Artemis; and such "nudism" is quite frequent in Egypt even where there is strong reason to think that the god in question had once borne an Egyptian name.[67]

Another possibility was the combination of names from the two cultures, the Zeus Ammon form.[68] This type was of uneven popularity in the Greek East. Syria offers, for instance, Zeus Marnas and Zeus Manaphos; many more examples are added if we include combinations such as Zeus Beelbaaros (the alternative is to treat Beelbaaros not as a name but as the title that it in fact was: Baal, "lord," of Baaros).[69] Some Egyptian examples were quoted in chapter 2, and there is even one compression into a single name, Hermanoubis. Instances in Anatolia and Thrace are harder to detect because of our almost complete ignorance of indigenous theonyms. Several instances may lurk where a god's name, usually that of Zeus or Apollo, is linked with an epithet that means nothing to us and might be a divine name; even so, the type is much less common in these regions than the "god plus toponym" combination.[70] It is also typically unstable (both in East and West): the god named second in the combination will normally also appear as a distinct entity, in a different context[71] or even in the same.

Another grouping looks like this but obeys a different logic, though the distinction may not always remain clear: this is the combination of Zeus's name with that of another major deity, Sabazios or Dionysus or Serapis or Asclepius. The point

67. And not just in dedications by Greeks, such as *OGIS* 98 (where see the note) and 142 (where see the note in *IGPhilae* 17): see, e.g., Otto, *Priester und Tempel,* 1:8–9. Similarly, Paz de Hoz, *Die lydischen Kulte,* 11–12, takes some plain Artemises as Anaitis. Amyzon: *IAmyzon,* e.g., nos. 2.10, 3.13, 10.4–5, 16.9, 21.13–14.

68. Cf. p. 43.

69. See Sourdel, *Cultes du Hauran,* 45; *IGLJ* 2.154 with note. There is now a Zeus Beelgalasos, *SEG* XLIX 2006–7. Names of sacred mountains juxtaposed to Zeus give Zeus Tourmasgades, "Zeus montagne de l'adoration" (J. T. Milik, *Biblica* 48, 1967: 578–80), and Zeus Tourbarachos, "Zeus de la montagne bénie" (J. Jarry, *Annales Islamologiques* 7, 1967: 164; cf. *SEG* XXXIII 1247; LI 2005–6). Novel is Azizos *theos* Hermes (Emesene), *BÉ* 2015, no. 697.

70. In Anatolia note the alternations *theos*/Zeus Olybris, *IAnazarbos* 44–47 (on this god, cf. B. Isaac, *ZPE* 117, 1992: 126–28); Tyrimnos/Apollo Tyrimn(ai)os, Paz de Hoz, *Die lydischen Kulte,* 280–86 (a god who eventually becomes Helios Pythian Apollo Tyrimnaios, *TAM* V.2.956); for other possibilities see appendix D below, p. 191. Note too combinations of Mother with possible (Atimis, Hipta, Phileis) or certain (Leto, Anaitis) theonyms (Paz de Hoz, 31–32); but in such combinations Mother becomes more title than name. In one case the theonym may be changed into an adjective: contrast "invincible god Asbameus/ei" (Marek, *Stadt, Ära und Territorium,* 165, no. 27, Amastris region) with the Zeus Asbameios of Tyana (*ITyana,* 2:317–19): cf. Mitchell, "Persians in the Sanctuaries," 167–68. Thrace: e.g., Zeus Zbelsourdos, Apollo Karabasmos, Hephaistos Dabatopeios. Several probable theonyms also combine with Ἥρως: see Kazarow, *RE* s.v. Thrake (Religion), 476, a list which could doubtless be extended.

71. The Arab god D(o)usares is just that at home and in Pozzuoli (Tran Tam Tinh, *Divinités orientales en Campanie,* 145–46), but Zeus Dousares in Miletus: *Milet* 1.3.165. For the West, see p. 45.

here is not to identify the gods of two different cultures, because in the combination Zeus Dionysus, for instance, the gods are both Greek. Instead, Zeus seems here to have lost specificity and to mean something like "great god." The dominant identity is that of the god named second, not of Zeus: a Zeus Herakles—an extraordinary conception—has recently emerged in Phrygia, and in the relief accompanying the dedication this Zeus Herakles is shown unambiguously as Herakles. This fading of "Zeus" to "great god" may also occur in some instances of the "Zeus Ammon" type (whence the qualification above that the distinction of the two types does not remain clear). It is unfortunately difficult to trace and explain the emergence of this weakened sense of Zeus. The type becomes common in the Roman imperial period, but there are isolated Hellenistic forerunners: on the mainland a single reference to Zeus Amphiaraos and a few to Zeus Trophonios; in Asia Minor the cult of Zeus Sabazios brought to Pergamum from Cappadocia by Stratonike, mother of Attalos III; Zeus Sarapis soon follows. The particular circumstances are not known of any of these namings; the trail leads into obscurity.[72]

Another category has only one member, but a remarkable one. This is the composite divine name Zenoposeidon, a blend of Zeus and Poseidon, sometimes added to the name of the chief god of Mylasa in inland Caria, Zeus Osogollis (probably a "Zeus Ammon" combination), to give as full name Zeus Osogollis Zenoposeidon. Already in antiquity people found this name bizarre: how could a visitor expect to find lodgings in Mylasa, a wit asked, where even gods have to double up in one temple?[73] Coins too present a mixed image: in the time of Mausolus this Zeus holds a trident in one hand, while an eagle perches on the other; later he acquires a crab at his feet, and sometimes a double ax. So he combines crab and trident, symbols that suggest the sea and Poseidon, with double ax and eagle, symbols that suggest Zeus and sovereignty. Such a unicum defies confident explanation. Zeus became Zenoposeidon, it has been suggested, when Mylasa acquired maritime interests, as happened at an identifiable moment in the fourth century.[74]

72. On all this, see Parker, "Zeus Plus." Zeus Herakles: *SEG* LIII 1460 (? Appia, in Phrygia; somewhat comparable is the dedication by a man of Askalon in Egypt in 228 A.D. to "my ancestral god Herakles Belos *aniketos*," Bernand, *Delta*, 1:242.14). Whether Zeus Ares (Pella in the Dekapolis, *SEG* XLI 1566) is simply of this type or influenced by assimilation to Dousares is unclear: see the note ad loc. in *SEG*.

73. Machon 53–63 Gow ap. Ath. 8.18, 337C-D. Here and in the neutral allusion by Theophr. ap. Ath. 2.15, 42A the god is simply Zenoposeidon without Osogollis. For the form Osogollis, not Osogoa, see *SEG* XL 991.2.

74. So Debord, "Zeus Cariens," 22–23; but K. Konuk, "Coinage and Identities under the Hekatomnids," in O. Henry, ed., *4th Century Karia: Defining a Karian Identity under the Hekatomnids* (Paris, 2013), 101–21, at 107 n. 35, points out that the crucial coin (which he illustrates) probably comes too early in Mausolus's reign to fit the suggestion. For the crab he points to "the crater lake of Deniczik where crabs are found in numbers and which is situated about 6 km southeast of Milas"; this leaves the trident unexplained. For the later coins, see Laumonier, *Cultes indigènes*, 117–20; A. Akarca, *Les monnaies grecques de Mylasa* (Paris, 1959), 46–50.

But such a rebranding of a god would be very unusual, and we have no way of knowing how young or old the title Zenoposeidon may be. Another possibility would be that the normal principle of identifying one native with one Greek god has proved unsatisfactory: *interpretatio* buckles under the strain of assimilating Osogollis, chief god of Mylasa, to Zeus, chief god of the Greeks, because Osogollis has a maritime aspect alien to Zeus which comes to be expressed in the unique new compound. Such may be the explanation for compound Latin names found in *interpretatio,* such as Mercurius Silvanus.[75] But it is hard to see why a god of inland Caria should originally have had such a pronounced concern with the sea.

GOD PLUS TOPONYM OR ETHNONYM

Much the commonest form in all regions is the combination of a theonym with a local specification. Most commonly this is a topographic epithet (typically formed with the suffix *-enos*[76] or less commonly *-eites*), but there are other possibilities: in Syria we find, for instance, "God Zeus of the Village of Ornea," " Zeus the God at Adatha," "Mercury Lord of the Village of Chamon," and simple juxtaposition of the god's name with a toponym (Zeus Bakathsaphrein);[77] there can also be reference to features of a local landscape—Zeus of the Twin Trees, Mother from the Cave (both Anatolian).[78] Such epithets are so common in Anatolia as to become a prime source for knowledge of Anatolian place-names: sometimes the corresponding place-name is independently attested, in which case the origin of the epithet is certain; at other times the place has to be inferred from the epithet, but can be done so with great confidence.[79] These epithets were applied in Anatolia both to

75. Mercury-Silvanus: Cadotte, *Romanisation des dieux,* 14, 123–29 (cf. 58–62 for Juppiter Saturn). A Roman woman dedicates at Akoris in Middle Egypt in 29 A.D. to a Hera Aphrodite who must be Hathor (*SEG* XXXVIII 1678, with J. Bingen, *BÉ* 1990, no. 824). An iconographic equivalent might be the varying depiction of Zeus Marnas as resembling Zeus or Apollo, reflecting a god seen as both young and supreme: Belayche, *Iudaea-Palaestina,* 240–41. For Balmarcod hovering between Zeus and Dionysus, see J. Rey-Coquais, "Deir el-Qalaa," *Topoi* 9 (1999.2): 607–28, at 612. It is widely held that certain Phoenician Baals were both celestial and maritime, and might emerge either as Poseidon, as at Berytus, or Zeus, as at Baitokaike: see, e.g., J. P. Rey-Coquais, in *IGLS* VII p. 74; for further references and some reservations, see Bonnet, *Enfants de Cadmos,* 150.

76. Cf. in brief A. Chaniotis, *AJA* 102 (1998): 249, who cites I. Duridanov, "Probleme der thrakischen Sprache III," *Linguistique Balkanique* 32 (1989): 85–112, for similar toponymic epithets in Thrace, Moesia, and Dacia.

77. *IGLS* XI.42; *Dura* V, 308, no. 610; Hajjar, *Triade d' Héliopolis-Baalbek,* no. 168 (Μερκουρίῳ δωμίνῳ κώμης Χάμωνος); *BÉ* 2004, no. 384 (cf. Zeus Baitokaike).

78. Twin trees: Petzl, *Beichtinschriften,* 9–12; from the cave *SEG* XV 816; but Mother ἀπὸ Κρανος Μεγάλου is not now thought to relate to κρήνη (*SEG* LVIII 1506–7).

79. Thus such epithets are a main source for Zgusta, *Ortsnamen.* For some demonstrable cases, see Schuler, *Ländliche Siedlungen,* 248.

gods with indigenous and to gods with Greek names: the long lists that can be assembled of local epithets of Men and Mother on the local side are matched by those of Zeus, Apollo, and Artemis on the imported.[80] The situation is just the same in Thrace, where variants of the "Thracian hero" earn innumerable toponyms, and so, for instance, does Asclepius, many more than in Greece.

As we have seen, such designations were common to all the ancient polytheisms, so the application of the Greek usage to indigenous deities was easy and natural; the exact form of expression may sometimes closely reflect a non-Greek original. The use of such topographic double names was an extension beyond the Greek world of a naming practice already characteristic within it, and was of central importance for the relation between originally Greek and originally non-Greek gods. The renaming of Amun-Re of Egyptian Thebes as Zeus Thebaieus added this profoundly foreign god to the long list of Zeus plus local epithet combinations already existent. To some extent it acknowledged his singularity through the epithet, but it laid no stress on the difference between the Greek god and the Egyptian: as was noted in chapter 2, though we might regard the difference between the Zeuses of two Attic mountains as one of small variation on a common model, and that between either of these and Zeus Thebaieus as drastic and complete, as a form of expression the cultic double name puts them all on the same level.

The example is a little misleading, however, because Egypt was much less productive of such combinations than Anatolia and Thrace; even "Zeus Thebaieus" became Ammon in later dedications.[81] The difference was not that Egyptian gods were less tied to localities than those of other regions; perhaps more strongly even than elsewhere, every place belonged to a god (whence the many names like Hephaistopolis and Letopolis). The difference probably lay rather in the plethora of existing named gods who would not be pushed aside to become merely yet another "Zeus plus toponym." The commonest recipients of topographic designations in Egypt are Egyptian gods, Isis above all.[82]

80. On the epithets of Men, see Labarre, *Dieu Mèn,* 36–44 (for a new one, Matrios, from Bithynia, see *BÉ* 2015, no. 665). The general assumption is that the god of a "Greek theonym plus local epithet" combination, when found outside Greece, will be an indigenous god principally worshipped by the local population, but Greek settlers seem sometimes to have adopted the local usage. See, e.g., the relations of the "settlers at Daphnous" (in Hellespontine Phrygia) with an Apollo Daphnousios, *SEG* XLIII 879–84; cf. Schuler, *Ländliche Siedlungen,* 33, 191, 255.

81. Cf. Ronchi, *Lexicon theonymon,* 109–18, and *SEG* XXXVI 1405–54; but Thebes remained "Diospolis."

82. E.g., Isis at Philai, Pharian Isis, "the gods at Abydos" (Perdrizet, *Memnonion,* 80), Tithoes god of Kellis (*SEG* XLV 2091), Petesouchos crocodile god of the village (*PTeb.* 63.25). Some demotic epithets appearing transcribed in Greek are apparently topographic, though were not necessarily so understood:

A topic on which research has barely touched is that of the kind of locality from which the toponymic adjectives derive in different cultures. The apparently shared naming practice may hide substantial differences. The typical source of a local divine epithet in Anatolia is a small-scale human community, a village or small town.[83] It was different in mainland Greece: the Attic deme of Erchia worshipped Apollo under six different epithets but not under the epithet Erchieus (*LSCG* 18); among the six epithets that were used, the only toponymic one was Pythios, which refers not to a settlement but to the prestigious sanctuary at Delphi. The main local epithets of Artemis in Greece, Brauronia and Mounychia and Amarynthia, come from sanctuaries set in isolated seaside spots, not from communities she protected. Probably the largest class of local epithets of Zeus in Greece derives from mountains: mountain epithets are common in Anatolia for Mother—she who is sometimes called simply "of the Mountains," *Oreia*—but rare for Zeus.[84] The Levantine pattern appears to be mixed: there are male gods both of mountains[85] and of communities. Both in Anatolia[86] (if not in all its regions) and in Syria we encounter, repeatedly, the idea of the god who rules over or is lord of a village; the same conception is likely to underlie the village-derived local epithets. Near Byzantium there is even Zeus Komatikos, "of the Village" or "of the Villagers"; dedications to him are explicitly made by or for the villagers.[87] The gods of a village are sometimes spoken of as an anonymous collective, the *theoi* Tabalenoi (Lydia)

e.g., Isis Esenchubis = Isis of Chebis (E. Bernand, *Inscriptions grecques d'Egypte et de Nubie au Musée du Louvre,* Paris 1992, no. 20); Aphrodite Tazbes (Ronchi, *Lexicon theonymon,* 724). Von Lieven, "Translating Gods," 78–79, suggests that animal form may have discouraged *interpretatio* of some gods (Khnum, Sobek, Tutu, and Apopis).

83. More usually the former, though note Strabo 13.3.2, C 620: Apollo Larisenos, of the town Larisa in the Kayster valley; *New Documents from Lydia,* no. 83, three gods who "rule Kollyda." On ancient Near Eastern toponyms, see n. 15 above.

84. Zeus Androklas (*Westkilikische Inschriften,* Kestros b) is taken as an oronym by G. Bean and T.B. Mitford, *AnatSt* 12 (1962): 215, and Zeus is Ὀρείτης in northeastern Lydia (*New Documents from Lydia,* no. 86, with note). Some Anatolian epithets are thought to derive from rivers (a type also found in Greece: e.g., Athena Krathia, Hdt. 5.45.1; Artemis Alpheiaia, Paus. 6.22. 8–10; Hera Imbrasie, Ap. Rhod. *Argon.* 1.187 with schol.); see Zgusta, *Ortsnamen,* ¶ 109, Artemis Asteleane (Akmonia); ¶ 353.6, Apollo Thymbraios (Troad); ¶ 797, Apollo Mekastenos (near Kyzikos). Meter Oreia: S. Şahin, *EpigAnat* 17 (1991): 127; *MAMA* X 307, with references; *SEG* XLIV 1028.

85. Zeus Kasios is a famous example; on the epithet Libanos used in relation to M. Hermon, see Aliquot, *Vie religieuse au Liban,* 17–23.

86. See N. Belayche, "Au(x) dieu(x) qui règnent sur . . .," in A. Vigourt et al., eds., *Pouvoir et religion dans le monde Romain* (Paris, 2006), 257–69, with works cited in her n. 9. Schuler, "Rural Population Groups," 81–87 and 95, argues, against the usual "theocratic" view of these formulae, that calling a god king is a way to stress his power to do one good, not a form of self-abasement before him. In *RC* 70.6 we encounter a reverse conception: the power of Zeus Baitokaike depends on his sacred village.

87. Robert, *Hellenica* X, 38–46 (*SEG* XV 423–25).

or the *theoi* of the Motaleis (Phrygia) or the gods at Thebnein (near Tyre, Phoenicia), for instance.[88] The presumption here (sometimes supported by iconography) is that identifiable gods are referred to, but it is the association with the place that is stressed. In old Greek naming, the addition of a toponym to a god's name will often reflect an outside perspective (an offering brought by a nonresident; a cult that has moved from its place of origin); but it is the Motaleis who speak of the gods of the Motaleis, "as a mark of cohesion between the god and his worshippers."[89]

Formally similar to the toponymic adjective, but in some ways differently used, is the adjective deriving from a people, what one might call the ethnonymic adjective: Persian Artemis, Carian Zeus, Macedonian Dionysus, Syrian Aphrodite, or, without the god's name, the Syrian goddess, the Rhodian god, the Arabian god.[90] It works in more than one way. It can give an outsider's view of a god particularly associated with a people, especially if that god has been introduced to the outsider's own territory. The "Syrian goddess" is normally[91] Atargatis seen from outside Syria; the "Rhodian god," known from a dedication in Phrygia, must be Helios, because Helios was more honoured in Rhodes than in any other place; the "Arab god" is attested chiefly in dedications made in the region of Gerasa in the Dekapolis (see map 2), which would not have considered itself an "Arab" town.[92] Artemis

88. See Paz de Hoz, *Die lydischen Kulte,* 84; Schuler, *Ländliche Siedlungen,* 248–49; Rey Coquais, *Inscriptions de Tyr,* 6 (for Syria, cf., e.g., the gods of Kiboreia, *IGLS* XI.4). In some cases there can be doubt with such expressions whether a specific narrow group (e.g., a triad) or all the gods known to the community are envisaged: see M. Wörrle, *Chiron* 21 (1991): 227–28, on the "gods of the Pernitai" (an eastern Lycian *koinon: SEG* XLI 1379); and cf. Schuler, *Ländliche Siedlungen,* 249 n. 205. Schuler suggests that such teams of gods will often have shared one temple. On the great gods of Keramos, see p. 144.

89. Robert, *Hellenica* IX, 22 n. 1. But there are exceptions in both directions: we meet Zeus Phainesios at Damet et-'Alya, whereas at Phaena itself he is "greatest, highest": Sourdel, *Cultes du Hauran,* 24; on Greece itself, see p. 17n63.

90. Somewhat different are generalised references to the gods of Greeks, Romans, and Persians, for which see Versnel, *Coping with the Gods,* 106.

91. Not exclusively: cf. Baslez, *Religions orientales à Délos,* 94–95; Lightfoot, *Syrian Goddess,* 38; Aliquot, *Vie religieuse au Liban,* 146. Syrians sometimes took up the external appellation, particularly Syrian slaves abroad, to mark a national cult (cf. n. 102). For the uncommon "Syrian Aphrodite," see Lightfoot, *Syrian Goddess,* 537; her understandable doubts about the presence of this cult in the Black Sea region in the fifth century B.C. are apparently met by a new text, *SEG* LIV 630.

92. θεῷ (ἁγίῳ) Ἀραβικῷ, *IGerasa* 19–21; cf. 22; *SEG* XIV 830; XXXII 1540; cf. XLI 1567, the "*ouranios* Arab god," from Pella of the Decapolis; on Gerasa's claim to a Greek origin, see P. L. Gatier, "La présence arabe à Gérasa et en Décapole," in H. Lochazmeur, ed., *Présence arabe dans le croissant fertile avant l' Hégire* (Paris, 1995), 109–18, at 112–13, 118. Much stranger is the solitary dedication to "Atargatis of the Arabs" from Faqra in the Lebanon, *SEG* XLIX 2011: cf. A. Kropp, in M. Blömer et al., ed., *Lokale Identität im römischen Nahen Osten* (Stuttgart, 2009), 112–17; T. Kaizer, in *Religious Identities in the Levant,* 27. Rhodian god: *MAMA* X 447. The single attested dedication to "Armenian god" appears to come from the Lebanon. *SEG* LV 1870; cf. *MonPiot* 87 (2008): 102, no. 15 with p. 77.

Persike[93] is likewise found not in Persia but in Lydia; she is the goddess also known as Artemis Anaitis or Mother Anaitis, Artemis in her Persian guise, a foreign goddess imported into Lydia (or at all events a goddess imagined to be such) and there identified with the Greek goddess Artemis, but also kept separate by the ethnic adjective.[94] So here the epithet not merely identifies an origin but also distinguishes between this and other forms of Artemis. "Phrygian Mother of the Gods" works similarly;[95] the title is attested exclusively outside Phrygia and points back to Phrygia as the home par excellence of the Mother, though also hinting at a distinction between her and any other Mother. "Tyrian Herakles" makes a slightly different distinction, between Melqart and the Herakles of Greece. A trio of Egyptian healing gods admitted to Epidaurus are marked off from their local equivalents, unusually, by being designated "Egyptian Hygieia, Asclepius, and Apollo." It was as a way of differentiating themselves from their neighbours in Jerusalem that the Samaritans sought permission to dedicate their temple at Gerizim to Zeus Hellenios.[96]

93. The variant Artemis Medeia ("Median") was detected by S. M. Sherwin-White, *ZPE* 49 (1982): 30, in *SEG* XXXII 1612.14. For Persian Zeus or Zeus of the Persians, see *TAM* V.1.267; *SEG* XXVIII 1186 (Nakoleia): very commonplace dedications, their recipient aside.

94. M. Brosius, "Artemis Persike and Artemis Anaitis," *Achaemenid History* 11 (1998): 227–38, denies that Artemis Persike is a "hellenised" Anahita; she sees her rather as a "persianized" Artemis. But in determining the reference of the term Artemis Persike that polarity is not the issue; those who used it certainly meant by it Anahita. On the Persian cult (probably of this goddess: aliter Brosius) at Hypaipa and Hierakome/Hierocaesarea, see L. Robert, "Types monétaires à Hypaipa de Lydie," *RN*, 1976, 25–48 (*OMS*, 6:137–60, without the plates) remains basic. Some later literary texts similarly add an ethnonymic adjective where Herodotus would have engaged in unadorned *interpretatio:* Assyrian Hera, Lucian, *Syr. D.* 1; Assyrian Aphrodite, Nonnus, *Dion.* 3.111. Cf. Ceres Africana, Tert. *Ad uxorem* 1.6.

95. For attestations from six places, all outside Phrygia, see Robert, *Documents,* 323. For a priesthood sale prescribing that one priestess in Iasos is to serve both the Mother of the Gods and the Phrygian Mother, see G. Maddoli, *SCO* 61.2 (2015): 101–18, and the discussion by M. Nafissi, ibid., 119–36; note the probably Phrygian names of the persons who contribute for the Phrygian mother in *IIasos* 229. "Sikel Hera" represents a Syracusan adoption of an indigenous figure interpreted as Hera, if B. Bravo's interpretation (B. Bravo, " 'Hera dei Siceli,' 'Dea di Hybla,' e 'Demeter Signora di Enna,'" *Annali dell' Istituto Italiano per gli Studi Storici* 12, 1991–94: 141–79) of Aelian *VH* 6.11, ἐν τῶι τῆς Σικελίας Ἥρας ναῶι, is correct. I venture no opinion about Lakaina Artemis (*IGerasa* 43) or Men Italikos (*MAMA* V 150 = Lane, *CMRDM,* 1:93, from Phrygia), both slightly supplemented. Other apparent ethnonyms too ill attested to analyse are Demeter Βοιωτάη (Graf, *Nordionische Kulte,* 69–70 [Chios]; a product of Theban diplomacy in the Aegean in the 360s according to A. Schachter, in N. Papazarkadas, ed., *The Epigraphy and History of Boeotia,* Leiden 2014, 327); Demeter Λίβυσσα of Argos (Polemon fr. 12 *FHG* III p. 119; Festus s.v. Libycus p. 108 Lindsay; speculation in F. Colin, *BIFAO* 97, 1997: 101); Molossian Apollo (schol. Lyc. *Alex.* 426); Demeter Achaia/Panachaia (for the former, see Schachter, *Cults of Boiotia,* 1:162; for the latter, *SEG* LIII 565.10–11, Phthiotic Thebes). On the problematic Thasian Herakles, see appendix E.

96. Epidaurus: Paus. 2.27.6. An unnamed "Egyptian god" follows Greatest Heliopolitan Zeus in *IGLS* VI.2731 (cf. *SEG* LII 1573). Tyrian Herakles: *RIB* 1129; cf. the dedication to Astarte from the same site (Corbridge), *RIB* 1124. M. Gerizim: p. 71 above. Note too, e.g., Aitolian Artemis in Ozolian Locri

Less often, such an ethnonym is used within the ethnos concerned. Sometimes it has a clear contrastive force. The unique dedication to "Lydian Mother of Gods" comes from Sardis, within Lydia; the point here is to separate a supposedly native goddess from a more famous foreign goddess of the same name, a point that becomes explicit in Macedonia with a goddess called Autochthonous Mother (of the Gods). These rivals to the famous Phrygian Mother have been joined by a Pontic Mother of the Gods whose temple was discovered at Dionysopolis in Bulgaria in 2007.[97]

In other cases the force of the epithet is less clear. "Lydian Zeus," the "Pisidian goddess," and "Pisidian gods" appear in a region on the borders of Phrygia, Caria, and Pisidia which had once been under Lydian control and where, according to Strabo,[98] four languages were spoken (Pisidia, Solymian, Greek, Lydian): here the ethnic adjectives given to gods may reflect the uncertainties and perhaps tensions of life in ethnically ambiguous territory (but Lydian Zeus occurs also on coins of Sardis).[99] Perhaps the ethnic adjective sometimes indicated a cult which members of an ethnic group felt to be a distinctive part of their identity. The temple of

(Paus. 10.38.12) and among the Eneti (Strabo 5.1.9, C 215); a dedication to "Macedonian" Dionysus by three women of Larisa in Thessaly (*SEG* XXVII 206, late 1st c. B.C.). Apollo Doreios from two sites in Perrhaebia (*SEG* XXXV 542, 657: late Hellenistic/Roman) appears to preserve/revive traditions of Dorian presence in that region (B. Helly, "Apollon Doreios: Recherches sur les Doriens de Thessalie," available at http://www.hisoma.mom.fr/bhelly). An "Ancestral and Argive Apollo" was honoured in Tarsus in the third century A.D. (*SEG* XXVII 947) as a way of asserting the Argive origins of Tarsus.

97. Lydian: Robert, *Documents*, 321–23. Macedonian: P.M. Petsas, M.B. Hatzopoulos, L. Gounaropoulou, and P. Paschidis, *Inscriptions du sanctuaire de la mère des dieux autochthone de Leukopetra (Macédoine)* (Athens, 2000). Pontic: *SEG* LX 758–79; cf. I. Lazarenko et al., *The Temple of the Pontic Mother of Gods in Dionysopolis* (Varna, 2013) (*non vidi*). Whether the claim is that there was just one "Mother," born in the place in question, or that plural mothers existed, is not clear; *autochthon* applied to Hera in the Samian poem *IG* XII.6.1.285.10 is clearly claiming Samian origin for the famous Hera (cf. Paus. 7.4.40).

98. Strabo 13.4.17, C 631; cf. Robert and Robert, *Carie*, 76. Pisidian gods: known from curses from the Acıpayam plain, J.H.M. Strubbe, *ΑΡΑΙ ΕΠΙΤΥΜΒΙΟΙ* (Bonn, 1997), nos. 129–39, with pp. 296–97; new examples, *EpigAnat* 45 (2012): 89–92. Pisidian goddess: Robert and Robert, *Carie*, 75 n. 7; for what is probably the same goddess without inscription on coins of Sebastopolis and Kidrama, Robert and Robert, 76. But the "Pisidian gods" of the curses may arise from the desire to evoke a collectivity in curses: "Lycian gods" similarly appear, uniquely, in a grave curse from Limyra, within Lycia (*SEG* XLV 1802).

99. Cf. the Zeus Pratomysios of *IApameia* 114, from a region originally Mysian but later Bithynian. On Lydian Zeus (coins of Sardis and Kidrama), see Robert, *Villes d'Asie Mineure*, 214 n. 4, 298, 301–2; Robert and Robert, *Carie*, 76, 367; note too the legend in Nonnus, *Dion.* 13.474–97 by which a priest of Zeus Lydos left his temple and stayed the devastating onset of Typhoeus by a word of power. Eumelus supposedly located Zeus's birth in Lydia: fr. 18 Kinkel and Bernabé; fr. dub. 4 Davies, ap. Lyd. *Mens.* 4.71, p. 123 Wünsch. I merely note Pamphylian Apollo, who appears just once, within Pamphylia, in a rather florid dedication (*IPerge* 56): I suppose the epithet expressed regional pride; so too "Carian god," known only from a rider relief of unknown provenance showing an Apollo-like figure (Delemen, *Rider-Gods*, no. 312).

Cataonian Apollo was imitated throughout Cappadocia, according to Strabo, so this was perhaps a national cult, Apollo of all the Cataonians/Cappadocians.[100] Herodotus speaks of what he implies is a pan-Carian sanctuary of Zeus Karios at Mylasa (1.171.6), which might suggest that the force of the epithet was again collective, "common to all the Carians." (But nothing else that we know about Zeus Karios supports Herodotus.[101]) And, finally, a group of co-nationals away from home can discover a national god who never existed at home: an association of Phrygians in Alexandria dedicated an image of Phrygian Zeus, much as Greek traders in Naucratis seem to have honoured "gods of the Greeks."[102]

THE SPEAKING EPITHET

The commonest category of epithet after the toponym is the speaking epithet of traditional type. When a god plus epithet combination of a type familiar from old Greece—such as Apollo Lykeios ("of Light"? "of Wolves"?) or Zeus Ktesios ("of Property") or Athena Polias ("of the City")—appears in the East, it will often have been imported by settlers. But there is no reason or principle why a whole "cultic double name" should not have been used in *interpretatio* just like a single name; and some will have fitted their recipients like a glove, Zeus Keraunios, "of the Thunderbolt," for the many storm- and weather-gods of Anatolia and Syria, for instance. A few less obvious cases can be identified. Within Herodotus's application of Greek names to foreign deities, his use of Aphrodite Ourania, "Heavenly," or plain Ourania stands out, since here he adds an epithet to the simple theonym which usually suffices him in *interpretatio*:[103] evidently (Aphrodite) Ourania identified a non-Greek Aphrodite of particular characteristics. Apollo Archegetes, "Leader," is a god traditionally associated with the early colonisation movement, but when we meet him far from the sea at Komana in Cappadocia he must cover a local deity. Zeus Epikarpios, "of Fruits," of Bostra and Gerasa (see map 2) will be

100. Strabo 12.2.5, C 537; on the relation of Cataonia to Cappadocia, 12.1.2, C 533. The epithet might, however, have begun as separative: Zeus of Cataonia as opposed to Cappadocia.

101. Hdt. 1.171.6. Herodotus's sanctuary disappears from the record; the epithet reappears at a different site, Panamara near Stratonicea: see Debord, "Zeus cariens," 31–34.

102. Phrygian Zeus: *OGIS* 658; F. Kayser, *Recueil des inscriptions grecques et latines (non funéraires) d'Alexandrie impériale* (Cairo, 1994), no. 74. Naucratis: Polinskaya, "Shared Sanctuaries," 53–57 (though note her doubts). A dedication to "the Thracian gods" from the region of Nicomedia in Bithynia may be comparable, *TAM* IV.1.84.

103. See V. Pirenne-Delforge, in *Nommer les dieux*, 274–79, on 1.105.2, 4.59.1 and 2, and (plain Ourania) 1.131.3, 3.8.1 and 3. She argues that the Egyptian Aphrodite mentioned by Hdt. without epithet in 2.41.5 and 2.181.4 was seen by him as a different figure, whence the characterisation of the Phoenician shrine of Aphrodite in Memphis (2.112.2), who will have been (cf. 1.105.2) an Ourania, as (in Egyptian eyes) ξείνη. Hdt. 4.33.5 speaks of Thracian and Paionian women sacrificing τῇ Ἀρτέμιδι τῇ βασιληίῃ, perhaps an epithet invented to cover a foreign cult (Lerat, *Locriens de l' Ouest*, 2:159).

heir to a god of the region (which god is not certain) responsible for the crops; Zeus Arotesios, "of Ploughing," of Hippos in the same region is said in a literary source to represent the West Semitic god Dagon. Apollo Lykeios, in whose epithet some heard the noun λύκος, "wolf," makes a startling appearance at Luxor in Thebes as recipient of a statue of a jackal.[104] A traditional epithet could also be applied to an untraditional recipient. Egypt offers, for instance, Boubastis Soteira, Anoubis Hegemon, and even (but not before the Roman period) Serapis Polieus—three of the commonest epithets for Greek gods now applied to Egyptians.[105]

Occasionally an epithet was taken over from a different language.[106] Zeus Pharnaouas and Anaeitis Barzochara in Cappadocia bear epithets transliterated from old Persian, thought to mean something like "Possessing Glory" and "Belonging to High Harā/Mount Harā"; a Persian origin, "of Horses," has also been suggested for Zeus Asbameios.[107] More such cases from different languages may well hide among

104. Archegetes: *BÉ* 1971, no. 668 (*SB* V 7747, greatest god Apollo archegete, is taken as an indigenous god too). Epikarpios: *IGerasa* 42; *IGLS* XIII.9104 (Sartre ad loc. argues for Baalshamin). Arotesios: H. Seyrig, *Syria* 36 (1959): 70, 77–78; cf. Philo of Byblos, *FGrH* 790 F 2 ap. Euseb. *Praep. evang.* 1.10.25 (Zeus Arotrios); Paz de Hoz, *Die lydischen Kulte,* 49, supposes the same for a Lydian Zeus Aithrios. Lykeios: *SB* I 683 (written Lykaios); contrast a priest of "the dog-headed god" at Abydos (?) (*IGR* I 1161). The Apollo Propylaios of Eumeneia (Greek epithet, Anatolian iconography) may rather reflect adoption of a local iconographic convention: see p. 222n88.

105. *SB* I 1162; *IG* XI.4.1253 (the epithet Hegemon belongs to Hermes, whom Anoubis sometimes "translated"; the same text offers Sarapis Soter and Isis Soteira); *OGIS* 708.11–12 (cf. Ronchi, *Lexicon theonymon,* 901–2). Cf. n. 125 below on Isis.

106. It has often been suggested that the Semitic root MLK underlies the epithet *meilichios,* which would then have been part-Hellenised by popular etymology: see, e.g., M.H. Jameson, D.R. Jordan, and R.D. Kotansky, *A* Lex Sacra *from Selinous,* GRB Monographs 11 (1993), 139–41; T. Petit, *Cahiers du Centre d' Études Chypriotes* 37 (2007): 283–98, which I know only from *SEG* LVII 1737. Saturn Balcaranensis = Baal Qarnaïm is a Latin example: Cadotte, *Romanisation des dieux,* 38. For a Greek epithet entering Etruscan cult, see S. Haynes, *Studi Etruschi* 69 (2003): 73, with references, on (Fufluns) Pachie = Βακχεῖος. For an epithet of El acquiring autonomy as the god Κόνναρος/Connarus, see p. 45n49.

107. C.P. Jones, *Epig Anat* 37 (2004): 98–99. For Old Persian *Farna-vant, "possessing glory," Elizabeth Tucker refers me to J. Tavernier, *Iranica in the Achaemenid Period* (Leuven, 2007), 179, and for "the exactly corresponding Avestan adjective" *xᵛarənahvan,* frequently used as "an epithet for various gods, including Ahura Mazda, in the Younger Avesta," to Yasna 1.1, 12.1 (ed. K.F.Geldner, 1888–96), Yašt 8.1 (ed. A. Panaino, 1990). She wonders whether "the Achaemenid winged disk figure on the Bisitun and Naqsh-i-Rustam reliefs *might* have been described in Old Persian as *Auramazdā farnavant-, "Ahura Mazda possessing/in possession of the farnah-." She notes that Yašt 5, a sacrifice to Anahita at the foot of Harā, a mountain elsewhere described (e.g., in the "hymn" to Mithra, Yašt 10) as "high Harā" (harā- bərəzaitī- or harā- bərəz-), "probably justifies seeing an Old Iranian compound adjective "belonging to high Harā/Mount Harā" in Barzochara, although it is not attested as such anywhere else." Zeus Asbameios: Mitchell, "Persians in the Sanctuaries," 167–68, citing predecessors: from Old Persian asp/b, "horse." Athena's Theban epithet Ὄγκα was taken in later antiquity as Phoenician (e.g., Paus. 9.12.2; more in Schachter, *Cults of Boiotia,* 1:130), a view not supported to my knowledge by Phoenician evidence.

the many epithets from Anatolia which we cannot interpret. The reference to Anaitis "from the Sacred Water" in a text from Lydia might, by contrast, be a translation; no exactly equivalent old Persian epithet has been quoted, but Anahita's close association with springs and water is well established.[108] There is one certain case of an Aramaic word used as an epithet in Greek both transcribed and translated: MDBḤ means "altar," and in northern Syria we encounter both a Zeus Madbachos (Μάδβαχος) and a Zeus Bomos, "Altar."[109] (Zeus Βέτυλος, "betyl," is closely comparable, except that here the Semitic original Bethel had been seminaturalised in Greek; in both cases we have the un-Greek juxtaposition of theonym with physical object.) An honorific certainly translated from Aramaic is "Pitying" found in a single Greek dedication of gratitude from Palmyra:[110] ἐλεήμων translates the RḤMN very common in Palmyrene texts. But it failed to enter broader Greek usage.

A plausible and much earlier instance of translation is Strat(e)ios, "of the Army," attested for Zeus in Caria in the fifth century by Herodotus (5.119.2) and later in Pontos and elsewhere in Asia Minor, applied also in western Asia Minor and the islands to Aphrodite and Hecate.[111] Thus its origin appears to be Anatolian. A. B. Cook rather charmingly refers to Zeus Stratios as "Lord of Hosts," and one might wonder whether there is any connection between the Greek epithet and that familiar phrase from the Old Testament; geographically more tempting is the strong case that has been made for associating it with the very frequently attested Hittite thunder-god or weather-god "of the army," an association that might help to explain its occurrence both in Caria and in Pontos but not in places between.[112] Ourania, "Heavenly," used of Aphrodite is another instance, one assimilated into mainstream Greek religious language, if we accept that it reflects the title "Queen

108. See L. Robert, *RN* 18 (1976): 43–46 (*OMS*, 6:155–58), using Yašt 5 to illuminate *TAM* V.1.64, τὴν Ἀναεῖτιν τὴν ἀπὸ τοῦ ἱεροῦ ὕδατος.

109. *IGLS* II.465–73, with commentary on 465; *IGLS* II.569; cf. Schwabl, "Epiklesen," 293, 331. Zeus Βέτυλος: *SEG* VII 341 (Dura-Europos).

110. εὐχαριστεῖ Μαλχος Βαρεα τοῦ Μαλιχου ἑνὶ μόνῳ ἐλεήμονι θεῷ, *IGLS* XVII.I.32, with commentary. For pity/mercy as a divine characteristic also in Ugaritic texts, the OT, and predicated of Allah, see Day, *Yahweh*, 26.

111. Zeus: see V. Gebhard, *RE* s.v. Stratios, 256–62; Schwabl, "Zeus, Nachträge," 1474; add now *SEG* LVII 1150 B 4–5; *TAM* V.1.681; *MAMA* XI 311, and for instances in Athens and the islands (brought from Asia Minor) the note to *SEG* LIX 1913. Aphrodite: Graf, *Nordionische Kulte*, 177–78, 262–64 (Erythrai); Hekate: *IG* XII.4.358.19–20; 624–32 (Cos). One would very much like to know whether the god was in fact already known as Zeus Stratios, and not some Carian equivalent, at the time (the Ionian revolt) referred to by Hdt.

112. See A. Teffeteller, "The Lineage and Role of Zeus Stratios," in G. R. Tsetskhladze, *The Black Sea, Paphlagonia, Pontus, and Phrygia in Antiquity* (Oxford, 2012), 223–27. Cook: *Zeus*, 2:591. On Lord of Hosts in the Old Testament, a contested term, see Mettinger, *In Search of God*, 123–57, esp. the survey, 154–57; Allen, *Splintered Divine*, 290–94.

of Heaven" that characterised several great goddesses of the ancient Near East.[113] The treatment of epithets in bilingual inscriptions is a special case, since the attempt to create a bilingual posed a special challenge, and an equivalence sought there would not necessarily be carried over into monolingual usage: it is discussed in appendix F. On the whole, evidence for translated epithets remains sporadic,[114] except for translation between Greek and Latin; and even here translated forms remain largely confined to bilinguals or learned contexts and do not enter common monolingual usage.

A god whose name is not translated but, remarkably, alluded to is Balmarkod (and variants), worshipped at Deir El-Qalaa inland from Beirut in present-day Lebanon. Balmarkod remains just that in Greek, not Zeus Balmarkod, though in Latin he becomes Iuppiter Optimus Maximus Balmarcod; when he is given an alternative name by the ὁ καί formula it is apparently again Semitic, Μηγριν, perhaps a participle meaning "beating grain." But his name can be broken down as "lord" (Baal), plus the Semitic root RQD, to "leap, dance," thus "lord of dances," and Greek dedications from the site refer to this sense: he is addressed as κοίρανε κώμων, "lord of revels," described as χορῶν ἀρχηγέτης, "leader of choruses," and apparently associated with ivy of Nysa (that Dionysiac place).[115] Perhaps the tension between his status, which would suggest Zeus or Poseidon, and his activity, which would suggest Dionysus, discouraged *interpretatio*.

GRECO-EGYPTIAN

Egypt needs to be treated as a special case, because here, uniquely, contemporaneous evidence on a grand scale is available from two languages. No culture generated more elaborate divine titulature than ancient Egypt: the eight stately volumes of the *Lexikon der ägyptischen Götter und Götterbezeichnungen* stand as witness.[116] Greek epithets are laid out comprehensively in the laborious work of Giulia Ronchi, *Lexicon theonymon rerumque sacrarum et divinarum ad Aegyptum pertinentium, quae in papyris ostracis titulis Graecis Latinisque in Aegypto repertis laudantur* (Milan, 1974–77)—this too is multivolumed, though its size was vastly inflated by the decision to list, for example, every occurrence of Ptolemy in a dedication or

113. See Day, *Yahweh*, 144–50 (the OT attestations are Jeremiah 7.18 and 44.17–19, 25); C. Houtman, in *DDD*, 678–80; C. Metcalf, *The Gods Rich in Praise: Early Greek and Mesopotamian Religious Poetry* (Oxford, 2015), 177–78; Allen, *Splintered Divine*, 4.

114. For the epithet Tiamou of Men possibly translated by καταχθόνιος, see p. 117n19.

115. On all this, see J. Rey-Coquais, "Deir el-Qalaa," *Topoi* 9 (1999.2): 607–28 (*SEG* XLIX 1990).

116. Ed. C. Leitz (Leuven, 2002–3). Useful pointers on Egyptian naming in Bricault, "Isis Myrionyme," 68; for the titles of Isis from one site, see O. E. Kaper, "Isis in Roman Dakhleh," in L. Bricault and M. J. Versluys, eds., *Isis on the Nile: Egyptian Gods in Hellenistic and Roman Egypt* (Leiden, 2010), 149–80, at 157–61.

Soter or the ubiquitous "great/greatest." These two corpora of material, one colossal, the other substantial, need to be compared by a person expert in all three relevant languages (hieroglyphic, demotic, Greek). But in default of such a study a few observations can be made.

An unmistakable first point of contrast is the relative simplicity of the Greek. Though the baroque indigenous naming of monarchs was sometimes reproduced in Greek, that of gods is unrecognisably pared down. In a trilingual of 12 B.C. from Dendera in Upper Egypt, "Hathor, Mistress of On, Eye of Rê, Mistress of the sky, Lady of all gods" of the demotic (and presumably the largely lost hieroglyphic) comes out in Greek as "Aphrodite."[117] Dedications to groups of gods, none of whom bear an epithet ("to Sarapis, Isis, Anubis, and the gods who share their temple"), are extremely frequent. It is a commonplace of scholarship that in Egypt an indigenous god may underlie an unadorned Greek theonym—this in contrast to Anatolia, where the assumption is that a local god will usually be represented in Greek by a theonym plus toponym combination. Translation of Egyptian epithets into Greek appears to occur only occasionally (some are found transcribed, not translated, mostly in Egypt itself).[118] The main exception is the pervasive application of *megas,* "Great," or *megistos,* "Greatest," or multiples of the same (culminating in *megistos kai megistos kai megistos,* i.e., *trismegistos,* "Thrice Greatest") to a great variety of gods. "Great" here directly reproduces the hieroglyph *ʿaʿa,*[119] and the examples cover pages in Ronchi's compendium. Min "traveller," *nʿy,* may live on in (Egyptian) Pan εὔοδος, "of Good Journeys," an epithet perhaps not found outside Egypt.[120] A certain number of other epithets are used—honorific rather than strictly functional—that are not standard in other regions of the Greek world: "Good" (ἀγαθός/χρηστός), "Grateful," "Gracious" (εὐχάριστος), "Strong" (κραταιός), "Living Forever" (ἀείζωος), "Giving Abundant Returns" (πολύφορος), and (once each) "Glorious" (εὔδοξος), "Hospitable" (φιλόξενος), "Victorious" (ἔκνικος), "Full of Excellence" (ἐνάρετος), "Most Happy" (εὐτυχέστατος).[121] These may be Hellenisations of Egyptian concepts, but their use is spasmodic. At Koptos in the Thebais, Isis is "of the Hair" (Ἶσις τριχώματος), with an

117. Cairo Museum 50044; *SEG* VIII 653; Bernand, *Portes du désert,* no. 24; Vleeming, *Short Texts in Demotic,* no. 163 (whose translation I use).

118. See, e.g., the entries Ἀθερνεβθφι, Ἑρμουθις, Ἐσενχηβις, Νεφερσης, Νεφρεμμις in Bricault, *Myrionymi.* On divine names without epithets, see n. 66 above. In *SEG* VIII 653 (cf. previous note) "Isis as Thermuthis, Mistress of [—]" of the demotic becomes "Isis greatest goddess called (ἐπικαλουμένη) Thermuthis" in Greek.

119. Cf. "Osiris-Apis, the great god" (demotic)/"Serapis, greatest god" (Greek) in the bilingual Vleeming, *Short Texts in Demotic,* no. 260. See too pp. 142–43 below.

120. J. Yoyotte, "Une épithète de Min comme explorateur des régions orientales," *RÉG* 9 (1952): 125–37, at 133 n. 6.

121. For the first seven: Ronchi, *Lexicon theonymon* (with Wagner, *Les oasis,* 331–32, for χρηστός); for the next three: *SEG* XLV 2091, 2123; XLVI 2102. But for a case of εὐχάριστος in Anatolia, see below, n. 135. For the special link of εὐεργέτης with Sarapis, see p. 147n86.

unusual genitive referring to the great tress that she deposited there in mourning for Osiris;[122] but no Egyptian parallel is quoted.

The conclusion may appear surprising, given that Isis was famously "she of myriad names,"[123] and her cult was propagated with the aid of a unique rhetoric that stressed her multiple capacities; all the goddesses worshipped anywhere were in truth, it was claimed, the one Isis differently named.[124] It was on the basis of all these supposed names that, apparently in the first century A.D., her priests created for Isis the new title, μυριώνυμος, "Thousand-Named," which spread rapidly; a papyrus of the second century A.D. then lists over several pages the names, some theonyms, some epithets, under which she is honoured throughout the world.[125] But she is not especially rich in Greek epithets; Athena could have become just as myriad-named, if her priests had been similarly ambitious. The primary claim was that the actual names of innumerable foreign goddesses referred to her, not that she possessed a multitude of epithets. Isis has both topographic and functional epithets (some of the latter borrowed from other gods, such as Lochia, "of Birth," Pelagia, "of the Sea," Soteira, "Saviour," and Nikephoros, "Victory-Bringer"),[126] but no more than other major powers have. What is more distinctive about her is the breaking down of her multiple powers by the juxtaposition of a second noun, whether a theonym or an abstract noun: thus on Delos in the Hellenistic period we meet Isis Aphrodite, Isis Aphrodite Dikaia, "Just," Isis Dikaiosyne, "Justice," Isis Euploia, "Fair Voyage," Isis Euphrosyne, "Good Cheer," Isis Nemesis, "Righteous Anger," Isis Nike, "Victory," Isis Tyche Protogeneia, "First-Born Fortune," Isis Hygieia, "Health."[127] The type of naming here is Greek (compare Athena Hygieia), its adoption on this scale a response to the perceived omnicompetence of this goddess; some details are perhaps Egyptian,[128] and there is a touch too of local fashion,

122. See E. Bernand, *ZPE* 45 (1982): 103–4 (Totti, *Texte der Isis- und Sarapis-Religion,* no. 67).

123. In one inscription also πολύμορφος and πανεπίσκοπος (*RICIS* 501/0154, Rome).

124. So already the first hymn of Isidoros, Totti, *Texte der Isis- und Sarapis-Religion,* 21.1–3, 14–24, and, less developed, in the Maroneia aretalogy of the second century B.C. (?) (Totti 19.19–20).

125. See Bricault, "Isis Myrionyme"; the papyrus is *POxy.* 1380 (Totti, *Texte der Isis- und Sarapis-Religion,* 20). The rather late aretalogy of Karpokrates similarly engages in some epithet accumulation in asyndeton, Totti 6.9–10. For lists of epithets of Isis, see Tran Tam Tinh, *Divinités orientales en Campanie,* 214–29; Bricault, *Myrionymi;* the indices to *RICIS.*

126. Cf. n. 105 above.

127. Aphrodite: *RICIS* 202/0322 (*ID* 2080); Aphrodite Dikaia 202/0350 (*ID* 2158), probably 202/0346 (*ID* 2040: N.B. names of dedicators!); Dikaiosyne 202/0282 and 0288 (*ID* 2079 and 2103); Euploia 202/0329 (*ID* 2153); Euphrosyne 202/0357 (*ID* 2107); Nemesis 202/0310–12 (*ID* 2038, 2062–63); Nike 202/0287 (*ID* 2059); Tyche Protogeneia 202/0283–84 (*ID* 2072–73); Hygieia 202/0307 (*ID* 2060). On Isis Dikaiosyne, known also in Athens, see E. Matricon-Thomas, in L. Bricault and M. J. Versluys, *Egyptian Gods in the Hellenistic and Roman Mediterranean,* Mythos Suppl. 3 (Palermo, 2012), 52.

128. On Isis Dikaiosyne as a possible reflection of Isis Ma'at, see Plut. *De Is. et Os.* 3, 352B with the note of J. G. Griffith ad loc. Note too the unusual Isis χρηστή, "good," *RICIS* 202/0262 (*ID* 2149).

the type being much commoner on Delos than elsewhere. It fades away after the Hellenistic period, perhaps, it has been suggested, because juxtaposing Isis's name with that of a single partner was felt to weaken her claim to encompass all other goddesses.[129]

NEW EPITHETS

The Greek epithet stock was not then greatly increased, in Egypt or elsewhere, by translation. But this does not mean that it ceased to grow. Commoner than translation was the creation of new epithets, a never-ending process in Greece itself which continued in some regions of the Greek East. Aphrodite acquired Hagne, "Reverend," as a fixed epithet as a way of rendering the Syrian goddess Atargatis in Greek. Another banal-seeming but important instance is that of Zeus Megistos, "Greatest." Cults of Zeus Megistos are very common in Anatolia, and the use of *megistos* as a fixed epithet of Zeus seems to originate there and not in Greece. The earliest attestations of *megistos* in this use are at a sanctuary at Iasos in Caria at which dedications in Carian to trqδ, a descendant of the Hittite storm-god Tarhunt, have also been found;[130] it looks as if in this sanctuary frequented by both Greek and Carian speakers Zeus Megistos was the Greek for trqδ. But the creation of new epithets was not just a phenomenon of first contacts between indigenous gods and Greek naming practices; on the contrary, it continued throughout pagan antiquity and in some regions apparently gained in momentum. An exemplary late case from Lycaonia is Meter Kouadatrene, if we accept the attractive view that the epithet derives from an attested Roman imperial estate known as Quadrata, that by transcription gives Kouadrata and then by metathesis and the addition of an ending Kouadatrene; an epithet unquestionably of the Roman period is *eparcheios,* applied to Zeus in Cilicia, deriving from *eparchia,* "province."[131]

In Roman Phrygia and Bithynia Zeus acquired a plethora of epithets; an article publishing material from a single provincial museum in Phrygia recently added some ten to the list.[132] Many of this ever-growing array are topographic, but many too have an obvious Greek meaning, and of the latter a good number are only

129. M. Malaise, *Les conditions de pénétration et de diffusion des cultes égyptiens en Italie* (Leiden, 1972), 189–91.

130. *IIasos* 1.14, 220.1, 233–34; cf. Laumonier, *Cultes indigènes,* 599, and on later evidence, the note to *MAMA* XI 313. The appeals to Zeus Megistos put in Cyrus's mouth by Xenophon (*Cyr.* 5.1.29, 6.3.11) do not prove a cult epithet. The dedication: D. Piras, in F. Rumscheid, ed., *Die Karer und die Anderen* (Bonn, 2009), 233–35. Aphrodite Hagne: see pp. 163–64. For Latin epithets apparently used only of interpreted indigenous gods, see Carter, *De deorum romanorum cognominibus,* 6–7.

131. Kouadatrene: *Konya Museum,* no. 6, accepted by McLean ad loc., with references; contra Zgusta, *Ortsnamen,* ¶ 597. For ἐπάρχειος, see *ICilicie* 109.

132. M. Ricl, *ZivaAnt* 44 (1994): 157–74 (the museum at Eskişehir) (*SEG* XLIV 1034–68).

attested in Anatolia. A study promised by Louis Robert in a characteristically eloquent passage was unfortunately never written:

> I will draw up a list of these epithets elsewhere. Most are functional and relate to the concerns of peasants, expressed in Greek: fertility of the soil, abundance of harvests, vines, milk, threshing floor, stable, farm, mountain, sprouting of trees, riches, sky, thunder and lightning and the justice of Zeus, the goodness of Zeus.[133]

The commonest concerns expressed through his epithets old and new are weather, particularly in its relation to agriculture, and agriculture itself;[134] herding is less represented but Zeus Phatnios, "of the Stable," is a clear instance. A taste has been noted in the region for epithet pairings, where the second gives an optimistic twist to the ambiguous or threatening implications of the first: Zeus Chalazios Sozon, "of Hail, Protector" (i.e., "Protector against Hail"); Zeus Bronton Karpodotes, "Thunderer, Crop-Giver"; or the different aspects can be spread out into a dedication to Zeus Bronton and Zeus Karpodotes and Zeus Eucharistos ("Thundering" and "Crop-Giver," and "Paying Back").[135] As for newly created epithets, Zeus Agathios, "Goody," is known from several dedications in Bithynia and a few elsewhere, but we also have one addressed to Zeus Archagathos by the supervisor of his newly established temple: Archagathos, not "Goody" but "Source of Good," looks like a variation to give the new temple a slightly distinctive mark.[136] On the Phrygo-Pisidian border there is Zeus Kalakagathios, a nice formation based presumably on the ideal of *kalokagathia*, "Gentlemanly" Zeus;[137] back in Bithynia we also find a Zeus Aretarchos, "Origin of *Aretē*"or "Controller of *Aretē*"; whether the *aretē* here is moral virtue or the physical strength of, for instance, a gladiator is unclear.[138] This tendency to stress the god as a source of good things is also found

133. *Documents de l'Asie mineure,* 340. The passage goes on: "The images correspond to the needs of the peasants: oxen under the yoke, ploughs, saddled donkeys, horses, women, children in cloaks, ears of corn, grapes. It is naturally in Greek that they proclaim their need for justice and holiness through the abstractions of the Divine (τὸ Θεῖον), the Just and Sacred (Ὅσιος καὶ Δίκαιος, Ὅσιον καὶ Δίκαιον, Ὁσία)."

134. For abundant references on Zeus as Bronton (his commonest epithet), Astrapton, Keraunios, Karpodotes, Epikarpios, and variants, see Schuler, "Rural Population Groups," 74–76. (Schuler notes, 74 n. 41, citing *TAM* V.1.360, a dedication after two deaths by lightning, that there was a self-protective aspect in the worship of Keraunios alongside the agricultural.) He stresses the unique (though cf. Ἐπιδώτης) but clearly agricultural epithet Ἀναδότης, "sender up," borne by Zeus in the polis of Aizanoi, *SEG* XLV 1719; note too Ἁλωνίτης, "of the threshing-floor" (twice attested in Anatolia, *TAM* V.1.166a, *ISide* 17; not in Greece). Phatnios: *MAMA* 1.7; for Zeus Enaulios as "de la cour des bestiaux," see L. Robert, *Hellenica* X, 33–37 (cf. H. Malay, *Researches in Lydia, Mysia, and Aiolis,* Vienna 1999, no. 84).

135. *JHS* 24 (1904): 21, no. 4 (Thrakia Kome near Kyzikos); *INikaia* 1084–85: all this from Schuler, "Rural Population Groups," 76.

136. Agathos: *INikaia* 1061–67, with note on 1061; *MAMA* IX 52 (?). Archagathos: *INikaia* 1071.

137. *SEG* VI 550. On the impulse to variation, see Chaniotis, "Megatheism," 131.

138. *INikaia* 1076. Cf. *IKyz* 2, no. 4: a wrestler gives thanks for victory achieved Διὸς ἀρετῇ.

in Phrygian titles, these too new on present evidence, such as χρύσεος, "Golden," εὐχοδότης, "Granter of Prayers," ὠφέλιος, "Beneficial."[139]

In Egypt, a god of the desert identified as Pan becomes, repeatedly, εὔοδος, "of Good Journeys" (possibly reflecting an Egyptian epithet of Min), also εὔαγρος, "of Good Hunting," and, from a site en route for gold mines, χρυσοδότης, "Gold-Giver"; these three epithets may be exclusive to Egypt (whereas those relating him to mountains, which he also receives in Egypt, are more widespread).[140] Egyptian gods are also πλουτοδόται, "Wealth-Givers," and πλούσιοι, "Wealthy," epithets not unknown but rare in other places.[141] A personally tailored medley of gods is seen in a late Hellenistic dedication made by a huntsman as a "setting out vow" to "Mother of the Gods of the Army, Ares Ally, Zeus Olympios, Pan Fellow Campaigner."[142] Thus in Egypt too the epithet system appears to be finding new applications in new situations.

This chapter has revealed a broad spectrum of naming possibilities. Some remain close to non-Greek originals, presumptive or even occasionally attested; as Greek, they may appear somewhat stiff or unidiomatic. The intrusion of actual Greek theonyms is uneven. Many forms of naming are situated in shared territory: local and functional epithets for gods were as familiar in all the ancient Near Eastern cultures as in Greece itself; avoidance of the actual divine name by periphrasis was less common in Greece, but far from completely bizarre to a Greek ear. But the relation between divinity and locality expressed through the local epithet has emerged as an important disguised variable; and the proliferation of "ethnonymic adjectives" (Persian Artemis and the like) illustrates an awareness, among the inhabitants of this multicultural religious world, of the varied differentiations that persisted, and could be insisted on, within the "universal polytheism."

Large-scale importation of new religious values, via epithets, seems not to have occurred. But we have seen how, if in differing degrees, non-Greeks could generate

139. Robert, *Hellenica* X, 104–7; T. Drew-Bear, *Nouvelles inscriptions de Phrygie* (Zutphen, 1978), 48, no. 24; *IHadrian* 10. Chaniotis, *Megatheism*, 135–36, sees such epithets as a new development: benevolence is stressed, not mere protective power. The new epithet in *MAMA* XI 312 (Perta), Παπιας Μιθραδάτου Διὶ Ἀκουσέῳ εὐχήν, is taken by the editor as probably a variant on ἐπήκοος.

140. See the indices to Bernand, *Pan du désert;* Bernand, *Le Paneion d' El-Kanaïs* (Leiden, 1972); Bernand, *De Koptos à Kosseir* (Leiden, 1972); Wagner, *Oasis*, 331–32. Epithet of Min: see n. 120 above; note also *SEG* LIX 1771, Isis "mistress of the good start of a journey," in a hieroglyphic inscription. I think Ronchi, *Lexicon theonymon*, 491, is wrong to see an epithet of Pan, θηροφύλαξ, in *SB* I 294.3; see the note ad loc., and cf. *SB* I 286.

141. See Ronchi, *Lexicon theonymon*, 897–98. For a "hero Megon πλουτοδότης," see Graf, *Nordionische Kulte*, 453, no. 68 (Chios); for Zeus πλούσιος, Paus. 3.19.7

142. ὁμοῦ τιμὴ <Μη>τρὶ θεῶν Στρατείαι, Ἄρηι Συμμάχωι, Διὶ Ὀλυμπίωι, Πανὶ Συνστρατευομένωι, εὐχὴν ἐξοδίας: Bernand, *Pan du désert*, no. 3.

new religious concepts while using Greek. A little parodoxically, the phenomenon is most clearly observable in conservative Phrygia, where Hosios kai Dikaios emerges and where Zeus luxuriates in new epithets. There, Greek had become the vehicle for religious imagination, and new values, deities even, could be shaped within it to meet local needs.

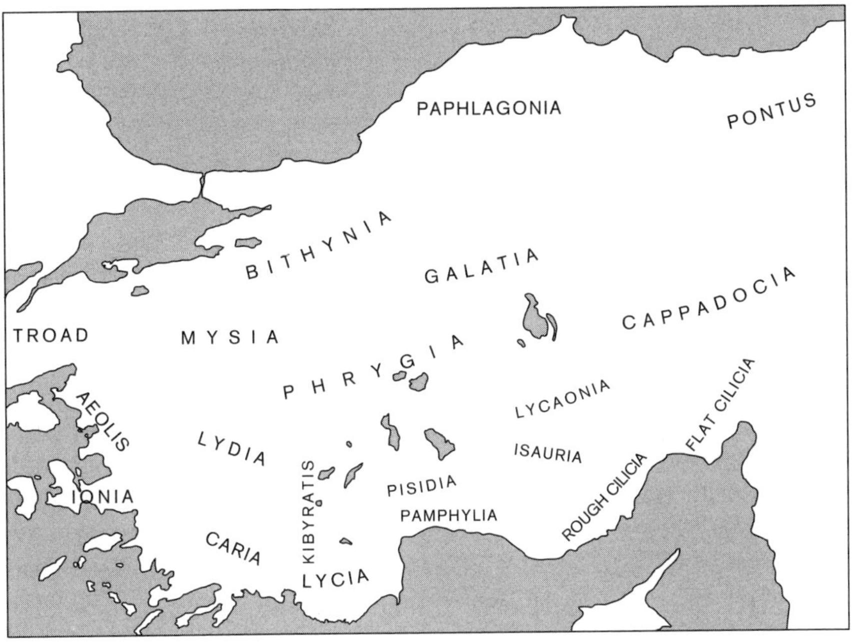

MAP 1. Anatolia

Hierapolis
Antiochia
Laodicea
Salamis
Apamea
Emesa
Palmyra
Tripolis
Heliopolis
Berytus
Sidon
Damascus
Tyrus
TRACHON
HAURAN
Scythopolis/
Nysa
Caesarea
Bostra
DEKAPOLIS
Gerasa
Philadelpheia
Jerusalem
Gaza
Petra

MAP 2. Syria

4

Supreme, Ancestral, and Personal Gods

This chapter continues the theme of its predecessor, the interaction between Greek naming practices for gods and those of the other cultures which adopted Greek either exclusively or in parallel with an indigenous language that remained in use. The issue is not a merely formal one, but can involve fundamental assumptions about the sources of divine power, the limits to human knowledge of the divine, and the character of relations between men and gods. Those "theological" aspects of naming were implicit in the previous chapter and will come more to the surface in the present chapter, which studies three particular forms of referring to the divine.

GODS "OF" INDIVIDUALS

In traditional Greek usage the name of a god is seldom combined with a genitive or adjective indicating an individual or group specially associated with it: gods did not belong to individuals or groups. The "Aphrodite of Dexikreon" on Samos was sufficiently unusual in a Greek context to require explanation as one of the *Greek Questions* of Plutarch. The Herakles Diomedonteios of Cos is a perhaps unique case of a god deriving an epithet from the founder of his cult. (Other examples of theonyms accompanied by adjectives formed from a human name seem to work differently.) Though festivals might bear the name of a benefactor who financed their endowment, the Coan Pythokleia, for instance, did not generate a Zeus of Pythokles.[1] An expression such as "Artemis Orthosia of the Geleontes" (a phratry

1. Pythokleia: *IG* XII.4.281.33–34; cf. 350. The frequency even of this type is much reduced if J. D. Sosin, "Endowed Eponymous Festivals on Delos," *Kernos* 27 (2014): 127–57, is right to reinterpret the

on Thasos) does not indicate any exclusive claim of the Geleontes on the goddess: the reference is to a particular altar or cult-place at which the Geleontes worshipped Artemis Orthosia, just as other groups might worship her at other altars.[2] But a usage emerges in Anatolia whereby a god's name is accompanied by a man's name in the genitive: "Men of Artemidorus" is an often recurring example. The usage breaks the very familiar association of a god with a place and presents one with a person instead.

The two earliest instances both have a Persian aspect. Chronologically perhaps the oldest is that Men Pharnakou by whom according to Strabo (sole source for the cult) the kings of Pontos swore their royal oath. The Pharnakes in question must certainly be one of the figures of that name associated with the house of Pontos:[3] perhaps King Pharnakes I who reigned in the second century, perhaps rather one of the several bearers of that name in Achaemenid times from whom the dynasty traced its descent. At all events it is clear that the title is one with distinguished associations. The second early instance is a statue dedicated (on the most persuasive interpretation) to Zeus of Baradates, or perhaps rather of Bagadates, at Sardis by Droaphernes, hyparch of Lydia in the thirty-eighth year of Artaxerxes, that is, in either 428/7 or 366/5 depending on which Artaxerxes is referred to. Who this Bara-/Bagadates was, why he established a cult of Zeus in Sardis, and what the nature of this "Zeus" (who may in time have mutated into Zeus Polieus) was are all unknowns; the presumption must be that he was a Persian administrator of high status.[4] These cults named after eminent Persians have been seen as the family cults of the person named in the genitive, perpetuated in their memory by dependents and descendants.[5] But there is no evidence except questionable analogy that such was their nature, nor indeed that Persian aristocrats had family cults of this type at all.

Whether an "Apollo of Moundos" or "of Toumoundos" is to be recognised in Lydia in the 270s/260s B.C. is doubtful;[6] he may rather be Apollo Toumoundos/es.

---

many supposed examples on Delos. Dexikreon: Plut. *Quaest. Graec.* 54, 303C-D. Herakles Diomedonteios: *IG* XII.4.348 (ca. 300 B.C.). Other examples: see appendix G.

2. Similarly nonexclusive (except in relation to a particular cult site) are the Zeuses of the various tribes and tribal subgroups of Mylasa in Caria ("Zeus of the Otorkondeis" and so on). Versnel, *Coping with the Gods,* 63 n. 152, compares the form of expression "Zeus Meilichios in (the plot of) Myskos" found at Selinus (*NGSL* 27 A 9, 17) with the Anatolian "theonym plus personal name in the genitive." But formally the former is just a designation of place, even if in reality the (e.g.) Myskos in question may have been a figure significant in the history of the cult.

3. So Gschnitzer, "Persische Kultstiftung," favouring an early Pharnakes; Mitchell, "Persians in the Sanctuaries," 164, prefers Pharnakes I. Strabo: 12.3.31, C 557.

4. *SEG* XXIX 1205; cf. the overview of debate in P. Debord, *L'Asie mineure au IVe siècle* (Bordeaux, 1999), 367–74. Bagadates: so Mitchell, "Persians in the Sanctuaries," 157–58.

5. So Gschnitzer, "Persische Kultstiftung."

6. *TAM* V.3.1545. The editor prints as τοῦ Μουνδου, but the presence of the article is abnormal in expressions of this type.

Thereafter evidence for the type resumes after a long chronological gap and becomes so abundant in the Roman period that it is no longer in doubt that the analysis as "god's name plus genitive of a human name," first strongly advocated in 1911,[7] is in most cases correct. But the long interval must leave it uncertain whether there is any continuity between this and the earlier usage. Almost twenty certain or probable examples (if we include the equivalent form which uses a preposition, such as "Men from Diodotos") are now known from Lydia (above all), Phrygia, Bithynia, and Caria.[8] There is at least one example of a god plus two genitives (Zeus Bronton, "Thundering," of Loukios and Aristainetos), and one where the name is attested in both the singular and the plural (Zeus Goneos and Zeus Goneon).[9] In none of these cases do we have any way of identifying the individual in question; except perhaps with Mother Adrastou, and in contrast to Men Pharnakou, there is never any reason to suppose that the figure was one of high standing and perhaps to be set some way back in time. Once he is described as a priest.[10]

The ruling assumption is that the name in the genitive is that of the founder of the cult; the rare plurals will point to joint foundations or a refoundation. Never, however, is this founding role made explicit, and a new text has suggested a different possibility.[11] A sanctuary of Zeus Megas Menophilou is said to have been founded by the Daokometai (of a hitherto unknown village near Aizanoi in northern Phrygia) after Menophilos "was struck/smitten/amazed" (either ἐξεπλήχθη or κατεπλήχθη). The force of the verb is not completely clear: was Menophilos "smitten with amazement" in response to an epiphany, or did he perhaps feel the punitive hand of god?[12] On either view he is not presented as having founded the cult, which was apparently done

7. J. Keil and A. Von Premerstein, *Bericht über eine zweite Reise in Lydien,* DenkschrWien 54.2 (Vienna, 1911), 104.

8. To the examples collected by Gschnitzer, "Persische Kultstiftung," 47–48, add Men Karou (Lane, *CMRDM,* 3:68) and Meter Adrastou (R. van Bremen, in R. W. C. Catling and F. Marchand, eds., *Onomatologos,* Oxford 2010, 443–47) from Caria; Zeus Poteos from Phrygia (*EpigAnat* 17, 1991: 76–77); Zeus Bronton Baboudos and Zeus Bronton of Loukios and Aristainetos from Bithynia (*Olba* 20, 2012: 358–60, nos. 8–10); Meter Skepnou (*SEG* XXXVI 1101), Anaeitis ἐγ Μητρῷ (?) (*SEG* XXIX 1174), *thea* Andronikou (Malay, *Researches,* no. 149), Apollo Tolou (*New Documents from Lydia,* no. 27), Zeus Medou (Malay, *Researches,* no. 110) from Lydia; Meter Menandrou from Mysia (?) (*SEG* LIII 1378). Petzl, *Beichtinschriften,* 103, proves the equivalence of the formulas with and without ἐκ from the alternations between Μεὶς Ἀρτεμιδώρου and Μεὶς ἐξ Ἀρτεμιδώρου.

9. Two genitives: see previous note; the separately attested instances of Zeus Bronton of Apellinarios and Meilesias, *INikaia* 1509–10, are probably combined in 1104. Singular and plural: C. P. Jones, *AA* (1994): 458–59 (*SEG* XLIV 865 and 915), with evidence for use of the rare name Goneus in the region.

10. *INikaia* 1510.

11. C. Lehmler and M. Wörrle, *Chiron* 36 (2006): 76–78 (*SEG* LVI 1434): A Δαοκωμῆται κατ' [εὐχ|ήν· ἔτους—-—?] B Λώου μηνὸς ἐννεα|καιδεκάτῃ Μη|νόφιλος C [κ]ατεπλήχθη δε[ινῶς καὶ?] | ἐκτίσθη Ζεὺς | Μέγας Μη|νοφίλο|υ.

12. Epiphany: so the editors, citing Mark 7.36 and Lucian, *Alex.* 26 (not, however, close parallels); punished, A. Chaniotis, *EBGR* 2006 (2009): no. 73.

by his fellow villagers. We might think rather of the use that has been identified in Jewish and early Christian texts in which the X of "god of X" is a person to whom God has made a revelation or through whom he has displayed especial power or favour: "Blessed be the God of Shadrach, Meshach, and Abednego" (who has just saved them from the fiery furnace; Dan. 3.28); "Unique is the god of Thekla" (who has just performed a miracle).[13] It was through Menophilos that Zeus Megas communicated, by one means or another, with the villagers. This new example need not take all the others with it: the genitive may have expressed a variety of relations. But it is the only case where the relation between the god and the mortal is actually made explicit.

Though many cults of this type are attested once only, others appear in more than one location. Men Tiamou in particular is very widespread, though new evidence cited below may support the recently revived view that Tiamou is not in fact a genitive of this type.[14] Such diffusion is not in itself a problem, on any view of the origin of such cults: the god as experienced (or founded) by X in a particular location proves effective, and is worth exporting to another under his original name. But some particular manifestations of the phenomenon are difficult.[15] Men Axiottenos (a typical epithet indicating a locality) is both known on his own and associated in the genitive with three or four different mortals. Several individuals, it seems, had become associated with the one powerful regional god. One of those mortals, Artemidorus, also gave his name to a Men who ruled not over Axiotta but over "Dorou Kome." Was Dorou Kome a subdivision of Axiotta, or was Artemidorus associated with Men in more than one place?[16] Still worse is a newly published text with a double genitive, Μέγας Μὶς Τιαμου Ἀρτεμιδώρου Ἀξιοττα κατέχων, "Mighty (is) Men of Tiamos (or Tiamou) of Artemidorus who [Men] possesses Axiotta."[17] To suppose that Artemidorus was associated with two separate cults of Men Axiottenos, one more properly his own (the one mentioned above), one a renewal or reinforcement of that of Tiamos (that of the new text), strains credulity. Perhaps the dedicator was piling up epithets without interest in their original reference.[18] Or perhaps Tiamou is not a genitive

13. Peterson, *ΕΙΣ ΘΕΟΣ*, 210–12, esp. 212: "The person named in the genitive seems to me in every case to have been given a role as a kind of mediator of revelation" ("eine Art Offenbarungsmittlertum erhalten zu haben"); cf. the addendum to p. 189 in Peterson, *ΕΙΣ ΘΕΟΣ*², 562.

14. See n. 19 below.

15. See P. Herrmann, "Men, Herr von Axiotta," in *Studien zur Religion und Kultur Kleinasiens, FS Dörner* (Leiden, 1981), 415–23, at 422–23.

16. Separate from this problem is the problem that Men Axiottenos could rule places other than Axiotta: Koresa (Lane, *CMRDM*, 1:no. 47), Tarsi (no. 69); on this issue, see Lane, *CMRDM*, 3:32–33.

17. *New Documents from Lydia*, no. 52.

18. So the editors of the new text, citing Lane, *CMRDM*, 3:68 n. 11 and 72 n. 26 for the opacity of many Men epithets to worshippers in the Roman period.

at all, but a Luvian-derived epithet meaning something like "of the Earth/ Netherworld."[19]

An apparently similar genitive occurs on the plateau of Trachon and in the Hauran (see map 2), with the difference that it is normally attached not to a theonym but to plain *theos: theos* Aumou is the best-attested example.[20] A theonym sometimes follows, but this addition seems to be a secondary development: only in later texts does "the "god of Aumos" become "Zeus invincible sun, god of Aumos."[21] The cult of *theos* Aumou is attested at seven sites and over more than a century, and was celebrated not just within a family but by villages; it has been claimed as the region's most important cult.[22] The *theos* Maleikathou received two dated dedications separated by sixty years. No more than in Anatolia do the texts make explicit the relation of the figure named in the genitive to the cult.[23] But here the dependence on an indigenous naming practice is clear: in monolingual Nabataean texts we find "god" (occasionally with a theonym attached) plus a similar genitive relating to otherwise unknown figures (e.g., "god of Sa'idou," *CIS* II.176).

19. So M. Ricl, *EpigAnat* 44 (2011): 146 (developing G. Neumann, *Untersuchungen zum Weiterleben hethitischen und luwischen Sprachgutes,* Wiesbaden 1961, 71–72), who adduces tiyam-maššiš, "of the earth/underworld," epithet of the sun-goddess Tiwad. On this view Tiamou would be translated by Men's common epithet καταχθόνιος. Note "lord (κύριος) Tiamou," in Petzl, *Beichtinschriften,* no. 68.16. Evidence for Tiamos as a personal name is slight: *LGPN* VB has a fifth-century Tiaimos from Miletus, *LGPN* VC will contain one Tiamos: G. Laminger-Pascher, *Die kaiserzeitlichen Inschriften von Lykaoniens* (Vienna, 1992), no. 371 (*JHS* 25, 1905: 173, no. 54).

20. See above all Alt, "God of the Fathers," 30–45, and 65–76 (lists). Add now Tyche Samethou, *IGLS* XIII.2.9571, and "the goddess of Moithos" (below); more possibilities are suggested in *BÉ* 2015, no. 714. That Αυμου is a genitive cannot be proved (for the other view, see H. Petersmann, in T. Fahd, ed., *L'Arabie préislamique,* Leiden 1989, 410–11 [SEG XXXIX 1834]), but the many parallel cases (Alt, 71–76; cf. Sourdel, *Cultes du Hauran,* 54–56, 95–96; *IGLS* XV.1.92 with note: note especially "god of Hameros" and "god of Arkesilaos") and the existence of a name Αυμος make it plausible; cf. *IGLS* XV.1 p. 340. θεὸς Αυμου is apparently misunderstood as θεὸς Αυμος in *SEG* XLV 1991 [Διὶ Ἡλίῳ] ἀνικήτῳ Θεῷ Αὔμῳ and *LBW* 2393.3–4.

21. *LBW* 2392–95 (Alt, "God of the Fathers," 73, nos. 41–44; to appear as *IGLS* XVI.24, 26, 27). Similarly, the theonym Theandrios is added to two separate "gods of X": to the God of Rabbos, R. Donceel and M. Sartre, "Théandrios, dieu de Canatha," in E. Dąbrowa, ed., *Donum amicitiae* (= *Electrum* 1, 1997, Krakow), 21–34 (*SEG* XLVII 2069, 2075–78, to become *IGLS* XVI.1.142–45, 153; add now *SEG* L 1535/LII 1722); to the god of Ouaseathos, *LBW* 2374 and 2374a (*IGR* III 1238; Donceel and Sartre, 29–30, no. 11: to become *IGLS* XVI.1.114), from 'Atil, 3 km west of Kanatha. On Theandrios, see p. 92. Slightly differently, the "goddess Moithou" apparently loses the genitive when turning into "goddess Leukothea (of Rachla), *IGLS* XI.20–30, with commentary on 20.

22. Alt, "God of the Fathers," 37: "There is no other god in this region who can be called a serious rival to him." Seven sites: *IGLS* XV.1.92 with note. Theos Maleikathou: *IGLS* XVI.1.92 and 96.

23. Not much can be built on the appearance in the late (320 A.D.) *LBW* 2393.7 (*OGIS* 619) of an Aumos as a priest in the cult.

(In what looks like a distinct, respectful locution there are references to the "god of our master" the king, once "of our master Rabel."[24])

A few scattered clues to the origin of these cults exist. Once a dedication is made to the "god of Hotaisu" by, among others, the grandson of an Hotaisu; similarly, in Greek, a single village yields a dedication by Loaithemos (to his unnamed "ancestral god") and one by a group "to the god of Loaithemos"; in Nabataean again, two separate dedications are made in the Hauran to "the god of Qasiu" and "to their god Baal[shamin] by the tribe of Qasiu."[25] The first two of these examples may suggest, though they do not prove, the perpetuation within a kinship group of a cult decisively marked in some way by an ancestor. The third complicates the issue by the ambiguity that it introduces between Qasiu as an individual and as a tribal name. (There are also separate traces in Greek texts of gods named after tribes, Kronos Sokarathon[26] and Tyche Thaimeios, but via the tribal name, not that of an individual.) Thus it remains problematic to determine the worshippers of these "gods of X": were they kinship groups by origin?—the cult of *theos* Aumou eventually sprang those bounds—and on what scale? The type has been seen as an inheritance from the nomadic past of the Nabataeans,[27] and its distribution (primarily the Hauran, not the coastal strip) supports that view; so too might the reference to a person, not a locality. But the case of a "god of Arkesilaos" (N.B. the Greek name) still lending valued aid in the year 324 A.D.[28] complicates the attempt to explain the usage in terms of nomadic origins. Another problem lies in that vague reference to a cult "decisively marked" in some way by an ancestor. A prominent priest, the founder of the cult, the recipient of a revelation?[29] A matter of

24. Alt, "God of the Fathers," 67–70 (the royal instances are his nos. 5–11); cf. Healey, *Religion of the Nabataeans,* 151–52; and Healey, "The Nabataean 'God of the Fathers,'" in K. Dell et al., eds., *Genesis, Isaiah, and Psalms, FS John Emerton* (Leiden, 2010), 45–57, with addenda (*CIS* II.3.3991 may be another), including a goddess.

25. *CIS* II.1.354 (Healey, *Aramaic Inscriptions,* no. 4); *IGLS* XV.1.138–39; *CIS* II.1.174 with *RÉS* IV 2042; the tribe of Qasiu also in *CIS* II.1.165. Cf. in general M. Sartre, "Tribus et clans dans le Ḥawrān," *Syria* 59 (1982): 77–91; in brief *IGLS* XV.1 pp. 24–25.

26. *Musée Soueida,* no. 198, taken as a tribe name by Sourdel, *Cultes du Hauran,* 36; Τύχη Θαιμειος, "Gad of the Benou-Taîmaî" (*IGLS* XVII.1.306.6, with commentary). Dussaud, *Arabes en Syrie,* 124, cites also Gad-ʿAwîdh, taken as Gad of the Aouidenoi (a tribe known from *LBW* 2236; *OGIS* 617; cf. Alt, "God of the Fathers," 70); he takes the epithet Γοζμαίη of Athena in *LBW* 2345 as a similar ethnonym (Sourdel, 71, leaves its character open).

27. The central thesis of Alt, "God of the Fathers," who used it to illuminate the biblical "God of our Fathers" and "God of Abraham." For reactions to Alt, see Mettinger, *In Search of God,* 58; according to Vorländer, *Mein Gott,* 303–4, reinforced by R. Albertz, *Persönliche Frömmigkeit und offizielle Religion* (Stuttgart, 1978), 88–91, the "God of our Fathers" has nothing to do with tribalism or nomadism but is a special application of the personal god.

28. *LBW* 1834.

29. "The head of the priestly family," Dussaud, *Arabes en Syrie,* 124; Sourdel, *Cultes du Hauran,* 95: "The name of an individual or clan probably invested with the function of priests"; cf. 21 (but Alt,

institutional authority, or of charisma? The texts do not yet reveal. Nor is there any point of contact between the gods so named in Syria and in Anatolia; their origins might be different.

GODS ANCESTRAL AND PERSONAL

In a sense expressions such as "god of Aumos" are just a special application of a broader usage. References to my, your, our, their god(s) are regular in Semitic languages: in them, everybody has a god/gods.[30] Greek is different: the Phrygian in Timotheus who appeals to "Artimis, my great god" (Ἄρτιμις ἐμὸς μέγας θεός) marks himself out as a barbarian no less by the "my" than by the form of the goddess's name. In a Hellenistic Greek-Phoenician bilingual from Cos, the "to my lady, to Astarte" of the Phoenician is plain "to Aphrodite" in Greek. When the chorus of Athenian women in Euripides' *Ion*, sightseeing in Delphi, see a representation of Athena and cry, "I see Pallas, my goddess," it is the joy of encountering their city's goddess far from home that evokes the unusual cry, but the sentiment is more civic than personal.[31] The closest Greek equivalent in regular use lies in the concept of the *patrōos* (later also *patrios*), "ancestral," god,[32] since an ancestral god is necessarily also an ancestral god of somebody.

In old Greek usage *patrōos* is applied to gods broadly in two ways. On the one hand, in certain contexts such as patriotic appeals, a city's gods collectively can be spoken of as (its) *patrōoi;* this use is primarily literary, not epigraphic. On the other, individual gods typically receive the epithet in the context of smaller groups, phratries above all; the reference is to the cults one enters into by right of birth. Much of the evidence comes from altars belonging to such groups, often accompanied by a genitive specifying the group whose ancestral altar it is ("of Zeus Patroos

---

"God of the Fathers," 41, warns against supposing hereditary priesthoods). Revelation: Alt, 42 (citing Peterson, as p. 116n13 above): "The founder of this cult must have experienced a revelation from a hitherto unknown god, and been compelled by it to worship him."

30. See p. 83n26.

31. Timotheus, *PMG* 791.260; *IG* XII.4.2.546 (similarly, e.g., *KAI* 41 and 47, from Cyprus and Malta); Eur. *Ion* 211, λεύσσω Παλλάδ' ἐμὰν θεόν; cf. Ar. *Nub.* 601–2, ἥ τ' ἐπιχώριος ἡμετέρα θεὸς/αἰγίδος ἡνίοχος, πολιοῦχος Ἀθάνα. Cf. Burkert, "Mein Gott?" Quite distinct is the usage whereby persons of low status in Egypt could speak of social superiors as "my god": Bernand, *Inscriptions d' Alexandrie,* 36 (*OGIS* 195); for similar language in papyri, see Brashear, *The Archive of Athenodoros* (Berlin, 1995), no. 2600.3 with note; and cf. I. Gradel, *Emperor Worship and Roman Religion* (Oxford, 2002), 44–49 (on expressions such as *mi Iuppiter terrestris* in Plautus).

32. πατρῷος, πάτριος. Related but less common terms are προγονικός: *IMylasa* 902; *BÉ* 2015, no. 633; *IGB* IV 2217; γενι(α)κός: *IGB* III 1766, θεῷ Ἀπόλλωνι γενιακῷ Εστρακεηνῳ; also 1767 (?), 1845; cf. 1683, which spells the idea out: Φοίβ[ῳ]ἡμετέρ(ου) γένους ἀρχηγῷ; συγγενικός: *INikaia* II.1.1130; *TAM* IV.1.96; παππῷος (but the editor suggests misreading of πατρῷος): *INikaia* II.2.1513; μητρικός: *IGB* IV 1985, ἐπηκόῳ θεῷ Σεβαζίῳ μητρικῷ.

of the Etymobysiadai"); members of such groups can also make dedications to the god in question under the epithet *patrōos.*[33] Usage expands somewhat in the broader Greek world.[34] Epigraphic references to a city's ancestral gods become common; most often they occur as a group, but gods especially associated with a city such as Ephesian Artemis can be singled out and are even occasionally given an explicit possessive pronoun: thus the city of Stratonicea in Caria issues invitations to neighbouring communities in the name of "our ancestral god."[35] An Ephesian decree once even speaks of Artemis as "our," or perhaps "the city's," "own goddess."[36]

The added genitive is occasionally found too in relation to the ancestral gods of individuals, references to which multiply; in the second/third century A.D. there emerges a polite epistolary formula "Before all things I pray that you are healthy and I make a supplication (προσκύνημα) for you before the (our) ancestral gods."[37] Expatriates from at least the Hellenistic period onward dedicate to the gods of their home city as *patrioi,* Boeotians in Egypt to "Zeus Basileus and their other *patrioi theoi,*" Syrians in Delos, to "Hadat and Atargatis, *patrioi theoi,*" Milesians in Egypt rather later to "our *patrios* god Helios Apollo of Didyma." Two more examples from many: Cornelius Gallus at Philai in Egypt in 29 B.C. dedicates to his ancestral gods in both Greek and Latin (and also to the "Nile which has assisted him [συνλήπτορι/adiutori]"); men perhaps from the Hauran and probably serving as soldiers dedicate to their "ancestral gods" Theandrios and Mana(l)phos in Morocco (in Greek) and in Spain (in Latin).[38]

There also start to occur dedications such as "to ancestral god Zeus Bonitenos" or "to Lord Zeus and Hera, ancestral gods," or simply "to the ancestral god," where the

33. A dedication from Larisa τοῖ πα[τ]έρος Ποτειδο[ῦνι] is a rare variant, if correctly read (*IG* IX.2.475). See in general R. Parker, "πατρῷοι θεοί," in A. H. Rasmussen and S. W. Rasmussen, eds., *Religion and Society: Rituals, Resources, and Identity in the Ancient Graeco-Roman World* (Rome, 2008), 199–212.

34. But not under the influence of the Septuagint, which speaks of θεὸς τοῦ πατρός μου (e.g., Gen. 31.5), not πατρῷος θεός.

35. *IStrat* 23 and often, ὁ πάτριος ἡμῶν θεὸς; cf., e.g., *IEph* 2026.16; J. Reynolds, *Aphrodisias and Rome* (London, 1982), no. 48.13–14; *TAM* V.2.828 a 6–7; Robert and Robert, *Carie,* 278–79, no. 156; *SEG* L 1270. Slightly different is ἡ παρ' ἡμῖν Ἄρτεμις, *ISardis* 4.10–11.

36. ἐν [τῇ] ἡμετέρᾳ πόλει τῇ τροφῷ τῆς ἰδίας θεοῦ τῆς Ἐφ[εσί]ας: *Syll*³ 867.42–43 = *IEph* 24 B 22.

37. An individual's ancestral god: *IDidyma* 504.6–7 (J. Fontenrose, *Didyma,* Berkeley 1988, 204, no. 30); *Milet* 1.7.205a.2 (Fontenrose, 194–95, no. 20); *IGLS* IV.1410, ὁ πατρῷός μου θεὸς Ἀρκεσιλάου (a "god of Aumos" type who is also a πατρῷος); Bernand, *Delta,* I, 242.14 (228 A.D.), Διὶ Ἡ[λίῳ] μεγά[λῳ] Σαράπιδι ἐν Κα[νώβῳ] θεὸν πάτρι[όν] μου Ἡρ[ακ]λῆ Βῆλον ἀνείκητον Μ(άρκος) Α(ὐρήλιος) Μάξιμο[ς Σύρος] Ἀ[σ]καλωνείτη[ς. Cf. the advice of Sansnos, E. Bernand, *Inscriptions métriques de l'Égypte gréco-romaine* (Paris, 1969), 165.3, ἡγοῦ μάλιστα τοὺς πατρῴους. The epistolary formula: Ronchi, *Lexicon theonymon,* 884–86.

38. *SB* III 6664 (2nd c. B.C.?); *ID* 2226 and 2247; *SB* I 1530 (τὸν πάτριον θεὸν Διδυμέα Ἥλιον Ἀπόλλωνα); *IGPhilae* 128; L. Robert, *RÉG* 49 (1936): 1–8 (*OMS,* 2:939–46) (Theandrios and Manaphos).

resident of a place gives the epithet to a local god.[39] Here it apparently serves not to identify but to legitimate: it provides reassurance of the rooting of the cult in tradition. There are abundant examples from all regions. It might seem that such language reflects a non-Greek usage, but, in the several Palmyrene bilinguals where gods are so identified in the Greek, there is either no Palmyrene equivalent or a simple possessive suffix, "their god": it is just once, in a monolingual Palmyrene text,[40] that we meet a dedication to a "god of the house of their ancestors." The Greek usage reflects an idea rather than a fixed expression. With these later applications of the concept of "ancestral," the question arises of the group within which this inheritance has been passed down: it is clearly now often the city, doubtless also sometimes the village, but are smaller or differently organised groups also still in question (as in the earlier usage), and what groups if so ("the house of my ancestors"?)? The identifying genitive ("of the Etymobysiadai") is no longer found, and the question remains largely unanswerable.[41]

It is not in doubt that *patrioi* could be the focus for strong feelings. In a funerary epigram from Hadrianoi in Bithynia a husband grieves for his wife who had to travel much, leaving her (unnamed) *theos patrios;* a group of brothers in Thrace celebrate setting up a statue of a familiar *patrios* after a move to a new region; a papyrus letter declares touching faith that the *patrioi* will help in a present difficulty as they always do.[42] But "my *patrios theos*" very seldom becomes "my god" tout court. Some of the rare exceptions come from bilingual environments where such expressions were regular in the other language (but one might have expected much more seepage than in fact occurred),[43] one or two do not: a temple warden

39. θεῷ [π]ατρῴῳ Διὶ Βονιτηνῷ, Marek, *Stadt, Ära und Territorium,* 180, no. 95 (Amastris region); Διὶ [Κ]ῡρίῳ καὶ Ἥρᾳ θεοῖς πατρῴοις, *IGLS* XIII.1.9002; θεῷ [π]ατρῴῳ, *IGLS* XV.1.119a, with a note quoting several named and unnamed examples. An inscription from Moushennef in the Hauran speaking of an *oikos* of Zeus and ancestral Athena introduces an interesting differentiation in point of antiquity (Sourdel, *Cultes du Hauran,* 71).

40. *CIS* II.3.3978.6–7 = *PAT* 0324, LŠMŠ ʾLH BYT ʾBWHN, adduced by J. B. Yon in his note on *IGLS* XVII.1.320.

41. *Illion* 143.2 is a nice example of a city: Ἰλιεῖς τὸ[ν] πάτριον θε[ὸν] Αἰνείαν. Dedications to *patrioi* are made by smaller groups: Drew-Bear and Naour, *Divinités de Phrygie,* 1929, no. 5 (*SEG* XL 1192), a *symbiosis* to Alsenos *patrios megas* Zeus; *IGR* III 1172, the Maneinoi, a tribe (Sourdel, *Cultes du Hauran,* 34) to *patrikos theos* Herakles; *INicaea* II.2.1513, a group of συνγενῖς to Zeus παππῷος. But this does not in most cases prove that the god in question was exclusive to the group: it is not clear that the precinct of Zeus and ancestral Athena at Moushennef in the Hauran was exclusive to the *synodos* which constructed an *oikos* there (n. 39 above).

42. *SGO* 08/08/09 = *IHadrian* 76; *IGB* III 1794; *POxy.* 935.9–12 (*Select Papyri,* Loeb, 1:no. 136).

43. See, e.g., *IGLS* XIII.13.2.9005, Δ̣αλσουμ[ο]ς Χεειλου ε̣[ὐ]σεβῶν ἀνέθηκ̣εν ἐν ἱερῷ αὐ̣τ̣οῦ θεοῦ Εδ[.]γ̣ωνος; *OGIS* 199.31 (Adulis), πρὸς τὸν μέγιστον θεόν μου Ἄρην. Bernand, *Portes du désert,* 98, Πουὼν ϛʹ Ἁρπαῆσις ἀνέθ[η]κεν τῷ ἰδιῷ θεῷ μεγίστ<ῳ>, is problematic. Bernand writes ad loc.: "Hogarth thinks of the god of the Christians. Might it not rather be the god of the Jews? The formula is rare, not to say unique, in the epigraphy of Egypt. For a Jew to have taken an Egyptian name would not be surprising."

of Asclepius and gout sufferer in second-century A.D. Athens, for instance, supplicates the god, in verse, to allow him to walk to his temple "that I may see you, my god, you who are brighter (to me) than the earth in spring" (ὅπως σ' ἐσίδω, τὸν ἐμὸν θεόν, τὸν φαιδρότερον χθονὸς εἰαρινᾶς). A variant, again addressed to Asclepius, is a dedication "to my own Saviour and Telesphoros ("Fulfiller").[44] But these are exceptional; more commonly, the personal connection is routed via ancestors and tradition.

When emphasis that the god one worshipped was ancestral was so common, it is unexpected to find stress on innovation, but an example emerged in 2007: a dedication (from an unknown provenance in Anatolia) "to Apollo Ta—Newly established":[45] the god by his power (*dynamis*) had helped the dedicator recover lost gold and silver objects, and perhaps it was because the cult was established in gratitude for that service that the newness was underlined.

SUPREME GODS

A special case which crosses categories is that of designations in the Roman Near East for a supreme deity. Here a variety of epithets, such as *hagios,* "Sacred," *hypsistos,* "Highest," *megistos,* "Greatest," *ouranios,* "Heavenly," are deployed in various combinations. At Palmyra, a very common description of Zeus in both monolingual and bilingual texts is "Zeus Highest [Greatest] and Listening (*epēkoos*)."[46] It is used as an equivalent in bilinguals for the following:

1. The so-called anonymous god, known by the periphrasis "Blessed be his name forever" (BRYK ŠMH L'LM': *IGLS* XVII.1.378, perhaps 382), which can be extended to "Blessed be his name forever, good and merciful" (BRYK ŠMH L'LM' ṬB' WRḤMN': *IGLS* XVII.1.322, 342)
2. "Lord of the World" (MR' 'LM': *IGLS* XVII.1.344, probably also 152 and 343), who may be the anonymous god again, since there exists an extended form "Blessed be his name forever, lord of the whole world" ([BRYK Š]MH L'LM'

44. *IG* II² 4514.10–11; *IG* IV² I.570. Cf. Ἀρτέμιδος Σωτείρας ἡμετέρας on a small Hellenistic altar (?) from the Milesian island of Lepsia, *ASAtene* n.s. 25–26 (1965): 324–25, no. 24; the identity of the "we" is unknown. See too p. 126n65 below. The speaker of "my Diktynna" in *GVI* 1902 (*IG* II/III² 13163) is a dog. The several references to "our" gods in *SEG* XVIII 578 are different: they distinguish the gods of a particular Cypriot community from the common gods (κοινός) of the island (so J. B. Cayla, as cited by T. Fujii, *Imperial Cult and Imperial Representation in Imperial Cyprus,* Stuttgart 2013, 80). In the text from Myndos, *BCH* 12 (1888): 279–80, no. 3, a fine for violation of a tomb payable τῷ θεῷ ἡμῶν Κυμνίσσει Ἀπόλλωνι, it is unclear whether "our" refers to a community, or the husband and wife for whom the tomb was reserved. Kings have special *basileioi theoi* to swear by, apparently: App. *Syr.* 60 (317).

45. G. Petzl, "Göttlche Hilfe bei Verlust," in P. Brun, ed., *Scripta Anatolica: Hommages à Pierre Debord* (Bordeaux, 2007), 331–38 (*SEG* LVII 2034): Ἀπόλλωνι Τα[-—- ν]εοκτίστῳ.

46. Ζεὺς ὕψιστος [μέγιστος] καὶ ἐπήκοος.

MR' KL ['LM']: *IGLS* XVII.1.130), again represented in Greek as "Zeus Highest and Listening."

3. "Baalshamin the Great and Merciful" (B'L ŠMN RB' WRḤMN': *IGLS* XVII.1.324).

But Baalshamin is also once plain Zeus (*IGLS* XVII.1.145), once (in Dura-Europos) Zeus *kyrios,* "Master," once (at a site between Damascus and Palmyra) Zeus *megistos keraunios,* "Greatest, of the Thunderbolt" (here he has also been expanded in the Palmyrene to Baalshamin "Lord of the World": B'LŠMN MR 'LM').[47] Baalshamin is clearly distinguished from Bel, who in Greek is Belos or Zeus Belos or something similar,[48] but he has no fixed Greek designation, and one that he often receives he shares with "the anonymous god," who is probably distinct from him. In none of these cases do we find an attempt to render in Greek the specificity of the Nabataean titles such as "Blessed be his name," "Great and Merciful."[49]

The god of Baitokaike (inland from Arados, in northern Phoenicia), who is believed again to be Baalshamin,[50] appears under a variety of appellations: God/Zeus Baitokaike, often plus one or several of the epithets "Great/Greatest," "Sacred," "Listening," "Heavenly," "of the Thunderbolt" (*IGLS* VII.4028–41); not far away is found "Highest Heavenly God" (*IGLS* VII.4027). At various places in the Hauran Zeus is plain Zeus (frequently), Zeus "Ancestral God," "Great/Greatest" (frequently), "Greatest, Highest," "of the Thunderbolt," *kyrios,* "Master" (frequently: here used as an epithet, not a title), and also in response to particular circumstances "[Sacred] Listening," *teleios,* "of Fulfilment," *epikarpios,* "of Crops," *sōtēr,* "Saviour," *phōsphoros,* "Light-Bringer": in one town (Hebran) he is indistinguishably "Greatest," "of the Thunderbolt," "Master."[51] He also bears epithets referring to a locality. Baalshamin receives Nabataean dedications in the region, and it is reasonable to suppose that the author of any of the Greek dedications, if pressed to give the god a Nabataean name, would have chosen Baalshamin in each case; but in Greek, alongside various words just suggesting supreme power, more specific

47. Du Mesnil du Buisson, *Inscriptions palmyréniennes de Doura-Europos,* no. 23; *CIS* II.3.3912 (*OGIS* 631, Greek text only)

48. *IGLS* XVII.1 p. 424, index s.v. Bel. Thus the remark of T. Kaizer, *The Religious Life of Palmyra* (Stuttgart, 2002), 29, that both Baal Shamin and Bel are identified with Zeus needs qualification.

49. For a slightly different view, see Kaizer, "Religious Mentality," 176–79, who suggests that the Greek concept of "Highest God" is primary, and the Palmyrene expressions quoted are differing and innovative attempts to capture it. I do not know a Greek rendering in this period of the Aramaic title/name MRLH', a contraction of MR' LH', "lord of the gods" (for which see Healey, *Aramaic Inscriptions,* 229 on his no. 48; cf. his 49.8; 60.6; and below, p. 199); Mara (from MR) becomes a theonym in Seleuceia on the Eulaios (= Susa), *IEstrOriente* 213.3, 221.30.

50. Niehr, *Ba'alšamem,* 47–50; so too the god of *IGLS* VII.4027, 51.

51. Sourdel, *Cultes du Hauran,* 19–31; Niehr, *Ba'alšamem,* 230–39; cf. the indices to *IGLS* XIII.1 and 2, XV.1 and 2. Hebran: *Musée Soueida,* nos. 175–79; Sourdel, 25 n. 6.

terms from the Greek repertory are sometimes chosen, and no fixed equivalent emerges. Elsewhere in Syria we find, for instance, "Sacred Listening God," "Greatest and Sacred God," "Sacred Heavenly God," "Heavenly Ancestral God the Master," "Heavenly Highest Zeus Who Listens of Saarna."[52] Isolated and unusual is a late dedication from Byblos in Phoenicia to "Sacred Master and Master of the Whole World"; whether this title stands alone or adjectivally with a named god is unclear.[53] Editors debate from case to case what god is referred to,[54] and it seems that the question has to be answered from contextual probability; the name alone provides no certain indication. A much earlier attestation of what may be a similar Greco-Oriental designation for a high god is "Greatest Heavenly Zeus" or "Greatest Heavenly Ares" (the reading is problematic) from Delos. A "God of Gods, Greatest Zeus" who supposedly received a dedication in East Cilicia is so isolated as to raise great doubt.[55]

HIGHEST GOD

An important controversy surrounds *theos hypsistos,* "highest god," or Zeus *hypsistos* or simple *Hypsistos,* designations particularly frequent in Anatolia. A recent count gives 220 epigraphic instances of *theos hypsistos,* 121 of Zeus *hypsistos,* 34 of plain *Hypsistos;* there are also two instances of a feminine *thea hypsiste.*[56] These are significant numbers: a typical handful of dedications from Phrygia in the second century A.D. will combine a good number addressed to Zeus, Apollo, Mother, and

52. θεὸς ἅγιος ἐπήκοος, *IGLJ* 4.20; cf. *IGLJ* 5.100; θεὸς μέγιστος καὶ ἅγιος, *IGLS* XI.40, Mount Hermon; θεὸς ἅγιος οὐράνιος, *Tel Aviv* 13–14 (1986–87): 61 no. 2, Kedesh in Upper Galilee; θεὸς οὐράνιος πατρῷος ὁ κύριος, *Syria* 6 (1925): 354–56, no. 38, Damascus; Ζεὺς οὐράνιος ὕψιστος Σααρναῖος ἐπήκοος, *IGR* III 1060, near Byblos. On titles of Zeus at Gerasa ("Heavenly," "of Crops"), see Niehr, *Ba'alšamem,* 245. At Petra a Ζεὺς οὐράν[ιος]/Βεελ[ ]/ὁ ἐν τόπῳ Μωθω has recently been detected: P. L. Gatier, *Syria* 84 (2007): 180–81 (*SEG* LVII 1931).

53. τῷ κυρίῳ ἁγίῳ κὲ κυρίῳ ὅλου τοῦ κόσμου, *SEG* XLVI 1781 B; cf. Bonnet, *Enfants de Cadmos,* 194–95.

54. Editors suggest that *IGLJ* 4.20 may be Dousares, *IGLS* XI.40 Baal of Hermon; *Tel Aviv* 13–14 (1986–87): 61 n. 2; and *Syria* 6 (1925): 354–56, no. 38, Baalshamin. On the θεὸς ἅγιος of Petra, see Niehr, *Ba'alšamem,* 270–71.

55. Delos: *ID* 2312; cf. Baslez, *Religions orientales à Délos,* 103. Cilicia: *JHS* 18 (1898): 310, no. 7, from Yarsowat, "a small town about twelve miles on the road from Missis to Osmanieh," read as θεῷ θεῶν Διὶ Μεγίστῳ. W. D. Ramsay, *The Social Basis of Roman Power in Asia Minor* (Aberdeen, 1941), 113, reprints the inscription and speaks of "the Semitic (Aramaic?) and oriental 'god of gods,'" without citing Semitic or oriental evidence.

56. Mitchell, "Further Thoughts," 16. I cite attestations from Mitchell's very useful catalogues, in his "Theos Hypsistos," 128–47, and "Further Thoughts," 198–208 (numbers in the latter are preceded by an A). For *thea hypsiste,* see Mitchell, "Theos Hypsistos," no. 167, and "Further Thoughts," no. A 44. For a differently arranged catalogue (by titles, not regions), see Wischmeyer, "ΘΕΟΣ ΥΨΙΣΤΟΣ," at 158–66.

Men under a wide variety of epithets with not a few to *theos hypsistos;* the phenomenon is far from marginal. The common cult title of Zeus expressing elevation in classical Greece had been Hypatos, again normally translated "Highest"; though Zeus is occasionally *hypsistos* in poetry,[57] as a cult title *hypsistos* is attested for the classical period only if we allow three monuments mentioned by Pausanias long to antedate his own day.[58] *Hypsistos* then emerges into prominence in the Septuagint as a standard designation of the Jewish god, standing on its own or accompanied by *theos* or *kyrios,* "lord, master"; it retains that reference in a good number of dedications unquestionably made by Jews,[59] and is avoided by almost all Greek literary sources of the first century B.C. to the second century A.D.

The earliest secure pagan attestation as a cult epithet is perhaps a priest of Zeus *hypsistos* at Mylasa in Caria in the early second century B.C., which like some other early instances may represent an *interpretatio* of a non-Greek god; but there is no indication that the Zeus *hypsistos* who receives a dedication at Anthemous in Macedonia ca. 100 B.C. is not a Greek.[60] Whether or not the pagan title derived from the Jewish,[61] it separated itself from it, as the application to Zeus makes clear. But in some regions the dividing line marked by the presence of the name of Zeus was much weakened. Mainland Greece provides almost half the attestations of *hypsistos* linked to Zeus (ca. 55) but less than a quarter of the whole corpus of *hypsistos* dedications (88 of 375); in other words, the anonymous form (*theos*) *hypsistos* is much commoner outside mainland Greece, and is all but invariable in the almost

57. Schwabl, "Epiklesen," 370; poetry: Aesch. *Eum.* 28; Pind. *Nem.* 1.60; Soph. *Trach.* 1191, *Phil.* 1289; Theocr. 25.159. *Trag. Adesp.* 617.12 is a blatant Jewish forgery. Poseidon gets the epithet in *PMG* 939.1.

58. Paus. 2.2.8, 5.15.5, 9.8.5: emphasised by Belayche, "Polysémie des épiclèses," 433–34.

59. See, e.g., *OGIS* 96 with Dittenberger's note 5; *IJO* II 215 (Sidibounda), to θεὸς Ὕψιστος and Ἁγεία Καταφυγή, which has acquired a partner *BÉ* 2013, no. 425 (Eurymedon valley, territory of Selge in Pisidia [?]). Septuagint: it renders Hebrew elyôn, on which, see Mettinger, *In Search of God,* 122. In *SB* V 8934 (36 B.C.: *OGIS* 742) a synagogue (προσευχή) is apparently dedicated θεῶι [με]γάλωι ἐ[πηκόωι (?); but Wilamowitz, *SBBerl* 1902, 1094–95, thought a more specific epithet was needed; M. L. Strack's doubts about line length and thus supplements (*ArchPF* 2, 1902, 559, no. 41) are partly based on a running together of two lines that has occurred in reprintings of this text after the editio princeps. Plain *theos* in a Jewish context: e.g., *OGIS* 73.

60. Mylasa: *IMylasa* 212.2; for the dating of the set of documents to which this belongs see R. Descat and I. Pernin, *Studi Ellenistici* 20 (2008): 285–314; they argue for ca. 210–145. There are then the guild of Zeus *hypsistos* in the Fayum ca. 69–57 B.C., Mitchell, "Theos Hypsistos," no. 28; *ID* 2306, "1st c. B.C." (Mitchell, no. 110a), which is probably (Bruneau, *Cultes de Délos,* 240–41) oriental. Anthemous: *SEG* XLII 562. Mitchell, "Theos Hypsistos," 130, tentatively dated his no. 41 from Edessa in Macedonia to the Hellenistic period, but *LGPN* IV s.v. Χάρης (2) has it early first century A.D. *Theos hypsistos* (as opposed to Zeus *hypsistos*) emerges outside demonstrably Jewish contexts in the first century A.D. (Mitchell, "Theos Hypsistos," 109).

61. For cautious remarks on this subject, see Wischmeyer, "ΘΕΟΣ ΥΨΙΣΤΟΣ," at 156–58.

50 instances from Phrygia and Lydia.[62] There is no way of telling, except sometimes from context, whether a dedication made to "highest god" from such regions is made by a Jew or Gentile.

So much is not controversial. Beyond that point, two views compete. For one, subtly expounded by Nicole Belayche,[63] such dedications, when made by pagans, are addressed to particular gods of traditional type who are taken for granted, though not named, and differ from case to case. The Zeus *hypsistos* Brontaios of Miletupolis[64] in Mysia ("Highest, of Thunder") would be halfway down the road from a traditional thundering Zeus to a new-style *theos hypsistos;* but even had Brontaios dropped out of the title, the old god would still have been there. An anonymous *theos hypsistos* will often be identical with a known Zeus of the region in question.[65] On this view, not merely different gods but different types of god bear the title. A *hypsistos* (commonly in this case a Zeus *hypsistos*) may be a traditional member of a Greek civic pantheon, worshipped by a community. Or the superlative title may reflect the more intense and intimate relation sought by an individual worshipper, perhaps influenced by Semitic conceptions, seeking direct access to divine power outside the parameters of civic cult;[66] or the traditional dominance of an Anatolian god who "rules" a community. But sometimes the philosophical conception of a supreme deity to whom the others are subordinate is present. We can call proponents of this view separatists, even if they stress that these conceptions of the divine also interact.[67]

62. For the comparison, see Mitchell, "Theos Hypsistos," 100–102; Mitchell, "Further Thoughts," 170–71 (though his denial in the latter place of the occurrence of *theos hypsistos* in Thessaly and Macedonia looks like a slip). For cases where both forms occur in the same sanctuary, see Mitchell, "Further Thoughts," 171. Phrygia and Lydia: the only exception is Zeus πανύψιστος in a verse dedication; Mitchell, "Further Thoughts," A 53l.

63. Above all in "Hypsistos"; cf. Belayche, "Polysémie des épiclèses." Mitchell's views are also criticised by G. W. Bowersock, *Hyperboreus* 8 (2002): 353–63; and in A. Bresson et al., eds., *Une koinè pontique* (Bordeaux, 2007), 251–54; and M. Stein, *EpigAnat* 33 (2001): 119–26; but they are more concerned with the link he postulates between "reverers of god" (*theosebeis*) and worshippers of *theos hypsistos* (cf. n. 70 below).

64. *IKyz* 2, no. 5 (Mitchell, "Theos Hypsistos," no. 185).

65. So Belayche, "Ritual Expressions of Distinction," 163, against Mitchell, "Further Thoughts," 173–74. Marek, "Allmächtige Gott," 129–35, associates the Bithynian dedication that he publishes, Θεῷ ἀρίστῳ μεγίστῳ ἐπηκόῳ σωτῆρι ἑαυτοῦ ̣καὶ τῶν τέκνων αὐτοῦ καὶ πάντων τῶν [βο]ῶν Κάττιọς Τέργος εὐχήν (*SEG* L 1222; there is also a poem, now *SGO* II, 09/12/01), with a nearby sanctuary of Zeus Sarnendenos (*SEG* L 1223–24). This would be conclusive, but is not certain, though plausible.

66. "Intimacy with a divinity both exalted and close to man because welcoming and helpful": Belayche, "Polysémie des épiclèses," 439; "the expression of a relationship with a privileged divinity . . . frequently as a result of a direct personal experience": Belayche "Ritual Expressions of Distinction," 146.

67. Belayche, "Polysémie des épiclèses," 442: "In the pagan East, the diffusion in the imperial period of the title ὕψιστος clearly benefited from receptiveness to Semitic, Jewish, and Syrian theologies. But it encountered there indigenous traditions and a philosophical conception of an elevated god."

The other, advocated first by Schürer and more recently by Stephen Mitchell, recognises here a distinct cult of what is in effect a new god significantly different from any member of the traditional Greco-Roman pantheon, but, though anonymous, as individual as them:[68] a cult that eschewed animal sacrifice, elaborate cult buildings, and anthropomorphic representation of a deity who was conceived as supreme, though not unique, and who in Anatolia at least was served by intermediate beings sometimes termed "angels" (ἄγγελοι). Its diffusion through the Greco-Roman world would parallel that of, for instance, Juppiter Dolichenus. Worshippers would meet in modest house chapels[69] or in the open air; the main ritual instruments would be lamps that evoked the light of the highest god, whom they associated with the sun. The coincidence in naming mentioned above between pagan and Jewish "highest gods" is one of the many areas where the two approaches diverge. For Mitchell, it marks a point of convergence between Jew and pagan (and subsequently Christian),[70] for the separatists rather a point of competition, with deities of radically different character disputing the same title of "highest."

Much here is difficult to judge. When an individual in Lydia makes a dedication to "Highest goddess who protects [a community whose name is now lost]," is it more important that he has a particular local goddess in view,[71] or that he transforms her into an anonymous supreme being? As with the majority of *hypsistos* monuments, the forms of cult by which he worshipped her escape us, and hence come most of our difficulties.[72] On one point, the separatist case goes too far. They argue that *hypsistos* is merely a glorifying title like many others that emerged in the imperial period, applicable not merely to many different entities disguised under the anonymous "god" but also to many different named deities. But this is

68. Though still answering traditional concerns (healing, crop fertility, etc.): Mitchell, "Theos Hypsistos," 106.

69. Cf. S. Mitchell, *AnatSt* 53 (2003): 151–54.

70. Mitchell, "Theos Hypsistos," 92–97, indeed identifies worshippers of *Hypsistos* with the Jewish sympathizers known as *theosebeis*, "reverers of god," on the basis above all of a report of Cyril of Alexandria speaking of self-styled *theosebeis* in Palestine and Phoenicia who worshipped *hypsistos theos;* he associates Cyril's report with those of Epiphanius and Gregory of Nazianzus on a pagan-Jewish mixed sect called Hypsistarioi by Gregory; Gregory of Nyssa also knows of Hypsistianoi. But, though these late reports speak of independent sects, the allusions in Acts treat the *theosebeis* in Mitchell's own words (115) as "Jewish sympathizers, who attended the synagogues without being members of the Jewish community"; they do not hint at a separate cult of *Hypsistos* (a point made by M. Stein, *EpigAnat* 33, 2001: 125). Tanais offers societies of those "revering (σεβόμενοι) *theos hypsistos* (Mitchell, "Theos Hypsistos," nos. 96, 98, 100, 101), but Ustinova, *Supreme Gods*, 203–39, has shown how difficult it is to see them as Jewish sympathizers. In favour of convergence, however, note Bickerman, *Studies*, 2:611: "A God-fearing gentile remained *theosebes* and did not embrace the faith of Abraham, Isaac and Jacob, precisely for the reason that the contradiction between Jewish monotheism and his polytheistic ideas was not evident to him. He just wanted to add the Eternal God, or let us say another Eternal Deity, to his pantheon."

71. So Chaniotis, "Megatheism," 120, on the dedication; Mitchell, "Further Thoughts," A 44.

72. Cf. Belayche's counter to Mitchell, "Hypsistos," 156–59.

exaggerated:[73] the only god to receive it with any frequency is Zeus, and the few countercases can be interpreted as predictable spillovers in a world where naming was not controlled. Thus its usage is different from that of epithets such as "Saviour" (Soter) or "Manifest" (*epiphanēs*) or several others that were indeed "transgod."

But the unifying case faces worse difficulties, at least in its extreme form, which postulates a distinct cult with distinct forms of worship. Mitchell himself has acknowledged that the prominent Macedonian cult of Zeus *hypsistos* falls outside his interpretative framework.[74] The bilingual evidence from Palmyra noted above illustrates what the separatists claim to be the general rule: "Highest Zeus" is there used as an equivalent for different local deities, without there being any reason to think that their cult practices have been assimilated one to another or to an international norm of *hypsistos* worship. Even more damaging is the observation that even in Anatolia, the supposed heartland of a distinctive cult addressed to *hypsistos,* he could receive dedications from communities;[75] are we to imagine whole villages going over to a nonsacrificial cult? What must be preserved from the case is the widespread tendency to honour an often anonymised "highest god": the convergence in naming should indicate some convergence in understanding (at least among pagans), even if forms of cult remained different.[76]

A point that no one denies is that the cult of the highest god was sometimes in interaction with the new theology, expressed to our amazement in oracles, of the second century A.D. (But for separatists this is an occasional convergence, whereas for Mitchell the two things are essentially the same.) When in Amastris in Paphlagonia "by the voice of the Unshorn One" an altar is dedicated "of the Highest God who rules over all and is not seen,"[77] a hierarchy is clearly being established between a lower god of sensible qualities, long-haired Apollo, and a supreme and invisible power who is unnamed; "Highest" here is not mere rhetoric but embodies a theological vision. We are close to the figure "who cannot be named, many-named, living in fire" and identical with all-seeing aither, whom Apollo saluted as god in a famous oracular response inscribed at Oinoanda in northern Lycia, in a sanctuary

73. See Mitchell, "Further Thoughts," 171–74.

74. Mitchell, "Further Thoughts," 170–71.

75. Mitchell, "Theos Hypsistos," nos. 189 and 224, noted (with a Syrian example) by Belayche, "Polysémie des épiclèses," 436. Very difficult too is the case of Tanais, where Ustinova argues (n. 70 above) that almost the whole population belonged to societies worshipping *theos hypsistos.*

76. So, persuasively, Wischmeyer, "ΘΕΟΣ ΥΨΙΣΤΟΣ": existing gods are given an injection of new power by "Anonymisierung" and "Abstraktion" within an "ecumenical Koine of religious language"; no new cult is created. In the prayer for vengeance from Alexandria, Mitchell, "Theos Hypsistos," no. 284, "highest god" is expanded to ὕψιστος καὶ πάντων ἐπόπτης, "highest and overseer of all."

77. Marek, "Allmächtige Gott," 135–37 (*SEG* L 1225; *SGO* II, 10/03/01; Mitchell, "Further Thoughts," A 51).

that also hosted a dedication to *theos hypsistos.*[78] Several other striking anonymised designations for high gods appear more fleetingly in Anatolia in this period, though their relation to philosophical speculation is seldom clear. Already late in the first century A.D. (?) a Roman citizen made a dedication in the Cibyratis north of Lycia to κυρίῳ ἄνω θεῷ, "the lord god above" (not a sole god, however: the same man also dedicated to a cluster of "great temple-sharing gods");[79] from roughly the same period from Aspendos in Pamphylia comes a dedication, whether pagan, Jewish, or pagano-Jewish is not clear, "to the unlying god not made by human hands";[80] Mylasa in Caria may yield one to "a good hidden god who hears";[81] much later in Lydia a priest serves the "one and only god"—an acclamation, "One God in Heaven," transformed into a deity—as well as Hosios kai Dikaios.[82]

More frequently attested, if within a restricted area, is a power called *to th(e) ion,* "the divine." A large number of small altars dedicated to Zeus *hypsistos* have been found in diverse places in and around Stratonicea in Caria.[83] In almost every case Zeus *hypsistos* (in the dative, as recipient of a dedication) is accompanied by *to th(e)ion,* sometimes standing on its own (seven times), sometimes accompanied by an adjective or noun: "royal/king divine" (four times), "great divine" (once), "good divine" (once), "divine good messenger" (once; once also just "good messenger"), "divine (heavenly) messenger" (four times; the same figure without Zeus two more times). When *th(e)io-* in such a text is accompanied by a noun, it can easily be understood as an adjective, to give, for instance, "to Zeus *hypsistos* and the divine messenger" as a rendering of Διὶ ὑψίστῳ καὶ θείῳ ἀνγέλῳ. But a dedication such as "to Zeus *hypsistos* and *to thion*" (Διὶ ὑψίστῳ καὶ τῷ Θίῳ, *IStrat* 1113) shows that in that case at least we have to recognise a neuter deity *to theion,* in fact known also in Lydia;[84] possibly different worshippers understand the grammar

78. *SGO* 4, 17/06/01.

79. *Burdur Museum,* nos. 103 and 184 = *IKibyra* 93 and 94. Probably no connection with "the gods above" of Delos, *ID* 2309; see below, p. 163n49.

80. Θεῷ ἀψευδ[εῖ καὶ] | ἀχειροποιήτῳ, *SEG* XXXVIII 1335 = *IJO* II 218, where see the commentary; and cf. P. W. van der Horst, *Journal of Jewish Studies* 43 (1992): 32–37; Chaniotis, "Megatheism," 137 n. 109; Mitchell, "Further Thoughts," 176. Though ἀψευδής is widely used of pagan oracular gods, and Clem. Al. *Protr.* 4.48.1 says "they have ventured to call Sarapis ἀχειροποίητος," the extremely emphatic decision to *address* the god in these terms, and these only, seems to me unlikely for a pagan.

81. *SEG* LIV 1109: published as Ἄρχη ἀγαθῷ | κρυπτῷ ἐπη|κόῳ εὐχήν, but since the name Ἄρχη is not known, A. Chaniotis, *EBGR* 2004 (2007): 249, no. 26, moots Archeagathos as the name of a heroized young soldier.

82. Εἷς καὶ Μόνος θεός, *TAM* V.1.246, 256/7 A.D.; on this text, see Belayche, "Ritual Expressions of Distinction," 153. For Εἷς θεὸς ἐν οὐρανῷ (followed by an acclamation μέγα τὸ Ὅσιον, μέγα τὸ Δίκεον), see *SEG* LVI 1432, and several among the acclamations from Delphi, *SEG* LI 613–31.

83. *IStrat* 519, 552 (in II.2 p. 1: *SEG* XXXVIII 1091), 814, 1110–20, 1306–10; *SEG* LIX 1255, 1257.

84. *TAM* V.1.186.1–2, 434.13, 524, *TAM* V.3.1635–36; cf. 1636; *Manisa Museum,* no. 184. I have not seen N. Belayche, "Des dieux au divin? Sur des appellations impersonnelles de divinités

differently.[85] One dedication at Stratonicea is made simply "to the many-formed (πολύμορφον) *theion*" (*IStrat* 1542).[86]

The standing of *to theion* is problematic.[87] In a confession inscription from Kotiaion in northern Phrygia it punishes an offender,[88] and in Lydia "the great *theion*" is invoked to protect a tomb (*TAM* V.1.434.13). Still in Lydia it is once associated with Men (*TAM* V.1.524); dedications are made to "*theos hypsistos* and the great manifest *theion*" (*TAM* V.1.186) and to "*theos hypsistos* [and] the great *theion*" (*TAM* V.3.1635: a probable reading), the latter by his priest. In a verse inscription from Nakoleia in Phrygia it is *pantodynastes,* "master of all" (*SGO* 3, 16/35/01). In none of this is there any suggestion that *to theion* has a subordinate status; the pairing with *theos hypsistos* can be seen as a kind of hendiadys, divinity seen in the absolute and as manifested in the human sphere ("great manifest *theion*"), rather than as implying a hierarchy. More problematic is the common designation of *to th(e)ion* at Stratonicea as a messenger (though there too it is sometimes "kingly" or "great"), as most explicitly in a recently published dedication: "Diodotos and Ammias having been saved by *to thion* as Messenger (or the Divine Messenger) with their whole household, thank-offering to Zeus *hypsistos*."[89] It has been suggested that in such expressions "*thion* as Messenger" is not so much an entity distinct from Zeus *hypsistos* as a manifestation of his power,[90] but that interpretation places some strain on the language used; a split between Zeus *hypsistos* and the messenger is hard to dispute here. But this certainly does not prove Zeus to be a transcendent being beyond direct reach of mortals; one could think rather of his traditional relation to Hermes. How high a high god is, whether transcendent or merely (like the traditional Zeus) higher than others, is often impossible to say.

This survey of renderings of the idea of a supreme god has produced partly different results in the two main regions studied. In the Near East, we can observe a process whereby different great gods (sometimes identifiable in their original form through bilinguals) receive the same designation in Greek. Differentiation does

gréco-romaines," in D. Hue, ed., *Le divin: Discours encyclopédiques* (Caen, 1994), 79–96, nor her "Les dédicaces 'au Divin (τῷ Θείῳ)' dans l'Anatolie impériale," in M. de Souza et al., eds., *Le sacré dans tous ses états* (Saint-Étienne, 2013), 181–94.

85. Belayche, "*Angeloi,*" 58.

86. The epithet πολύμορφος is elsewhere applied to Isis (*RICIS* 501/0154).

87. See Belayche, "*Angeloi,*" 57–65, for a spirited and largely successful attempt to dissociate pagan "messengers" both from Judeo-Christian angels and also from a neo-Platonic "transcendent god served by intermediaries" schema; cf. Belayche, "Hypsistos," 148–51; "Ritual Expressions of Distinction," 153–54, followed by Bonnet, *Enfants de Cadmos,* 322–27.

88. M. Ricl, *EpigAnat* 29 (1997): 35–43.

89. Διὶ Ὑψίστῳ Διόδοτος καὶ Ἀμμίας σωθέντες ὑπὸ τοῦ Θίου Ἀγγέλου πανοίκιοι χαριστήριον (*SEG* LIX 1255).

90. Bonnet, *Enfants de Cadmos,* 325 (following Belayche): "The messenger *represents* the god's divine power . . . it is neither separated nor different from the god."

not vanish in every case, however; these "greatest heavenly Zeuses" coexist with Zeuses who bear a more specific epithet (topographic, or even a retained theonym such as Marnas). In Anatolia, we lack any indigenous tradition with which to compare the Greek names; it is not even clear to what extent one existed. Innumerable gods retain separate designations. But there emerges also a tendency to blur (though probably not to efface) distinctions by describing particular gods without reference to their specific traditions simply as "highest." There emerges also, if in a limited region, an abstraction "the divine" which sometimes is coordinated with a highest god but sometimes, rather strangely, serves him as a messenger. The tendency is not in either region toward monotheism. A dedication from Lydia brings this out: "For *theos hypistos* and the great manifest *theion* Demo daughter of Tyrannos set up the goddess Larmene (or Larmenian goddess) in the year 256" (*TAM* V.1.186: 171–72 A.D.). It neatly illustrates the coexistence of powers described in untraditional ways such as *theos hypsistos* and *to theion* (which yet are not completely revolutionary: there have always been top gods and amorphous divine power) with a figure such as Larmene, so local and specific that we can say nothing confident about her history or nature.

5

Ad Maiorem Deorum Gloriam

The Growth of Praise Epithets

Chapter 3 sketched the developments in the naming of gods in Greek that occurred with the spread of the Greek language throughout the eastern Mediterranean, and chapter 4 treated three particular aspects of that phenomenon. Changes that occurred through extension in space and through extension in time inevitably became intermingled, because the evidence for many of the regions concerned only begins in the Roman imperial period. This chapter will pursue the issue of the changes that occurred over time. To start with an extreme example we can take a low pillar found in the entrance to the baths of Caracalla in Rome which bears a dedication Διὶ Ἡλίῳ μεγάλῳ Σαράπιδι σωτῆρι πλουτοδοτῃ ἐπηκόῳ εὐεργέτῃ ἀνεικήτῳ Μίθρᾳ χαριστήριον, "to Zeus Helios Great Sarapis Saviour Wealth-Giver Listener Benefactor Invincible Mithras, a thank-offering." Or possibly the object was dedicated by one Mithra (reading Μίθρα for Μίθρᾳ), in which case just nine words and not ten are used to identify the recipient.[1] It is not just the conflation of three or four gods into one that marks the text as a product of the high Roman empire, but also the accumulation of epithets. No less unclassical, if in a different way, is an altar from Athens inscribed "Of Apollo of the Alleys, Protector, Ancestral, Pythian, Clarian, Panionian."[2] We saw in chapter 1 that the primary function of the cult epithet in Greece was not to glorify gods but to identify them. The epithets discussed in chapter 3 mostly still worked in the same way. But epithets were increasingly used to show respect to gods and to acclaim their powers: the first of

1. *RICIS* 501/0126.

2. Ἀπόλλωνος Ἀγυιέως Προστατηρί[ου] Πατρώου Πυθίου Κλαρίου Πανιωνίου, *IG* II² 4995; cf., e.g., 4852, Apollo Prostropaios Aguieus Alexikakos.

these two texts elevates the god by amassing vague general epithets, the second by creating a casserole of what had originally been rather specific titles, some pointing to particular cult sites, some to functions. A dedication from Aphrodisias in eastern Caria "to the Manifest Heavenly Goddess of All the People Aphrodite" blithely effaces the distinction between Heavenly Aphrodite and Aphrodite of All the People set up by Plato in the *Symposium.*[3]

I turn in this chapter to these new usages and in particular to the new epithets they bring with them. They share the negative feature that their role is not primarily to identify. Their positive functions differ slightly from epithet to epithet and from one application to another. To call a god "Greatest" may be to praise him or to express the hope that the god indeed has the power to do what is asked, or a mixture of both. To call a god "Saviour" may be an appeal for help, or a grateful acknowledgment of help received. To call him "Lord" or "Holy" stresses rather the mortal's inferiority to the superior or more sanctified being. An entirely satisfactory general term for them cannot be found, but "praise epithets" may be the best available. Some of these praise epithets arose, as we shall see, by absorption into Greek of honorific terms from other languages, but then spread (with some restrictions) outside the regions where the donor language was spoken. As a result, as noted above, the distinction between change owed to cultural contact and change occurring over time loses clarity: both processes are at work, and in interaction.

Before turning to the praise epithets I note a phenomenon that is separate but coincides chronologically and geographically with the emergence of several of that group: the addition of *theos* to the actual theonym, to give expressions such as *theos* Apollo or *thea* Hera (less often *theos/thea* follows the name) which lie quite outside the norms of Greek naming. The usage is found throughout Anatolia, Syria, and Thrace, and spreads back to Macedonia but apparently not to Greece itself or the islands.[4] Just the same usage emerges in Latin with *deus* and *dea;* it is much commoner in the northern and western provinces and Africa than in Italy, rarer still in Rome, and much commoner with some gods than others; *deus* Juppiter in particular is all but unknown.[5] The earliest datable Latin instances are from 124 A.D. (with the order theonym-*deus*) and 135 A.D. (*deus*-theonym); examples

3. τῇ ἐπιφανεῖ οὐ[ρα]νίᾳ πανδήμῳ θεᾷ Ἀφροδίτῃ, *IAph* 12.26 d (*MAMA* 8.413d) 15–16; Pl. *Symp.* 180D-182A—a real distinction in cult, though not of the kind claimed by Plato.

4. Apparent instances in Greece are explicable. In *IG* V.1.312 Lycurgus is called *theos* because he is functioning as *patronomos* and must be distinguished from mortals; in the expression Β̣ορθίῃ θ̣εῷ in *IG* V.1.1376 (cf. 260 [?]) Β̣ορθίῃ is adjectival. Cf. n. 7 below on *Syll.*[3] 747. In *SEG* LIV 323 θεῷ Ἡρακλεῖ ἐπηκόῳ κατὰ κέλευσιν· Νεικητίων ἀνέθηκα (Attica), θεῷ probably links with ἐπηκόῳ, not with Ἡρακλεῖ. But for late instances in Macedonia, see *SEG* XXVIII 546, XLVII 884, and often.

5. θεὸς Ζεύς is rare enough for Schwabl, "Epiklesen," 315, to capitalise Θεός and treat it as an epithet, but to the cases he cites add two from Tralleis (*ITrall* 4; *EpigAnat* 41, 2008: 88–89; cf. Cook, *Zeus,* 2:958) and *SEG* LVIII 1186 (Aphrodisias).

abound thereafter.[6] In Greek, we meet Nanaia *theos* at Susa in the second century B.C. and, under a Seleucid Antiochos, *theos* Zeus Baitokaike;[7] but the vast bulk of the Greek material, which has been much less studied, seems to fall within the same time frame as the Roman, and there may have been a break and a new start within Greek. If so, it is not clear in which direction the influence went within the Greco-Roman *Sprachbund,* nor what impelled the usage.

An old view is that *theos/deus* indicated a foreign god,[8] the added noun perhaps serving to mark out the unfamiliar theonym as a theonym, as *oppidum* could mark out a town. That view has some plausibility for the Hellenistic Greek instances but none for the later Greek or Roman, given that the same gods had been identified by simple theonyms for a long period before the *deus/theos* convention arose. A variant makes *deus/theos* the marker of an interpreted god, a reminder that the Apollo or Juno in question is not the authentic Greek or Roman deity but a local equivalent. But here too it is hard to see why such a marker should have emerged so long after *interpretatio* began; nor do the facts of usage fit the hypothesis.[9] One may wonder whether the Semitic languages in which a theonym is regularly preceded or followed by "god" exerted pressure: a Nabataean dedicates to *thea* Aphrodite, corresponding to "Al-'uzza, goddess" (LL'Z' 'LHT'), in a bilingual from Cos in the year 9 A.D.[10] The movement would be from east to west if so. For the moment the phenomenon remains unexplained. In usage the added *theos* often seems equivalent to *kyrios,* "lord," an imprecise title of respect.[11]

6. See D. Vaglieri, in E. de Ruggiero, *Dizionario epigrafico di antichità romane* II.2 (Spoleto, 1910), 1716–21 (lists); and especially M.T. Raepsaet-Charlier, *Diis deabusque sacrum* (Paris, ca. 1993), 11–17; Van Andringa, *Religion en Gaule romaine,* 134, speaks of "a specific, but not exclusive, marker of local pantheons." For the very limited attestations from Rome itself, see R. Mancini, "Deo-deae nelle iscrizioni di Roma," *Tituli* 2 (1980): 173–78. The plural use (*dei Penates,* etc.: *TLL* V.1.888.55–889.4; Vaglieri, 1721–26) is earlier, and apparently unrelated. For the idea of a Greek-Latin *Sprachbund,* see, e.g., J. Kramer, *Von der Papyrologie zur Romanistik* (Berlin, 2011), 57–80. I am grateful to J.N. Adams for correspondence on this usage.

7. *IEstrOriente* 192, 193, 197, all from Susa, the last dated to 142/1 B.C.; *RC* 70.4 (but the four parallels adduced by Welles ad loc. are all to be explained differently; in particular, there are obvious reasons why Amphiaraos is repeatedly called "god" in *Syll.*[3] 747, given that his status as a god was precisely the point at issue in the dispute between the sanctuary and Roman tax-collectors there resolved in favour of the sanctuary). Robert, *OMS,* 1:405 n. 9, notes that *theos* Herakles is scarcely conceivable before the Common Era. Two perhaps Hellenistic inscriptions from Alexandria look anomalous: Bernand, *Inscriptions d' Alexandrie,* 55:"Ισιος, Σαράπιος, Ἄνουβιος θεῶν; and 55b: Εἴσει, Σαράπιδι, "Ερμῆι θεοῖς; the reader for the press suggests plausibly that we should understand "[and other] gods [of the sanctuary]."

8. So, e.g., Welles on *RC* 70.4, and by implication *TLL* V.1.888.42–54. In a dedication such as *IG* II[2] 4591 (4th c.) to "hero Bouthon and heroine Eudosia," "hero" probably has the function of explaining the category to which the recipient belongs; distinct is the later application to dead mortals, where the point is to elevate their status. Use of *oppidum: TLL* IX.2.758.35–57.

9. For critique, see Raepsaet-Charlier (n. 6 above), 16–17; Aliquot, *Vie religieuse au Liban,* 151 n. 147.

10. *IG* XII.4.558.

11. See, e.g., the index entries to the two words in the volumes of *IGB.*

A different development is exclusive to Thrace. Thrace has yielded thousands of examples, beginning perhaps in the second century B.C., of the iconographic type of the "Thracian rider." A majority of these are dedicated to figures described initially as Hero (Ἥρως) or Heron (Ἥρων), either just so or more often with a further specification which is typically a toponym in *-enos* or *-eites* but may also be a Greek functional adjective (e.g., Propylaios, "before the gate"),[12] or an unexplained Thracian word (theonym? epithet?). From the second century A.D., when the type really begins to proliferate, Hero or Heron can be replaced by *theos* or expanded into a unique compound *theos* Hero or *kyrios,* "lord," Hero, to give, for example, *theos* Hero Auloneites or *kyrios* Hero Pyroumeroulas. (In the same period the iconographic type begins to be applied to certain Greco-Roman gods, in particular Apollo, Asclepius, and Silvanus.)[13] Some suppose that Hero or Heron is a rendering of a phonetically similar Thracian term (appellative or name?), others that the Greek word has replaced a Thracian term or filled a gap in the Thracian vocabulary.[14] The iconography of the Greek hero-relief has evidently influenced that of the Thracian rider, and on any view that iconographic association must relate in some way to the adoption of the name. But the expansion to *theos* Hero or *kyrios* Hero shows that these "heroes" were or came to be perceived not as that but as gods; the change from Hero X to *theos* Hero X is exactly comparable to that discussed above from, for example, Apollo to *theos* Apollo, and may suggest that Hero had always been understood as in effect a theonym (as a mental experiment one can try substituting for it "Zeus").

I revert to the praise epithets. Greeks had always, obviously, celebrated the gods and humbled themselves before them. In classical poetry gods are constantly referred to or addressed or sworn by with reverential adjectives (*semnos, hagnos*, both "reverend"; *polytimētos,* "much-honoured"; *makar,* "blessed"; *megas,* "great")[15] and titles (*basileus, basilissa, despotēs, despoina, anax/anassa, potnia,* all variations on "king/lord" and their feminines). Most are applied to a wide range of powers (but *basileus* is almost reserved for Zeus), though some gods may particularly attract some titles (Despoina Cybele).[16]

12. For Hero Propylaios, see *IGB* I² 291bis, 350; III 1768, 1770; Hero Katoikadios, *IGB* III 1874; Hero Proastios, *CCET* 1:no. 77; for epithets found in Rumania, see *CCET* 4:16 -17; and cf. Oppermann, *Thrakische Reiter,* 278.

13. Oppermann, *Thrakische Reiter,* 277–78, 315–16. In old Greece, such an expression was a deliberate oxymoron: Pind. *Nem.* 3.22 (Herakles: Maas sought to remove it by changing θεός to θοός); *SEG* XXXVIII 947, Megara Hyblaea, Dioscuri?

14. For a thorough doxography, see Oppermann, *Thrakische Reiter,* 277 n. 2270.

15. But not ἰσχυρός, "strong." Everything about the dedication of Komosarue to "strong god Sanerges and Astara" at Phanagoria in the second half of the fourth century (*Syll.*³ 216; RO 65 D) is non-Greek except its language: deities, epithet, and (at this date) the use of a praise epithet of whatever type.

16. See Henrichs, "Despoina Kybele": he notes that in Callimachus and Catullus the title sometimes replaces the theonym. On "Sun first king" in a Hellenistic verse dedication, *SEG* XXXIV 492, from Atrax, see E. Voutiras, *ZPE* 61 (1985): 289, who suspects philosophical influence.

The examples are far too numerous to cite:[17] Socrates' address to his own special deities in *Clouds* (264–65) provides a rich selection of possibilities elsewhere applied to the traditional gods:

> ὦ **δέσποτ'** ἄναξ, ἀμέτρητ' Ἀήρ, ὃς ἔχεις τὴν γῆν μετέωρον
> λαμπρός τ' Αἰθήρ, **σεμναί** τε θεαὶ Νεφέλαι βροντησικέραυνοι,
> ἄρθητε, φάνητ', **ὦ δέσποιναι.**

> **Lord, master** (*despot' anax*), boundless Air, who holds the earth suspended, and gleaming Aither, and **reverend** (*semnai*) goddesses Clouds, thundering with lightning, arise, appear, **mistresses** (*despoinai*).

Some of these titles, as we saw, had served in Mycenaean Greece, or still served on the fringes of the Greek world or in special contexts (mysteries) in its heartland, in lieu of the actual names of the goddesses concerned.[18] We cannot doubt that worshippers regularly addressed their gods with such terms of respect. But they barely appear in surviving dedications of the classical period in prose, and when, later, reverential expressions become common in dedications the actual terms used are largely new. An inscription on an architrave from Thasos of the second century A.D. gives a flavour of a very different style of naming; it honours a Thracian god Heron, an otherwise unknown Gallia, and the "Syrian goddess":[19]

> κυρίου Ἥρωνος δεσπότου ἱερόν
> μεγάλης Γαλλίας ἀθανάτης ἱερόν
> μεγάλης Σύρων ἁγνῆς δεσποίνης ἱερόν.

> Sanctuary of lord master (*kyrios despotēs*) Heron. Sanctuary of great (*megalē*) immortal Gallia. Sanctuary of the great reverend mistress (*megalē hagnē despoina*) of Syrians.

Despotēs, despoina, kyrios, megas, hagnos, and *athanatos,* "immortal," all belong to the old Greek vocabulary, but their use here to glorify the deities (in an inscription the primary purpose of which is merely to identify the owners of a precinct) is decidedly postclassical.

The commonest praise epithets are *hagios,* "sacred," *kyrios,* "lord," and *epiphanēs,* "manifest," all entirely new in this sense, and *sōtēr,* "saviour," *megas,* "great," and *epēkoos,* "who listens," much extended. Instances are so frequent, and in the main so chronologically indeterminate, that to attempt a detailed account of their diffusion in space and time, even to attempt a ranking in terms of frequency, would be a huge and still frustrating endeavour. But some broad outlines are clear. I begin with *hagios.* In

17. For a few, see Williger, *RE* s.v. Kurios, 180.

18. See p. 5.

19. E. Will, *BCH* 64–65 (1940–41): 200–210; C. Dunant and J. Pouilloux, *Recherches sur l' histoire et les cultes de Thasos* (Paris, 1958), 2:182 and 223, no. 374. It is adduced as showing Syrian influence by Henrichs, "Despoina Kybele," 283.

classical Greek the adjective was used of things connected with the gods that demanded particular respect, such as especially venerable sanctuaries or mysteries, and under the influence of Herodotus this usage persists in, for instance, Strabo and Pausanias. It was not applied to gods as an epithet either regularly or occasionally.[20] It starts to appear in inscriptions as a divine epithet perhaps in the first century B.C. The first instance may be a dedication to Artemis Hagia by an Athenian, found in Sarapieion C of Delos and probably dating to the late second/early first century B.C. But this is not a certain example of the new usage, because the epithet comes after the theonym and may be functioning not as an honorific but as an identifier of a cult specifically dedicated to a goddess (otherwise unknown) entitled Artemis Hagia. Clearer is a dedication to an unnamed Eshmun as *theos hagios* at Sidon in Phoenicia, 48/7; then comes a dedication from Thrace datable ca. 21 A.D. to *theos hagios hypsistos*, "highest";[21] we see here, and in later instances, the tendency of worshippers of a non-Greek god when using Greek to prefer respectful periphrasis to naming.[22] The epithet is also applied, among many others, to Isis in a hymn of Isidorus, conventionally dated to the first century B.C. Later instances show a notable concentration (applied to male gods) in Syria,[23] but by the second century A.D. the epithet is formulaically given, in the superlative, to Artemis of Ephesos,[24] and other goddesses with Greek names receive it too in Anatolia, and eventually even on the mainland. Among male gods the only regular recipient with a Greek name is Zeus (there is one Herakles and one Kabeiros);[25] Sarapis gets it too, and it is often given in Syria to indigenous gods.[26]

When applied to a god the word has not changed in meaning but merely in application:[27] it still means "reverend," as it did when used of sanctuaries or rites, as *hagnos* and *semnos* had done in earlier poetic usage when attached to the names

20. The cases cited by Williger, *Hagios*, 83, are not epithets but predicates.

21. *ID* 2068; Stucky, *Eschmun-Heiligtum*, 321, no. 1, θεῶι ἁγίωι (cf. 324. no. 6, θεῶι ἁγίωι Ἀσκληπιῶι; by supplement in nos. 7–9); *IGR* I 777 (*ÖJhBeibl* 23, 1926: 128, no. 22), θεῶι ἁγίωι ὑψίστωι.

22. Cf. the probable application to the god of Doliche, *SEG* XXXII 1388. For an anonymous Cappadocian instance, see *EpigAnat* 28 (1997): 120, no. 6; cf. *IGLS* XV.2.495 (Hauran).

23. See, e.g., Sourdel, *Cultes du Hauran*, 97–98; and p. 124 above; examples can easily be multiplied. Isidorus: Totti, *Texte der Isis- und Sarapis-Religion*, no. 23.2.

24. Artemis of Ephesos: e.g., *SEG* XLV 1708, with the note. Other goddesses: Weinreich, *Ausgewählte Schriften*, 1:279–80. Add, e.g., *Ilion* 97.8, τῇ ἁγιωτάτῃ θε[ᾷ Ἀ]θηνᾷ; *IAph* 5.204.i, νεοποιοὶ τῆς ἁγιωτάτης θεοῦ Ἀφροδείτης.

25. *MélBeyrouth* 38 (1962): 17 (187/8 A.D., from Tyre); *SEG* XLIX 817.6, Thessalonike, 260 A.D., ὁ κρ(άτιστος) ἱεροφάντης τοῦ ἁγιωτάτου θεοῦ Καβείρου.

26. Cf. Williger, *Hagios*, 81–83. I know no Thracian example other than the one cited in n. 21 above, but ἁγνός seems occasionally to be so used: *IGB* 1² 306, 386, II 758. In Egypt it is rare: Ronchi, *Lexicon theonymon*, 6–7.

27. So rightly, Williger, *Hagios*, 83.

of gods. The new application, it has long been recognised,[28] apparently occurred under the influence of the Semitic QDYŠ, *qadoš*. The Septuagint shows how *hagios* and words related to it underwent multiple semantic extensions among Greek-speaking Jews seeking Greek equivalents for the Old Testament vocabulary of sacredness; it came to denote (like *hagnos*) the purity of worshippers, but could also be used to glorify God as in Isaiah 6.3: "*Hagios, hagios, hagios* is *kyrios* Sabaoth."[29] Some epigraphic examples refer to the Jewish god.[30] But the new pagan usage seems to derive rather, or also, from other Semitic languages which also came to render *qadoš* by *hagios*. In Phoenician building inscriptions ascribed to the fifth century B.C. from Sidon the inhabitant of the temple is probably described as "my god Eshmun, sacred (*qadoš*) prince" (though the expression has also been rendered "prince of the sanctuary"):[31] the expression "sacred god" recurs in neo-Punic. This is close to the Greek locution "*hagios* god X," and note that the first clear instance of the Greek usage came from the same temple at Sidon of Eshmun; it is in Syria that examples continue to be most frequent. It is, however, common even in contexts in Syria where the use of a Semitic language at the relevant time is not attested in inscriptions—Gerasa, for instance.[32] (That usage may support the case that a Semitic language in fact continued in spoken use.) And it spread beyond its probable region of origin; *hagnos* too, which had always been used of deities in verse, but in inscriptions only in one fixed combination (Hagne Aphrodite, for the Syrian goddess), now occasionally appeared in a sense indistinguishable from that of *hagios*.[33]

28. Since C. Clermont-Ganneau, *Études d' archéologie orientale* (Paris, 1880–95 [*sic*]), 1:104. *Sanctus* had been applied to gods earlier and independently of Semitic influence: a fragment quoted from Naevius, *Bellum Punicum* 2 (fr. 30 Morel), by Macr. *Sat.* 6.5.8, two poets quoted in Cic. *Div.* 2.115 and *Tusc.* 4.73; more in H. Delehaye, *Sanctus* (Brussels, 1927), 11–12. But that influence may have encouraged the extension of that usage later, when it becomes very common: for examples, see Delehaye, 12–20. The theory that *sanctus* acquired this application in early Hellenistic times as a calque on ἅγιος cannot stand (H. Fugier, *Recherches sur l'expression du sacré dans la langue latine,* Paris 1963, 270–92): *sanctus* was so used long before ἅγιος.

29. Ἅγιος ἅγιος ἅγιος κύριος Σαβαώθ: cf. Williger, *Hagios,* 84–108. For many further examples of ἅγιος applied to God, see A. Fridrichsen, *Hagios-Qadoš* (Kristiania, 1916), 33; on later Jewish usage, 62, mentioning θεὸς ὁ ἅγιος vel. sim. in 3 Macc. 5.13, 6.1, 6.29; on the OT usage in brief, Mettinger, *In Search of God,* 152–54. It enters hermetism dramatically with a ninefold repetition: *Corpus Hermeticum* 1.31.

30. It also continued to be regularly applied to synagogues: *BÉ* 1976, p. 558.

31. *KAI* 15 (see the commentary ad loc. for the ambiguity), 16; cf. 14.17. H. P. Mathys and R. Wachter, in Stucky, *Eschmun-Heiligtum,* 273–74 and 321, argue that the Greek inscription θεῶι ἁγίωι is decisive in favour of "sacred prince." Cf. gods called *qadoš* in *KAI* 4.4–5, 14.9, 22; El as sacred in Ugarit, W. H. Schmidt, *Königtum Gottes in Ugarit und Israel*[2] (Berlin, 1966), 28–29. Neo-Punic: *KAI* 145.4 = Jongeling, *Neo-Punic Inscriptions,* Hr. Maktar N 64.

32. *IGerasa* 17–18, 20.

33. *IAdramyt* II, 13; cf. n. 26 for Thrace.

Another title that becomes very common in Syria, Egypt, Thrace, and Anatolia[34] is *kyrios*, "master" (which acquires a Latin translation *dominus*). In old Greek *kyrios* had been applied to human persons in authority but not to gods; in the later usage it was applicable, like *anax* and *despotēs*, to both, an ambiguity which underlies a joke in *Philogelos* (76).[35] The new application achieved huge importance in the Jewish world as the rendering in the Septuagint for "our Lord" ('DNY, *adonay*, substitute for the tetragram YHWH), and subsequently among Christians when used of Christ. It has accordingly been much studied by theologians, above all in a four-volume monograph of 1926–29 by W. von Baudissin.[36] Noting the geographic distribution, Baudissin argued that it originated as a rendering in Greek of one of the various "lord" words used of gods in Semitic languages, such as Phoenician 'DN and RBT or Aramaic MR'; the use in the Septuagint would be a special variant of that usage.

The earliest attestations, however, come from Egypt and relate to Isis, who is dubbed *kyria* slightly more than fifty times in inscriptions of the Ptolemaic period from Philae, the first apparently dating from 142 B.C.[37] Baudissin knew of the existence of hieroglyphic signs *nb* and *nb.t*, "master," but denied their relevance to the new use of *kyrios* on the grounds that they, unlike *kyrios*, are always accompanied by a genitive indicating the sphere or place over which the god in question exercises mastery;[38] the Semitic usage, by contrast, (my/our) lord X, is closely

34. The fullest collection is still that of Drexler, in *RML* s.v. Kyria und Kyrios; see too Williger in *RE* s.v. Kurios; Baudissin, *Kyrios*, 2:263, 266; Sourdel, *Cultes du Hauran*, index sv. κύριος. For Egypt there is a comprehensive collection in Ronchi, *Lexicon theonymon*, 600–639; according to Bernand, *Fayoum*, vol. 1, note on no. 73.4, it indicates the top god of the locality. Older treatments (Drexler, Baudissin) play down its frequency in Anatolia, but there are many new instances. Instances from old Greece south of Macedonia (for which see now, e.g., *ILeukopetra* 73, 78; *EAM* 147; *SEG* XLVII 936) remain rare (I have only *IG* II[2] 4815 and *IC* II.XXI 1, both Sarapis). In *CIL* VIII 9020 (Mauretania), *Pluto Cyria et Ceres dii sanctissimi*, it has been thought that Cyria is used as a euphemism for Persephone (Drexler, 1758), but VIII 9021, *Pluto et Cyria Ceres dii sanctissimi*, may suggest a simple miswriting in 9020. Occasionally the word stands on its own, with reference to a deity identifiable from the context, as in Μελέαγρος Διομήδους ἀρχιερεὺς κατὰ κέλευσιν τῆς κυρίας Μητρὶ Ζιζιμμηνῇ εὐχήν (*JHS* 22, 1902: 341, no. 64, Iconium); *SEG* LIX 1694 (Tripolis in Syria); *RICIS* 504/0215; *Dura* IX, 118, no. 976; *ZPE* 194 (2015): 219–22, with *BÉ* 2015, no. 683 (a masculine instance); cf. Nock, *Essays*, 75 n. 100.

35. In the papyrus letter *SB* X 10278 (Trajan or Hadrian) the word appears twice applied to human masters/mistresses (once in direct address), twice to Hermes.

36. Baudissin, *Kyrios;* see especially 2:257–85.

37. See *IGPhilae*, I, p. 61 and (index) 404; first is *IGPhilae* 14.7, apparently dated to 142 B.C. She is ἡ Ἶσις ἡ κύρια even in the narrative of the life of Aesop, B. P. Perry, *Aesopica* (Urbana, Ill., 1952), 37, chap. 5. A "prayer for justice" from Arcesine on Amorgos (*IG* XII.7 p. 1 [no number]) which begins with an appeal to "κύρια Demeter" has been variously dated between the second centuries B.C. and A.D. (see H. W. Pleket, in *Faith, Hope, and Worship*, 189); the early date would be surprising in terms of the usage of the title.

38. Baudissin, *Kyrios*, 2:266–69.

paralleled by the Greek *kyrios* X.[39] The fixed title *kyria* Isis (normally without a genitive dependent on it) emerged, he supposed, under Syrian influence; the absence of earlier evidence from the Levant would simply be due to the dearth of inscriptions set up in Greek by persons whose first language was Semitic.[40] But a Phoenician dedication to Isis from the Ptolemaic period found at Memphis in Egypt (*KAI* 48.2) gives her the title "mighty" ('DR), not quite what Baudissin's hypothesis requires. And the sheer abundance of evidence for *kyria* Isis, not readily explicable as a simple borrowing, was not known to Baudissin; the precise interplay between Jewish, broader Semitic, and possible Egyptian influences must remain uncertain.[41]

At all events it is clear that *kyrios* emerged as a divine title through cultural and linguistic interaction, the attempt to render in Greek a characteristic title of honour from a different language and religious tradition. Why it was *kyrios* that came to be so used, with a transference from the human to the divine sphere, in preference to its near synonyms *anax* and *despotēs,* traditional elements in the religious as well as (in the case of *despotēs*) the secular vocabulary of respect, is a puzzle; applied to humans, the vocative *kyrie* is a less deferential form of address than *despota.*[42] *Despotēs/despoina* continue to appear in direct addresses to gods, but as praise epithets in dedications are much rarer than *kyrios.*[43]

Kyrios as an epithet is sometimes accompanied by the article—Διὶ μεγάλῳ τῷ κυρίωι[44]—and here Greek idiom suggests as a rendering not just "lord great Zeus" but "great Zeus, my master"; occasionally, particularly in direct address to a god, a genitive which makes that connection explicit is added.[45] In Egypt, the article is almost invariably present. In the Semitic-speaking regions, theophoric names in

39. He points out, *Kyrios,* 2:264, that *ba'al* and *ba'alat* by contrast are irrelevant, since they never introduce a theonym but stand alone. For *adon* with theonym, see, e.g., the trilingual *KAI* 66 (below, p. 198), where, however, the Greek omits it.

40. Baudissin, *Kyrios,* 2:268–69.

41. The curse from Delos published by Bruneau (*Cultes de Délos,* 650–55), which invokes Κύριοι θεοὶ Συκοναῖοι and Κύρια θεὰ Συρία ἡ Συκονα, is interesting; it is dated vaguely to the "Athenian period" on the island, i.e., after 166 B.C.

42. Dickey, *Greek Forms of Address,* 107; see 90–107 on all these terms.

43. Direct addresses: see Henrichs, "Despoina Kybele," 274 n. 58; in dedications, Baudissin, *Kyrios,* 2:156–57; L. Robert, *RPhil* 33 (1959): 222 (= *OMS,* 5:252 with instances from Macedonia, Bulgaria, Egypt, Ephesus, and Thrace); *BÉ* 1967, no. 645; *CRAI,* 1968, 583 n. 5; *SEG* LVII 1389 (Tyriaion), Zeus Despotes; *IEstrOriente* 16 (Armenia), Ἐμίλιος Οὐάλης Γῇ Μετρ[ὶ] Ὀλυβρι θεᾷ δεσποίνᾳ. Note *IGLS* VI.2728–29 (Hajjar, *Triade d' Héliopolis-Baalbek,* nos. 16 and 18), one addressed to "greatest Heliopolitan god" as δεσπότης, one as κύριος (cf. 2731). On ἄναξ/ἄνασσα, cf. Ronchi, *Lexicon theonymon,* 121–24.

44. So, e.g., *Musée Soueida,* no. 200; cf. the various dedications to Athena κύρια with and without the article in Sourdel, *Cultes du Hauran,* 71–72.

45. Κύριέ μου Σαρᾶπι῞Ηλιε εὐεργέτα, *POxy.* 8.1148, first century A.D., in a question to an oracle; cf. τοῦ κυρίου ἡμῶν "Ερμοῦ, *PPar.* 19.5, and in magical papyri, Baudissin, *Kyrios,* 2:281; τῆς κυρίας ἡμῶν Ἀρτέμιδος, *SEG* LX 1161 (Ephesus region).

Taim- and ʿAbd- ("slave of" a named god) remained very common,[46] and it has been argued that the title *kyrios* applied to a god suggests the master-slave relation whether or not the article is present.[47] But equals could address one another as *kyrie,* even if this was rather obsequious; the term was not confined to master-slave relations. The nature of the subordination understood to underlie *kyrios* (and comparable words) may have varied from region to region and even case to case; the other side of the supposed relation (mortal as slave of the god) is seldom made explicit in Greek.[48]

By late antiquity the glorification of gods as *megas,* "great," or *megistos,* "greatest," had become, in some regions, very general. Earlier, Zeus had regularly received both titles in poetry, in dedications influenced by poetry, and no doubt often in prayer; Kronos too is often "great," and other gods received the epithets occasionally when context made it appropriate.[49] Several more specific usages can also be distinguished. There was (the) Great Mother (of the Gods), sometimes also named Cybele, more often known by title alone: this use is already common in the fifth century.[50] There were Great Gods and Great Goddesses, the figures honoured in mysteries. The most frequently attested such Great Gods are the Kabeiroi, whose cult on Samothrace offered protection to seafarers. An inscription dated by letterforms to the fourth century from the related mystery cult on Lemnos provides what is apparently the first attestation of the term Great Gods;[51] the term spread with the spread of the cult, and came to be shared by those other protectors of seafarers, the Dioscuri.[52] (Lemnos

46. See, e.g., Sourdel, *Cultes du Hauran,* 61, 97.

47. Baudissin, *Kyrios,* 2:79–80.

48. *IGLS* XI.45 is often quoted, where a begging priest of κυρία Atarchates calls himself her slave; note too a distinctive use in the Sarapieion of Memphis, *SEG* XLIX 2260–2314. Gods and goddesses always signify masters, according to Artem. 2.36, p. 163.7–11 Pack. κύριε, deferentially, between equals: Dickey, *Greek Forms of Address,* 107.

49. Mueller, "Megas Theos," 396–99, with reference to Bruchmann, *Epitheta deorum.*

50. Mueller, "Megas Theos," chap. 2, p. 303, gives fifth-century instances, e.g., Pindar, *Dithyramb* II 7 Snell-Maehler; Eur. *Bacch.* 78; Ar. *Av.* 875.

51. S. Accame, *AsAtene* 3–5 (1941–43): 89, no. 11; θεοὶ μεγάλοι also in 99, no. 17 (1st c. B.C.), but cf. Kabeiroi or Kabeirion in p. 79, no. 3.5; p. 83, no. 5.7 (both dated to the 3rd c.), p. 84, no. 6.4 (1st c.), p. 105, nos. 23 and 24 (inscribed storage jars).

52. Mueller, "Megas Theos," chap. 1. In inscriptions of Samothrace itself (Mueller, 290–91), and often elsewhere (292–93, 298–99), the gods are almost without exception named simply as (the) Great Gods; cf. Paus. 1.31.1 and 8.21.4, who explains that gods locally described as Great Gods are in fact the Dioscuri. Herodotus, however (2.51.2), names the gods of Samothrace Kabeiroi. Outside Samothrace a further identifier is sometimes added: Great Gods Kabeiroi/Samothracian/Dioscuri (Mueller, 286–88); on Delos a cult of Great Gods and Dioscuri and Kabeiroi apparently collapses into one of Great Gods Dioscuri Kabeiroi (separate in *ID* 1898 but combined in 1562 and often: see Bruneau, *Cultes de Délos,* 397). *IG* XII.8.74 from Samothrace, Θεοὶ Μεγάλοι, Θεοὶ δυνατοί, ἰσχυρροὶ καὶ Κασμεῖλε ἄναξ πάτ[ρι]οι. Κοῖος, Κρεῖος, Ὑπερείων, Εἰαπετός, Κρόνος (so Fredrich: Πάτ[εκ]οι for πάτ[ρι]οι Keil ap. Mueller, 291 n. 2), is exceptional. The Great Gods to whom stoas and other buildings were dedicated at Kurmalar

was also inhabited by a Great Goddess, known to Aristophanes but not with any precision to us.)[53]) The Kabeiroi have plausibly been explained since Scaliger from Semitic KBYR as "powerful ones,"[54] of which "Great Gods" might be a calque. But there were also, in the Peloponnese, mystery cults of Great Goddesses,[55] and though these Peloponnesian Great Goddesses are not attested till later, we find Demeter and Kore already so referred to in Sophocles (*OC* 683); so this designation for mystic gods is not necessarily a translation from Phoenician. It is at all events a clearly circumscribed special usage. A figure known simply as Great God is attested at Istros and Odessos (both on the west coast of the Black Sea) in the third and second centuries B.C. respectively; in the Roman period he sheds his anonymity at Odessos and becomes Great God Derzelas (or something similar).[56] The general view is that he is a "chthonic god" (he is shown with a cornucopia) and liable to euphemism for that reason, the particular appellation perhaps being suggested by analogy with the cult on Samothrace. Perhaps, rather, Greek-speakers preferred to refer to a powerful indigenous figure by periphrasis instead of a non-Greek name, much as a goddess of the Aegean coast of Thrace became simply "the Maiden."[57]

Another circumscribed and very frequent usage is the addition in Egypt of "great (or greatest) god(dess)" to the name of an Egyptian god who receives (under his or her own name, or in *interpretatio Graeca*)[58] a dedication in Greek: "great(est)

---

in the Mount Olgassys region in Paphlagonia (*SEG* XXXIII 1113) are a puzzle: if the editor's Hellenistic date is sound (3rd/2nd c. B.C.; but doubts are expressed in *SEG*) they should be the Samothracian gods, but these would be out of place on a mountain far from the sea.

53. Photius M 162, μεγάλην θεόν· Ἀριστοφάνης ἐν Λημνίαις (fr. 384 K/A)· ἴσως τὴν Βενδῖν· Θρᾴκιος γάρ; Steph. Byz., Λῆμνος . . . ἀπὸ τῆς μεγάλης λεγομένης θεοῦ, ἣν Λῆμνόν φασι· ταύτῃ δὲ καὶ παρθένους θύεσθαι. Perhaps the recipient of *IG* XII.8.45, Halonnesus, Φιλόστρατος Μεγάλει Θεῶι.

54. Cf. W. Burkert, in N. Marinatos and R. Hägg, eds., *Greek Sanctuaries* (London, 1993), 178; "hard to resist," M. L. West, *The East Face of Helicon* (Oxford, 1997), 58.

55. At Andania: Paus. 4.33.5 and many other passages (Mueller, "Megas Theos," 295–96); at Megalopolis: Paus. 8.31.1 (identified by Pausanias as Demeter and Kore) and *IG* V.2.517. It is a standing puzzle that the Andania mystery inscription speaks of "(Karneian) Great Gods," not Goddesses (*Syll.*[3] 735. 24–25; *LSCG* 65.34, 68–69, 91), while mentioning separately Demeter and Hagna (= Kore, according to Paus. 4.33.4). This is the only occurrence of masculine Great Gods who are not associated with seafaring. On the problem, see N. Deshours, *Les Mystères d' Andania* (Bordeaux, 2006), 66–69 and 218; L. Gawlinski, *The Sacred Law of Andania* (Berlin, 2012), 17–22. *IG* V.2.510–11 attest a Great God at Thisoa in the territory of Megalopolis: Jost, *Arcadie*, 212.

56. Cf. p. 8n30. It is ambiguous whether the genitive Ὀδεσιτῶν of coins attaches to the god's name (so Chiekova), or just indicates the issuing polis (Mihailov in *IGB* I[2] p. 92). I speak for convenience as if the various attestations of a Great God in this region refer to a single figure; but this is sometimes contested. For full discussion, see Chiekova, *Pont gauche*, 179–200; cf. Oppermann, *Thrakische Reiter*, 292–93.

57. Cf. p. 7.

58. Mueller, "Megas Theos," 340, even takes it as a diagnostic sign, when applied to a Greek theonym, that the god is really Egyptian; but he allows that *OGIS* 50 (ca. 240 B.C.), *technitai* of great Dionysus, is an exception.

god" here translates a the hieroglyph *'a'a* of the same meaning traditionally applied to deities, as we see, for instance, from the expression "great great god," reflecting a repetition of that hieroglyph and even "great great great god" (whence the familiar Hermes Trismegistos, "Thrice Greatest").[59] This Egyptian use is a fixed part of the titulature of many gods, and we find, for instance, "Soknopaios great great god" in the driest administrative contexts. We can also note the use of Megistos (less often Megas) as a fixed cult title of Zeus (shown to be such by its use in administrative language, such as a job description for the "priest of Greatest Zeus") in Anatolia; it is first attested at Iasos in a context which has also yielded a Carian dedication to trqδ, and may have been coined to convey the idea of this great indigenous deity in Greek.[60] A single passage in Timotheus's *Persai* finally suggests that Artemis of the Ephesians was already regularly described as "great" in the late fifth century.[61]

Different from all these is the usage seen in the following small selection from a much larger total: "priesthood of the greatest and most manifest Zeus Panamaros"; "for great god Ares and Eleuthera the leader, manifest goddess"; "for the young god friendly to mankind, greatest, manifest Asclepius"; "priest of great Zeus Krezimos."[62] The function of the epithet here is simply to glorify; often it is added to a more specific epithet, and any god can receive it: a dedication from Amaseia in Pontos is made to "the great goddesses Nymphs," figures often viewed by moderns as rather minor.[63] Occasionally it is used of the Jewish or Christian god, though much less frequently than was *hypsistos*.[64] Very often it is combined with

59. Mueller, "Megas Theos," 339–81, i.e., a long list; Ronchi, *Lexicon theonymon*, 674–704, μέγας; 704–20, μέγας μέγας; 721–22, μέγας μέγας μέγας. The same usage in Aramaic from Egypt: Cowley, *Aramaic Papyri*, 72 B 15 (= *TADAE* 3, C.3.12.26). Examples come predominantly from Egypt itself, but spread beyond in the cult of Serapis and Zeus Helios Megas Serapis (though not, according to Mueller, "Megas Theos," 374, Isis). Unusual, but certainly Egyptian, is an unnamed Great God in *ID* 2180 = 2181, Θεῶι Μεγάλωι καὶ Διὶ Κασίωι καὶ Ταχνήψει, Ὧρος Ὥρου Κασιώτης κτλ. I would add Mueller's nos. 167–68 to this context and dissociate them from Thrace (cf. the note to *OGIS* 740).

60. Iasos: p. 107n30 above. As a fixed title at Pergamum: Ohlemutz, *Kulte in Pergamon*, 77–79. At Acraephia in Boeotia it apparently replaces Soter as a title of Zeus in the early first century A.D.: *IG* VII.2712.85–86; Schachter, *Cults of Boiotia*, 3:95. Whether dedications to Zeus Megistos echo a fixed cult title or are simply elative is often unclear, as, e.g., in the Galatian and Syrian instances in Mueller, "Megas Theos," 312–13. For Zeus Megas/Megistos of West Cilicia, see p. 88n42.

61. *PMG* 791.260. For later references to "the great(est) goddess Artemis" in Ephesos, see Mueller, "Megas Theos," 332.

62. ἱερωσύνην τοῦ μεγίστου κ(αὶ) ἐπιφανεστάτου Διὸς Παναμάρ[ου]: *IStrat* 704 (such expressions are pervasive in inscriptions of Stratonicea with reference to Zeus and Hekate); θεῷ μεγάλῳ Ἄρει καὶ Ἐλευθέρᾳ ἀρχηγέτιδι ἐπιφανεῖ θεᾷ: *IGR* 3.700 (Kyaneai); Δαίμονι φιλανθρώπῳ νέῳ Ἀσκληπιῷ ἐπιφανεῖ μεγίστῳ: *IErythrai* 223; ἱερεὺς μεγάλου Διὸς Κρηζίμου: *IEph* 3414.

63. *Studia Pontica* III, no. 96, ταῖς μεγάλαις θεαῖς Νύμφαις.

64. Mueller, "Megas Theos," 329–30; some of his cases are questionable, but, e.g., *OGIS* 742 is a clear Jewish instance.

one or more further exaltatory epithets such as *epēkoos,* "who listens," or *epiphanēs,* "manifest" (as in two instances above), or *hypsistos.*[65] The superlative, when used, exalts but does not compare: in an inscription from Stratonicea in Caria (*IStrat* 663) three different gods are each acclaimed as *megistos.* Instances of both "great" and "greatest" come thickest in Asia Minor and Syria, and occasional references to unnamed "great(est)" gods or goddesses in these regions may be spin-offs, influenced also by the long-standing tendency in those regions to avoid specific naming: context will have determined the god referred to. Keramos in Caria developed the habit in the Roman period of referring to what was presumably its civic pantheon as "great gods of Keramos," apparently a local idiosyncrasy.[66] Instances become rare further west, but one finds, for instance, "great Dionysus" on Andros and in Phocis and Rome, "greatest Herakles" in Macedonia (nos. 147, 148, 150, 157–58), "great goddess Artemis Thermia" on Lesbos (nos. 190–92), and probably a "greatest goddess Limnatis" in Messenia (no. 193).[67]

This usage coheres closely with the acclamatory use best known from the cry of the goldsmiths of Ephesus "Great is the Artemis of the Ephesians" (Acts 19.28) and now much more widely attested in relation to various gods; an important early example is the cry "Great is Zeus Panamaros" raised by that god's worshippers when he repelled Labienus and his Parthian troops in Caria in 39 B.C.[68] The use of *megas* to translate the hieroglyph *ʿaʾa* traditionally applied to Egyptian deities (see above) may also have had an effect; a text of 240 B.C. from Ptolemais Hermiou in Upper Egypt which speaks of "the artists of great Dionysus" (*OGIS* 50.7–8) already shows the traditional expression "artists of Dionysus" modified under this influence, and Serapis took his title "great" with him even when his cult moved outside Egypt.[69] There is, however, no sign of the use of *megas* as a general exaltatory epithet even in the late Hellenistic period; it is absent, for instance, from the many

65. For *IG* IX.2.1201, a singular product of the desire to enhance the power of a tomb protector, see n. 89 below. *IG* IX.2.106, again from Thessaly, invoked "the great wrath of great Zeus" in similar circumstances.

66. For Anatolia, see p. 88. For Syria, e.g., "the great god of Nazala," *IGLS* V.2697, 2700, 2702 (cf. note 70 below for a Palmyrene equivalent); θεῷ μεγάλῳ δε[σπότῃ] from Neocaisarea, *CIIP* II 1133; θεὸς μέγιστος καὶ ἅγιος on Mount Hermon, *IGLS* XI.40; a great goddess from the Ledja and a greatest god from Petra, Sourdel, *Cultes du Hauran,* 72 nn. 4 and 6; *CIG* III 4525; *IGLS* I.15. Keramos: θεοῖς μεγάλοις Κεραμιήταις, *IKeramos* 17; cf. 18, 21–23, 28. The designation of Astarte as "Greatest goddess" in a bilingual from Byblos may support the low dating of that controversial text (p. 85n31 above).

67. The numbers refer to the catalogue in Mueller, "Megas Theos."

68. *IStrat* 10.13; for other gods, see, e.g., Mueller, "Megas Theos," nos. 101, 102, 140, 146, 171–72, 185, 191, 205, 210.

69. See n. 59 above. Μέγιστος was readily available to render part of the titulature of Juppiter Optimus Maximus, as, e.g., in the bilingual *INikaia* 1141: Διὶ κρατίστῳ μεγίστῳ φροντιστῇ/*Iovi optimo maximo tutori,* but the Latin was not obviously an influence except in particular cases (e.g., Zeus Megistos Dolichenos/Helioupolitanos, Mueller, "Megas Theos," 314–15).

religious inscriptions emanating from worshippers of many nationalities on Delos, most of which antedate the first century B.C. The acclamatory formula "Great is X" is securely attested, we have just seen, in 39 B.C. (and at about the same time Antiochus of Commagene speaks generally of the worship of Great Gods, *OGIS* 383.62), but clear instances of "to great/est (god) X" do not occur so early, perhaps not even before the second century A.D. The popularity of new titles may often have multiple sources. In Syria, where *megas* and *megistos* were very frequent, they may have picked up on an indigenous RB.[70]

A very widespread epithet in postclassical dedications is *epēkoos,* "who listens" (rare variant *euēkoos*); the list of over 50 recipients of 138 instances of the epithet assembled by Weinreich in 1912[71] could now be hugely extended. The hope that gods would be *epēkooi* to prayers is expressed, not surprisingly, in classical texts;[72] but the word is absent from dedications, and its very frequent occurrence in that context later seems to indicate a new start. A dedication to Agdistis *epēkoos* made on behalf of King Ptolemy (probably from the letterforms Ptolemy II Philadelphos) and Queen Arsinoe (*OGIS* 28) may be the earliest datable instance;[73] a clutch of further Hellenistic examples come from Delos, a clear majority of them applied to Egyptian gods,[74] and from Egypt itself. Two of the Delian instances are on bronze small plaques to which are attached an ear or ears,[75] and it has rightly been concluded that image and epithet have the same referent, the power of the god to hear and answer prayers; such hearing ears could convey the same meaning by themselves in a dedication without the epithet. Clement of Alexandria describes the dedication of ears (and eyes) of costly material as an Egyptian practice[76] enigmatically hinting at the god's power to see and hear all, and hieroglyphic prayers

70. Baudissin, *Kyrios,* 3:71. We find in fact the "Great God of Nazala" attested *tel quel* in both Greek (above, n. 66) and Palmyrene (L'LH' RB' DNZLY), *CIS* II.3.3911 (on the findspot, see Millar, *Roman Near East,* 299).

71. Weinreich, "Θεοὶ ἐπήκοοι," still the standard treatment; cf. H.S. Versnel, in *Faith, Hope, and Worship,* 34–37. An extended list of instances with Zeus is given by Schwabl, "Epiklesen," 305–6.

72. E.g., Ar. *Thesm.* 1157; cf. Pl. *Menex.* 247D, *Phileb.* 25B, *Leg.* 931C-D.

73. The dating of *AM* 85 (1970): 105–7, no. 9 (*SEG* XXXVII 143), Σιμύλος ἥρωι ἐπηκόωι ὑπὲρ Λακράτου εὐχήν, to "c. 280 B.C." on the basis of a supposed prosopographical link is not compelling; that of *IOlbia* 72, a dedication of various σιτωνήσαντες . . . Ἥρωι Ἐπηκόωι to 200–150 is more plausible. Also claimed to be third/second century (letterforms) is εὐήκοος in *IG* XII.I.914 (verse: cf. n. 94 below). M. Avi-Yonah, *IEJ* 9 (1959): 1–12, proposes a second-century B.C. date for a dedication to Hadad and Atargatis, θεοὶ ἐπήκοοι, at Ptolemais in Egypt.

74. Bruneau, *Cultes de Délos,* 167–68.

75. *ID* 2173, 2365; cf. probably 2160; ears on dedications not made explicitly to a "hearing" god, e.g., *ID* 2394–98; for ears on dedications to Egyptian gods outside Delos, see Moyer, *Limits of Hellenism,* 199 n. 183.

76. Clem. Al. *Strom.* 5.7.42.2 p. 354 Staehlin: τά τε ὦτα καὶ τοὺς ὀφθαλμοὺς οἱ δημιουργοῦντες ἐξ ὕλης τιμίας καθιεροῦσιν τοῖς θεοῖς ἀνατιθέντες εἰς τοὺς νεώς, τοῦτο δήπου αἰνισσόμενοι ὡς πάντα θεὸς ὁρᾷ καὶ ἀκούει.

from the thirteenth/twelfth century addressed to "the god who hears prayers," sometimes accompanied by ears (which also occur separately), prove the influence to have gone from Egypt to Greece and not vice versa.[77] The early instances of *epēkoos* in dedications are so commonly made by non-Greeks or to non-Greek gods that doubt arises about the nature of the Apollo *epēkoos* of a Delian dedication, found though it was in Apollo's sanctuary.[78] But by the imperial period the use of the epithet has become much more general, though still rare in Greece itself.[79] Its spread in the Levant was doubtless encouraged by a convergence with the tradition, pervasive in Phoenician and still more in neo-Punic dedications, of thanking the gods for giving ear.[80]

The four praise epithets surveyed so far are all Greek words, but their usage calques, certainly or probably, usages in other languages. *Sōtēr,* "saviour," and *epiphanēs,* "manifest," by contrast are homegrown. The transition of *sōtēr* from a functional epithet applied to an ever-increasing range of particular gods into a praise epithet of almost universal application has yet to be traced in detail,[81] but the fact is clear. For the use of *epiphanēs* as a praise epithet, it may be possible to identify a point of origin. Greeks had always believed that gods could appear or become manifest to mortals, and the idea gained new solidity with the emergence in the third century of a noun designating the phenomenon, *epiphaneia.* Around the year 200, Cnidus announced to the Greek world that, because of the many *epiphaneiai* of its goddess Artemis Hyakinthotrophos, it had named her *epiphanēs,* and was now organising an international festival in her honour.[82] The Cnidians would not

77. For all this, see Weinreich, "Θεοὶ ἐπήκοοι," 175–87 (with much further evidence for depiction of hearing ears); for the Egyptological material, see, e.g., A. Erman, "Denksteine aus der thebanischen Gräberstadt," *SBBerl* 49 (1911): 1086–1110, at 1088 (Berlin 23077), 1097 (Brit. Mus. 276), 1108 (Turin, wooden *naos,* 913); G. Wagner and J. Quaegebeur, *BIFAO* 73 (1973): 54–58 (on the god Mestasytmis, "the ear hears," of the late Hellenistic/Roman period); G. Pinch, *Votive Offerings to Hathor* (Oxford, 1993), 246–64.

78. Bruneau, *Cultes de Délos,* 168, on *ID* 2365 (dedicated by Zabdion—a non-Greek name—and Antigonos); Baslez, *Religions orientales à Délos,* 121, speaks a little too categorically of the epithet as "given only to Egyptian and Ascalonian gods."

79. So Weinreich, "Θεοὶ ἐπήκοοι," 156, a restriction which newer evidence seems not to have removed.

80. See, e.g., *KAI* 10.3, 8; Jongeling, *Neo-Punic Inscriptions,* 407, index s.v. šmᶜ. The formula εὐξάμενος καὶ ἀκουσθείς (*IGLS* XVII.I.103, 323, 326, 379, and often elsewhere) echoes, and very probably derives from, Aramaic QRLH W 'NY (e.g., *CIS* II.3.4092 = *PAT* 0438; Kaizer, "Religious Mentality," 179; cf. Yon's note on *IGLS* XVII.I.323). God "listens" (ἐπακούει) in the LXX, but there is no secure use of the epithet in a Jewish dedication: see W. Ameling's note on *IJO* II, no. 19.

81. T. S. F. Jim is preparing a study of the topic.

82. *IKnidos* 220, contained within and partially supplemented from *IG* XII.4.166, the Coans' reply; from the latter (which reproduces the Knidian request) I quote 30, ποταγορεῦσαι τε διὰ ταῦτα τ[ὰ]ν θεὸν Ἐπιφανῆ, and 36–37, τᾶι Ἀρτάμιτι τᾶι Ἰακυνθοτρόφωι θεᾶ[ι] Ἐπιφανεῖ; Artemis is often thereafter Hyakinthotrophos (and) Epiphanes in Cnidian inscriptions.

have laid such emphasis on the title if it had already been banal; no earlier instance presents itself, but it soon afterward passes over into the sphere of royal titulature in the person of Ptolemy V Epiphanes Eucharistos. Succeeding instances come in a trickle[83] which changes to a flood in the Roman empire, when great figures such as Artemis of Ephesus, Aphrodite of Aphrodisias, and Hecate and Zeus of Stratonicea receive it repeatedly, many lesser figures occasionally; in 76 A.D. an Egyptian god is acclaimed as "the most *epiphanēs* of all gods, ever."[84] There emerges too the concept of the hero *epiphanēs,* sometimes in connection with actual supposed epiphanies of heroes, but becoming banal enough to be applied also to dead mortals.[85] As so often, mainland Greece stands somewhat aside.

These six were the main praise epithets, but others such as *euergetēs,* "benefactor,"[86] and *anikētos,* "invincible,"[87] also entered the repertoire. Some retain an element of specific meaning: *anikētos,* for instance, is most frequently applied to "invincible" powers such as the sun or gods of war, though we also meet (from an uncertain location) a "goddess Kore listening **invincible** heavenly." Among words expressing power, *kyrios* was dominant but did not exclude *anax* and *despotēs* and *basileus; tyrannos,* "tyrant," however, was all but reserved for Men,[88]and *pantokratōr,* "all-ruling," is almost always Judeo-Christian or under Judeo-Christian influence.[89]

83. I have noted *ID* 2102, of 122/1 B.C., to Sarapis ἐπιφανής; *ILampsakos* 7, "first third of 1st c B.C.," which gives in line 9 Priapos ἐπιφανής, in 28 Aphrodite ἐπιφανής (in both cases the god is standing in as eponym of the city).

84. Bernand, *De Thèbes à Syène*, no. 174: τὸν τῶν πώποτε θεῶν ἐπιφανέστατον Πέμσαον.

85. L. Robert, *RPhil* 13 (1939): 200–201 (*OMS* 2:1353–54), who for epiphanies of "real" heroes cites Ath. 6, 266D (Drimakos); Diod. Sic. 5.79. In divine dedications a sense of link with epiphany is still sometimes explicit, as in *IGR* I 1162 (Egypt, Roman period), ἀνέθηκεν βωμὸ[ν τοῖς ἐν ὀνεί]ροις ἐπιφανέσι θε[οῖς ἐπιστᾶσι]. For lists of gods bearing the epithet, see F. S. Steinleitner, *Die Beicht im Zusammenhange mit der sakralen Rechtspflege in der Antike* (Leipzig, 1913), 15–18; F. Pfister, *RE* Suppl. IV, 301; for kings, see now F. Muccioli, *Gli epiteti ufficiali dei re ellenistici* (Stuttgart, 2013), 281–309. For a dedication from Hierocaesarea to the Roman senate with this title, see Robert, *Hellenica* VI, 50–52.

86. See L. Robert, *Le sanctuaire de Sinuri près de Mylasa* (Paris, 1945), 23; L. Robert, *Hellenica* V, 21 n. 1 (where he notes a special link with Sarapis); *BÉ* 1968, no. 593 (cf. *OGIS* 233.29–30).

87. ἀν(ε)ίκητος/*invictus* is constantly applied to Mithras and to emperors, also to the sun (Helios), Zeus (particularly when linked with the sun as Zeus Helios), military deities (Ma in Macedonia, Herakles, Ares), and Nemesis. Literary use of ἀνίκητος goes back to Tyrtaeus fr. 11.1 W. Invincible Kore: θεὰ Κόρη ἐπήκοος ἀνείκητος οὐρανία, Graf, *Nordionische Kulte,* 70–73 and 461. The relation of the Greek and Latin epithets is inconstant: Isis is *invicta* but not, it seems, ἀνίκητος.

88. It is applied to Zeus Masphalatenos in *TAM* V.1.537; and there is Isis's claim in her aretalogy to be τύραννος πάσης χώρας, *RICIS* 202/1101.3a, 302/0204.4–5; cf. 701/0103.4–5. On ἄναξ and δεσπότης, see n. 42 above.

89. See, e.g., *IJO* I, nos. BS [Black Sea] 20–22; *IJO* II, pp. 262, 487–88; O. Montevecchi, "Pantokrator," in *Studi in onore di Aristide Calderini e Roberto Paribeni* (Milan, 1957), 2:401–32; J. P. Batut, *Pantocrator* (Paris, 2009). By pagans it is used almost exclusively (and that not often) in magical invocations or in verse (Chaniotis, "Megatheism," 137 n. 115; *RICIS* 102/1702); but *INikaia* 1121 and 1512 are prose dedications to Zeus Pantokrator. *IG* IX.2.1201, from Methone in Thessaly, ἕξει κεχολωμένον

Other words traditionally associated with gods or certain gods, such as *athanatos,* "immortal," or *ouranios,* "heavenly," sometimes now acquire a glorifying rather than defining function.[90] There is no point in trying to assemble a complete inventory of terms of praise; a recent study lists twenty-nine (some admittedly very rare), and adds "etc."[91] Terms such as *patrios* and *patrōos,* "ancestral," and the various expressions for "who guards the village/city," though not strictly praise epithets, also abound, and play their part in impressive accumulations such as a dedication "to our ancestral most manifest god Helios Apollo Larmenos, who protects the city of Motella" (in northwestern Phrygia).[92]

As that example shows, the transformation—for such it is reasonable to call it—in the language of dedication that occurred in the Hellenistic and, still more, the Roman periods was a matter both of the creation of new epithets and also of new ways of deploying them, a matter that has been well studied by Chaniotis. Epithets are enhanced by the universalising prefix *pan-*, or promoted to the superlative: the taste for the superlative mirrors and may be influenced by Latin practice, where Juppiter had always been Optimus Maximus. The world becomes bestrewn with best and greatest and most manifest gods (but whether "highest" is an equally promiscuous term is, as we have seen, controversial).[93] Multiple epithets are tumbled together in a single dedication, as we saw at the start of this chapter.[94]

Βασιλέα θεὸν μέγιστον παντοκράτορα κτίστην ὅλων καὶ θεοὺς πάντας καὶ θεοὺς ἥρωας καὶ αὐτὴν τὴν δέσποιναν Βασιλίδα, has apparently undergone Judeo-Christian influence (note κτίστην ὅλων); Marek, "Allmächtige Gott," 3:137–46 (*SEG* L 1233) (Paphlagonia), beginning Κύριε παντοκράτωρ, σύ με ἔκτισες, is probably Christian, despite its early date (237/8). On the dedication to a Zeus and a Great Mother, both said to "rule all things," in *IG* XI.4.1234, see p. 168n72. κοσμοκράτωρ (applied chiefly to Helios and Helios Sarapis and Roman emperors) emerges late under astrological influence (see works cited on *RICIS* 501/0126); an unnamed highest god is addressed as κοίρανε κόσμου in the verse section (*SGO* II, 09/12/01) of *SEG* L 1222; cf. too *SEG* XLVI 1781 (p. 124 above).

90. ἀθάνατος (ἄφθιτος is confined to verse): Robert, *Documents d' Asie Mineure,* 427–28 (citing *TAM* V.I.75); Robert, *Hellenica* XI–XII, 437 n. 5; Chaniotis, "Megatheism," 137 n. 112. A dedication δαίμονι ἀθανάτωι from Atrax in Thessaly is isolated if really second century B.C. as claimed (*ArchDelt* 46, 1991: *chron.* 223). "Immortal god" (singular) is Judeo-Christian (*IJO* II, no. 186.20–21, with note; no. 227.16; *MAMA* XI 211 with note); influence thence is conceivable. Also Judeo-Christian is "living god": R. Kotansky, *Greek Magical Amulets* (Opladen, 1994), 269. On οὐράνιος, see Chaniotis, "Megatheism," 125 n. 53 and appendix 1, where its extension to a wide new variety of deities is noted.

91. Chaniotis, "Megatheism," 130 n. 80 (cf. 137–38). Very rare: e.g., ἄλυτος, "inexorable," only in Petzl, *Beichtinschriften,* no. 4.

92. *SEG* L 1270: τῷ πατρίῳ ἡμῶν ἐπειφανεστάτῳ Ἡλίῳ Ἀπόλωνι Λαρμηνῷ, τῷ προεστῶτι τῆς Μοτεληνῶν πόλεως.

93. See on all this Chaniotis, "Megatheism," passim: on the superlative, 129; on παν, 138 (two in one epigram from Lesbos, *IG* XII supp. 30); multiplication of epithets, 138 n. 118. "Highest": pp. 124–29 above.

94. To those cited there we can add, exempli causa, Zeus Μέγιστος Καρποδότης Σωτὴρ Ὀλύμπιος (Drew-Bear and Naour, "Divinités de Phrygie,"1949–50); Θεὰ Ἐπήκοος Ἄρτεμις Αὐλίς Σώτειρα (*IGR* I 34 = *IGUR* I 145); Ζεὺς Πόρισος Κτήσιος καὶ Καθυπερδέξιος (L. Robert, *Hellenica* X, 62–66, no. 11,

(Gods are also sometimes tumbled together, as again we saw, though less regularly than is often supposed.[95]) For these reasons, though very simple formulae still predominate in some rural Phrygian shrines,[96] and many individual dedications remain straightforward even in contexts where others are expansive, it would seldom be possible to mistake the date of any sizable collection of dedications from the Roman imperial period. One can speak, if with notable local variations, of a shared Greek dedicatory vocabulary of the second century A.D. which is different even from that of the Hellenistic period. Or perhaps that should be qualified to "a shared dedicatory vocabulary of the Greek-speaking East." In Greece itself one can detect a different development, not the multiplication of superlatives but a taste for dainty new epithets or exquisite variations on old ones or poetic epithets now deployed in cult. From a single site, Epidaurus, one can cite Selene πολυώνυμος, "of Many Names," Apollo Ἑκατηβελέτης, "Far-Striker," Apollo and Asclepius συγγνώμονες, "Merciful," Asclepius εὔκολος, "Easy-Going," Asclepius and Hygieia and Telesphoroi ἀλεξίπονοι, "Averters of Pain," and Athana καλλίεργος, "of Fair Works."[97] The epithets so abundantly listed in Pausanias combine the traditional with this more recherché and antiquarian tendency.

The changes that occurred outside Greece had multiple causes. As we have seen in some detail, new praise epithets enter the Greek repertoire from other traditions. Praise itself acquires a new importance as a form of religious action. Hymns had always been a key component of Greek cult, but contexts and formulae for praise of the god in prose had not.[98] Here too the earlier parallels are to be found outside Greece. The verb *eulogein*, "to praise," occurs in Greek texts, but it is

Istanbul museum); Σωτὴρ Παμφύλιος Ἐπήκοος Ἀπόλλων, associated with Σώτειρα Ἄσυλος Περγαία Ἄρτεμις (*IPerge* 56); Ἐπήκοος Θεὸς Ἀσκληπιὸς Ἐπιδαύριος Περγαμηνὸς διώρυγα κατοικῶν (*IApameia* 5), where a humble toponymic designation follows the glorious "Epidaurios Pergamenos"; Ἑρμῆς Κώρυκιος Ἐπίνικιος Τροπαιοῦχος Ἐπικάρπιος (*ICilicie* 17); Διόσκοροι Σαμοθρᾴκων ἐπιφανεῖς θεοὶ ἀδαμεῖς ἀεί (Swoboda, *Denkmäler*, 17 no. 16); τὸν παντα[λ]ηθ[ῆ καὶ ὀνει]ρ[ο]δ[ότ]η[ν] κ[α]ὶ χρη[σ]μοδότ[ην] κ[α]ὶ ἄψευστον κα[ὶ] δι' [ὅ]λης οἰκουμένη[ς] μ[α]ρτυρ[ο]ύμενον οὐράνιον θεὸν [Βησᾶν] (Perdrizet, *Memnonion*, 500). An epigram from the territory of Lindos, *IG* XII.1.914, εὐξάμενος ἱερᾶι Σωτείραι τόνδε ἀνέ̣[θηκα]/τὸμ πίνακα Εὐή̣κωι Φωσφόρωι Ἐννοδ[ί]α̣[ι] (Ἐννοδ[ί]ο[ς] Wilamowitz, a name first attested in late antique Cilicia), looks like an unexpected Hellenistic (3rd/2nd c. B.C. [?]) anticipation of the type, but the verse form makes a difference.

95. Cf. Parker, "Zeus Plus." The matter is too complicated to discuss here.

96. Chiai, "Machtepitheta," 223, on the material in *Phrygian Votive Stelai.*

97. *IG* IV² 1.422, 427, 432, 469, 472, 485. The dedications to Athena under a variety of titles (Archagetis, Kalliergos, Polias, Stoicheia, Hypata: *IG* IV² 1.484–88) at Epidaurus by men who had served as πύρφορος also suggest a conscious engagement with the range of the epithet system. For further exquisite/antiquarian epithets, see, e.g., "playful" (παραπαίζων) Dionysus (*IG* II² 4787) and several in *IG* V.1.497 and 559. Such tendencies could even reach Moesia, *SEG* LIX 794–96 (dedications to Zeus Ombrimos, Hera Basilissa, Poseidon Kyanochaites).

98. They are absent from the summary of cultic "fair speech" in S. Gödde, *Euphêmia: Die gute Rede in Kult und Literatur der griechischen Antike* (Heidelberg, 2011), 353.

applied to gods so seldom[99] that praising god was manifestly not a fixed and familiar obligation. In the Septuagint by contrast, the instruction "Praise the lord" (εὐλογεῖτε τὸν κύριον, or small variants) occurs 114 times; there are 690 instances in all of words from the *eulog-* root.[100] In the cult of Egyptian gods, *aretalogos,* "teller of (the god's) powers," is an established office, and the inscribed aretalogies give us an idea of their work. Individuals joined in: alongside two proclamations by one Aischrion, a Thracian, near Antinoe that "I praise the god of fair journeys" and "I praise Isis," we can set Tibullus's exquisite vignette of his mistress Delia, obliged, in gratitude for a healing, to dress in linen and, hair unbound, "speak the praises" (*dicere laudes*) of Isis twice a day "amid the Pharian crowd."[101] Public praise of the god by individuals who have offended, confessed, and been pardoned then becomes a central part of the ritual process known from the "confession" inscriptions of eastern Lydia and western Phrygia in the second century A.D.[102] Whether this tradition is indigenous to Anatolia or externally influenced is unclear; what is obvious is that it did not come in from Greece.

The expression of gratitude in dedicatory formulae develops in a way that may reflect a related development in behaviour. Gratitude to a god was expressed in archaic and classical Greece by the fact of bringing a gift; in the Hellenistic period it became common to designate such a gift a "thank offering," *charistērion;* there then emerges a formula by which the dedicator states in the first person that "I give thanks to" the god.[103] This third stage might arise from a new insistence that persons believing themselves to have received divine aid should make a declaration of gratitude on a public ritual occasion; certainly the Maeonian confession inscriptions constantly link confession, praise, and thanksgiving.

A form of public praising that manifestly influenced the inscribed material was that seen in cries such as "Great is the Artemis of the Ephesians" or "Unique god Serapis"; crowds of worshippers joined in such acclamations, but confession inscriptions too often begin with, for example, a "Great is the Men of Axiotta," and celebrations of εἷς θεός, "unique god," are very frequent in graffiti and other media. The early history of such religious acclamation is uncertain. "Great is . . .," as we saw,

99. All I have found are Eur. *Ion* 1614, *HF* 356, and marginally *Hec.* 465.

100. The greeting εὐλογία (πᾶσιν) (e.g., *IJO* II, nos. 27, 155, 156) is purely Jewish: Robert, *Hellenica* XI-XII, 394–96.

101. Aischrion: see Dittenberger's note 1 to *OGIS* 73; Tibullus 1.3.31–32: *bique die resoluta comas tibi dicere laudes/insignis turba debeat in Pharia.*

102. See Petzl, *Beichtinschriften;* and Paz de Hoz, *Die lydischen Kulte,* indices s.v. εὐλογεῖν. L. Robert, *Nouvelles inscriptions de Sardes* (Paris, 1964), 30, used such evidence to criticise the statement of E. J. Bickerman, *RBibl* 69 (1962): 531 n. 31, that "εὐλογεῖν was not a term in the Greek religious vocabulary." That will depend on the period and region one is speaking of.

103. Robert, *Hellenica* X, 55–62. But cf. already, with Belayche, "Ritual Expressions of Distinction," 146, Plaut. *Trin.* 820–37 and other spoken expressions of gratitude in Roman comedy.

is first attested at Stratonicea in Caria in 39 B.C. and is soon followed, again in western Asia Minor, by the Ephesians' praise for their goddess; one isolated instance of "unique god Dionysus" occurs, on accepted datings, much earlier in the Gurôb papyrus, which, though found in Egypt, is thought to reflect Greek religious practice, if of an unorthodox Orphic kind.[104] Cries such as *iē Paian* had always been heard in Greek cult, and there are indications that gods may sometimes have acquired the title *sōtēr* as a formalisation of shouts of "Hail saviour" (*iē sōtēr*) made on a particular occasion; somewhat differently, acclamations addressed to worshippers, not gods, are likely to have been heard in mystery cults.[105] But it would be a surprise if cries of "Great is Athena of the Athenians" have completely evaded our sources. As a regular feature of public religion, at least, such celebration looks post-classical and probably not of mainland origin, though eventually adopted there: by the second century A.D. the cry of "Great is Pythian Apollo" was indeed raised at Delphi, amid a uniquely rich hubbub of further religious acclamations.[106]

External contact never provides more than a partial explanation for cultural change: there must be factors that make the receiving culture receptive to influence from another. The new language reflects, it has been argued,[107] a new subservience to the gods, a self-abasement of mortals before heavenly rulers with powers no less awesome than those of the absolute rulers now dominant on earth. On this view, the development in Greek religious vocabulary would occur in response to new political circumstances. But this approach may simplify a more complex reality. The divine epithet which most simply speaks of autocracy, *tyrannos*, was almost confined to a single god whose cult was largely restricted to a particular region, Men of Anatolia. Appeals, from the regions of Phrygia and Lydia where the cult of Men predominates, to "the gods who rule our village" or something similar are often taken to express the same conception of gods as absolute masters.[108] But they may derive, it has recently been suggested, from the civic language of the Greeks

104. Stratonicea and the Ephesians: see p. 144 above. Chaniotis, "Acclamations," 211–13, notes the predominance of imperial evidence but tentatively seeks antecedents. Belayche, "Ritual Expressions of Distinction," 157, speaks of "the tradition of glorifying the gods in Ptolemaic Egypt," adducing the Gurôb papyrus (*P.Gurôb* 1, reedited by J. Hordern, *ZPE* 129, 2000: 131–40) and, 152, a "gem of the second or first century B.C. from a Greek-speaking Egyptian environment" (S. Michel, *Die magischen Gemmen im Britischen Museum,* London 2001, no. 24) which acclaims Horos, Hathor, and Akori as "one."

105. Σωτήρ: Isyllus, in Powell, *Coll. Alex.* 132–35, line 82; *SEG* XXXVI 555.6. Mystery cults: to judge from the congratulations addressed to worshippers in some "Orphic" gold leaves, e.g., *SEG* XXXVII 497.

106. *SEG* LI 613–31. But, as already noted, the dedicatory formulae of the mainland remain for the most part restrained: there is no simple link between acclamatory formulae and dedicatory practice.

107. I paraphrase crudely a position associated in particular with the "Leiden School": see in particular H. W. Pleket, in *Faith, Hope, and Worship,* 152–92; H. S. Versnel, *Ter Unus* (Leiden, 1990), 39–95.

108. See especially N. Belayche, "Au(x) dieu(x) qui regnent sur . . . ," in A. Vigort et al., eds., *Pouvoir et religion dans le monde romain* (Paris, 2006), 257–69.

of the western coast which spoke of "gods who stand in front of" their cities: the relation would be one of protection, not domination.[109] That counter can in turn be countered by the observation that the behaviour of the gods of the regions in question, as described by their worshippers, is indeed oppressive and despotic; the evidence for "theocracy" is not confined to the vocabulary of "ruling." But one cannot generalise from the Lydia and Phrygia of the second century A.D. to earlier times and other regions. A more relevant effect of autocracy might be the effect that it had on the vocabulary of respect more generally. The "superlativism" with which gods are praised is matched, indeed abundantly outdone, by that applied not just to emperors but also increasingly to high officials.[110]

The cults which above all spread the ideas of unlimited divine power within the late Hellenistic and early imperial world were those of Isis and Serapis. The source of their appeal is not fully understood, but it is only very partially reducible to the modelling in heaven (primarily by a goddess, not a god!) of absolute monarchy on earth. Isis's power is not so much absolute like a monarch's as unlimited, in that it extends to every nook and cranny of the cosmos, which indeed she first set in order and now controls; and, much more emphatically than earlier gods, she can lend assistance in every situation of danger, at sea no less than in sickness. Paradoxically combined with this cosmic power is a new intimacy in her relation with worshippers: "Telethusa, you who are one of my people (*pars, o Telethusa, mearum*), lay aside your heavy cares" is how Ovid imagines the goddess addressing a devotee (*Met.* 9.696).[111] When she is appealed to as "mistress," we should think less of a subject's relation to a ruler, a relation irrelevant to the vast majority of her worshippers in their daily lives, and more of that of a social inferior to a superior with whom the inferior may nonetheless be on close terms; while displaying deference, "mistress" may also establish a claim. We cannot, however, explain the whole linguistic phenomenon by a growth in "personal religion" less mediated by civic structures. The language of praise becomes just as emphatic in civic as in individual dedications.

Another factor that has been much stressed of late is competition between cults: the goldsmiths of Ephesus were famously concerned to insist on the greatness of their Artemis in order to protect their own income; Lucian's *Alexander* depicts the promotion by an ambitious individual of a cult that he has founded. But the worshipper who proclaims the power of a god in a dedication does so in

109. Schuler, "Rural Population Groups," 80–88.

110. A. Zehetmair, *De appellationibus honorificis in papyris graecis obviis* (Marburg, 1912); O. Hornickel, "Ehre- und Rangprädikate in den Papyrusurkunden" (diss., Giessen, 1930); L. Dinneen, *Titles of Address in Christian Greek Epistolography to 527 A.D.* (Chicago, 1929). Most of the evidence they quote is Byzantine, but the tendency is already present earlier.

111. The blend of "inaccessible majesty" and attention to the humble is stressed by Versnel, *Faith, Hope, and Worship*, 35.

gratitude for the exercise of that power, not to provide free advertisement for a cult; competition is relevant only to the extent that the worshipper may have internalised the propaganda of priests and others who had a material interest in promoting the cult.

These phenomena defy easy synthesis; too many factors are in play, pushing in too many different directions. A Greek vocabulary of praise was greatly extended by (to a large extent) the calquing of terms from other languages. Praise itself acquired a new importance as a form of religious action, but was influenced also by new forms of exaltation in the secular sphere: "superlativism" and acclamation developed pari passu in cult and outside it. The disposition to address a god who was simply "Best Greatest Listening" (*aristos megistos epēkoos*)[112] coexisted with appeals to named gods. Intermingled with forms of expression that stressed power and grandeur were others that spoke more of mortals' dependency on the gods and even aspired to intimacy with them, a combination seen particularly in the cult of Isis and her associates. A geographical dimension remains important: some of the expressions discussed above continue to occur most frequently in regions where they picked up expressions or forms of thought of the indigenous tradition. The shared language was a mixing pot, but not a melting pot in which all cultural differences were effaced.

112. E.g., *SEG* L 1222.

6

Delos

Where God Meets God

The previous chapters have ranged widely in place and time. This chapter by contrast will seek to recover specificity by studying the religious culture of a small island over a finite period, that of Hellenistic Delos.[1] Delos was twice sacked in the first century B.C. and ceased to be important; evidence, so abundant in the earlier centuries, becomes scarce. As a result this case study fails to illustrate most of the phenomena discussed in the previous two chapters. But that disadvantage is more than outweighed by the superb illumination it provides of the broader theme of this book, the interaction of Greek and non-Greek religious cultures. No other site begins to rival Delos in richness of archaeological and epigraphic evidence.[2]

This abundance is in part due to the island's later decline, which saved the Hellenistic levels from overbuilding. But it also reflects a religious life that became in fact unusually varied. Since archaic times Delos had been the greatest religious centre of Greece outside the mainland, its sanctity as the "birthplace of two gods" supposedly known and respected even by the Persians (Hdt. 6.97).[3] Prestigious

1. My debt in this chapter to the works constantly cited below of Roussel, Plassart, Launey (in *ID*), Marcadé, Bruneau, and Baslez is great and manifest; see too now Bonnet, *Enfants de Cadmos*, 475–520. The many inscriptions involving Egyptian gods cited below are also available (and easily accessible via the concordance) in *RICIS;* to avoid overelaborate references I omit *RICIS* numbers in this chapter except in special cases.

2. The most extraordinary single document, however, is that just published from Marmarini in Thessaly: J. C. Decourt and A. Tziafalias, *Kernos* 28 (2015): 13–54, restudied by J. M. Carbon, *Kernos* 29 (2016): 185–208; R. Parker and S. Scullion, *Kernos* 29 (2016): 209–68.

3. On Datis's supposed dedication, *IG* XI.2.161 B 95–96, see M. F. Baslez, *RÉA* 87 (1985): 139: it originated as a dedication by a Greek woman Batesis or Patesis daughter of Batis, and was fraudulently ascribed to Datis at some time before 279 B.C.

and well-sited festivals had commercial potential, as Strabo describes for a later period (10.5.4, C 486): after the Roman destruction of Corinth in 146,

> traders transferred thither (to Delos), lured by the sanctuary's tax immunity and the excellent situation of the harbour; for it is well positioned for those sailing from Italy and Greece to Asia. The festival is a kind of commercial occasion, and the Romans frequented it more than any other people even while Corinth was still standing. When the Athenians acquired the island [in 166] they took good care of the sanctuaries and of the traders.

The crucial advantage of tax immunity, which drew huge quantities of trade away from Rhodes (Polyb. 30.31.10), had been granted by the Romans in 166, when the Athenians "acquired the island" by their gift as a reward for loyalty against Macedonia (at the cost of the native Delians, who were ruthlessly expelled). It was probably of this period that Pausanias was thinking when he spoke of Delos having been once (8.33.2) "the common marketplace of Greece." Notoriously, a commodity much traded in was slaves.[4]

Delos had long been an international centre, of diplomatic contacts and of trade; already in the fourth century a group of "sacred sailors from Tyre" had dedicated images (?) of Tyre and Sidon to Apollo, thus paying tribute to the greatest Delian god; an accompanying inscription in Phoenician mentions the name of the king Abdastart.[5] But it is during the Athenian occupation that the island's status as an international entrepôt, a place of contact between, above all, Italy and the Levant, really becomes visible; then too that elaborately organised cult associations emerge.[6] Most conspicuous is the place of Italians and Romans. Colleges of persons with Latin names known as Hermaists and Apolloniasts and Poseidoniasts in Greek, Magistreis (a nominative plural form) Mirquri Apollini(s) Neptuni in Latin, have left numerous dedications, singly or in collaboration; it is debated whether such Magistreis are the officials of three private associations, or rather of the three main cults of an informal *conventus* of Italians on Delos.[7] There are also Italian trade societies (wine sellers, oil sellers: *ID* 1711–14), and from about 100 even slaves and freedmen serving, as they might have done in Italy, as Competaliasts (*ID* 1760–71); the rites they organised in honour of the Lares Compitales

4. Strabo 14.5.2, C 668.

5. *ID* 50; cf. Bonnet, *Enfants de Cadmos*, 500–501, for Phoenician dedications before 166; 501–3, for the sacred sailors and the problem of what they dedicated (also A. Hermary, *RA*, 2014, 271–84); 518 n. 151, for early Carthaginian dedications. For foreigners, see Tréheux, *Étrangers*, with 93–113, lists by identifiable place of origin.

6. So Baslez, "Associations à Délos," an important study.

7. *ID* 1730–59; for references, see Bruneau, *Cultes de Délos*, 586–89, who favours the latter view and is followed by C. Hasenohr in several studies, most recently "Italiens et Phéniciens à Délos," in R. Compatangelo-Soussignan and C.G. Schwentzel, eds., *Étrangers dans la cité romaine* (Rennes, 2007), 77–90. The first such list is *ID* 1731, of ca. 140.

appear to be reflected in rough paintings on altars outside private houses or in niches in their walls.[8]

Two Phoenician associations have permanent establishments, the "Berytian Poseidoniast traders and shipowners and agents" (*ID* 1520) and the "Tyrian Herakleist traders and shipowners" (*ID* 1519); the "traders who sail to Bithynia" make a dedication (*ID* 1705), and the "society of senior agents (ἐγδοχεῖς)[9] in Alexandria" confers honours. There is a synagogue which has yielded dedications to *theos hypsistos,* "highest god,"[10] while quite early in the second century B.C. "the Israelites on Delos who pay tithes to the sacred Garizim" (i.e., Samaritans) already exist as a group; individual dedications by people originating from the Levant are very numerous, and there are a few from Arabia.[11] Honorary decrees are regularly voted by mixed groupings such as "the Athenians and Romans and other Greeks who reside on Delos or sail to the island as traders and shipmasters";[12] one which specifies that the honour is conferred by five "peoples (*demoi*)"—Athenians, Naxians, Andrians, Tenians, and Syrians—seems to show that Syrians, and by implication other Greek-speaking peoples of the Near East, are included among "other Greeks." The many slaves and freedmen have already been mentioned, predominantly Eastern in origin even though serving or having served Italian masters. Remarkable above all is the ethnic composition of the ephebate, which seems to have been open to all free persons from the beginning of the Athenian occupation:[13] the ephebes of 119/8 (*ID* 2598) were made up of six Athenians, eight Antiochenes, six Romans, four men of Tyre, two from each of Hierapolis, Sidon, Berytus, and Arados, and one from each of Karpasia, Nikopolis, Salamis, Naxos, Amphipolis, Hierapytna, Rhenaia, a Herakleia, Ptolemais, Laodicea, and Anthedon; other lists tell a similar story.[14] This may help to

8. Bruneau, *Cultes de Délos,* 589–620.

9. *ID* 1528–29, presumably linked to a similar grouping on Delos: for Alexandrians on Delos, see Roussel, *Délos, colonie athénienne,* 92–93; and Tréheux, *Étrangers,* 94.

10. *ID* 2328, 2330–32.

11. See n. 47 below for offerings of Temallatos of Gerra (cf. Roussel, *Délos, colonie athénienne,* 88); *ID* 2319–22. On the synagogue, see the very cautious study of P. Bruneau, "'Les Israélites de Délos' et la juiverie délienne," *BCH* 106.1 (1982): 465–504 = *Études,* 603–42, who publishes the new inscriptions of the Samaritans, *SEG* XXXII 809–10, allows as a theoretical possibility that the synagogue in fact belongs to the Samaritans, but points out that Jews are already attested for Delos in 139 B.C. by 1 Maccabees 15. For the identification as a synagogue, and building phases, see too M. Trümper, *Hesperia* 73 (2004): 513–98.

12. *ID* 1642; cf. 1643–82, and the analysis of the various formulae in Roussel, *Délos, colonie athénienne,* 51; Roussel argues, after Ferguson, that in the third quarter of the second century such decrees replace those issued by Athenian inhabitants on their own, and attest what was thereafter the sole decision-making body on Delos itself (one competent only to decree honours).

13. Roussel, *Délos, colonie athénienne,* 194.

14. E.g., the ephebic dedications *ID* 1922–40; the παρεύτακτοι (apparently, ephebes of the previous year) of *ID* 2598, the gymnasiarchs of the Hermaia in *ID* 2595.30–53; cf. Roussel, *Délos, colonie athénienne,* 85–87, stressing the predominance of Alexandrians, Syrians, Phoenicians, and Anatolians,

explain why Greekness was understood culturally, not ethnically; youths from all origins shared an education. Non-Greek names other than Roman are uncommon. All thirty Syrians listed in *ID* 2598, for instance, bear Greek names. But the phenomenon of non-Greeks "translating" their name into a Greek equivalent or simply adopting a second name for use in a Greek milieu is one attested elsewhere by bilinguals; and the predominance among Syro-Phoenicians of a small number of mostly theophoric names such as Dionysios, Apollodoros, and Herakleides suggests that it occurred here too.[15] Only a tiny number of dedications are made in languages other than Greek or Latin,[16] and even of these all but one are bilingual. The population was multifarious in origin, with Athenians and perhaps even ethnic Greeks in the minority; the common culture was Greco-Roman, even if, as we shall see, dramatically enlarged in the religious sphere. Even the establishment of the Poseidoniasts of Berytus, it has been argued, was physically a "thoroughly Greek style clubhouse."[17]

On taking control of the island in 166, the Athenians drastically reorganised the public cults on familiar Athenian lines: ten main priesthoods to be held for a year by Athenian citizens, with probably some element of rotation among tribes.[18] Some traditional cults lost out in the new system; most notably, Poseidon as a civic god disappears from view, though a mysterious new cult emerges of a Poseidon Aisios, "Auspicious," with (unusually) a life priesthood held by an Athenian.[19] It has been doubted whether, after the early years of the occupation, there were many

and the virtual absence of mainland Greeks and Macedonians. On the evidence of dedications, see the map in Baslez, *Religions orientales à Délos,* 144, and the chronological fine-tuning on p. 143: till 130 B.C. predominantly islanders; 130–120, "apogee of Syro-Phoenicians"; 120 onward, Italians becoming dominant population group.

15. Baslez, *Religions orientales à Délos,* 71–75, with statistics; cf. 92–93 for Theo- names, 130 for Helio- names; but, pp. 37–41, she doubts the same phenomenon for Egyptians. Cf. appendix B. For non-Greek names, see O. Masson, "Quelques noms sémitiques en transcription grecque à Délos et à Rhénée," in *Hommages à André Dupont-Sommer* (Paris, 1971), 61–73, who collects seventeen, more from Rheneia than Delos itself, mostly borne by just one individual; for Egyptian names, Roussel, *Cultes égyptiens,* 280 n. 3. The names in the long subscription list *ID* 2611 which has been associated with the Poseidoniasts of Berytus are of striking banality, with the single exception of Dionysios Theodorou Magabbon (B 6). Only one Hierapolitan priest of Atargatis has a perhaps Semitic name, Marion (*ID* 2257–58; Will, *Déesse syrienne,* 140—to be added to Masson's list); even he is son of an Achaios.

16. *ID* 2315–16, 2319–20, 2322.

17. Trümper, "Clubhouses in Delos," 122.

18. Mikalson, *Hellenistic Athens,* 216–18.

19. Disappearance: Bruneau, *Cultes de Délos,* 266. Poseidon Aisios: *ID* 1562, 1581, 1582. The standard account (Roussel, *Délos, colonie athénienne,* 71 and 274; Baslez, *Religions orientales à Délos,* 112) is that the untypical life priesthood proves Aisios to be an oriental god. But it should be noted that the holder of the priesthood was a powerful Athenian, dedicator of the Mithradates monument, and that he mentions his tenure of the life priesthood alongside earlier annual tenure of the civic priesthood of the Great Gods. This was an important office. Is a life priesthood in an Athenian cult established in we know not what circumstances inconceivable?

permanent Athenian residents on the island, though numerous Athenians divided time between there and Athens or passed through.[20] A great range of religious activity took place outside the public cults; the Poseidoniasts of Berytus and the Herakleists of Tyre, the various Italian societies, and the synagogue were already mentioned in the previous paragraph.

Even before 166, one cluster of foreign cults had become naturalised on the island, and retained their popularity apparently unaffected by the transformation of 166 by which the native Delians were expelled in favour of Athenian settlers. A famous inscription tells how one Apollonios, an "Egyptian of the priests," brought Sarapis to the island and maintained the cult "as was his inherited tradition" until his death at an estimated age of ninety-seven. The cult and priesthood then passed (after the Egyptian manner) to his son, to the delight of the "servants" (θέραπες) of the god; so a cult society already existed. The son lived to age sixty-one; his son Apollonios II in turn inherited the cult, and in obedience to a dream established a proper Sarapieion (what is known to scholarship as Sarapieion A) for the god, who had hitherto been lodged in hired premises. "Certain persons" moved by "wicked envy" then brought a public lawsuit against the founder, on grounds left deliberately vague which we can only guess; but Sarapis intervened miraculously and deprived the prosecutors of the power of speech at the trial (*IG* XI.4.1299).[21] The inscription recording the god's intervention is dated by letterforms to the late third or early second century; so Sarapis will have arrived at some time in the third.[22] Two further sanctuaries of Egyptian gods (conventionally termed Sarapieia) become archaeologically visible at about the same time, late in the third century, as Sarapieion A; while it was long held that their foundation was inspired by Apollonios II's successful legal defence, a newer suggestion is that the priests of Sarapieia B and C, which were perhaps already in existence when Apollonios II established Sarapieion A, may in fact from rivalry have been the "certain persons" who brought the lawsuit.[23] By 180 the largest of the three (Sarapieion C) has been adopted into the public cult of Delos and acquires a public Delian priest. That priesthood of Sarapis was subsequently taken over into the Athenian system as one of the main ten.[24] The three sanctuaries of the Egyptian gods also hosted a cluster

20. Roussel, *Délos, colonie athénienne,* 56–69. Even among "Athenians" he suspects cases of naturalisation (70).

21. On all this, see now Moyer, *Limits of Hellenism,* 142–207; on chronology (cautious, and disposed to down-date), 156–57; on the charge against Apollonios, 160 n. 51. For older views, see Bruneau, *Études,* 290–96 (with photo of the inscription), 310–313; Bruneau, *Cultes de Délos,* 457–66 (essentially a summary of Roussel's account in *Cultes égyptiens*).

22. On datings, see the cautious remarks of Moyer, *Limits of Hellenism,* 156–57.

23. Moyer, *Limits of Hellenism,* 194–207.

24. On dating of the Athenian priests, see L. Bricault, "Les prêtres du Sarapieion C de Délos," *BCH* 120 (1996): 597–616.

of small private societies, popular with non-Delians,[25] which have left quite numerous documents from the period before the Athenian occupation.

A later priest, but now a Delian, at Sarapieion A encountered further opposition in the Athenian period, for reasons that are not stated; but on appeal to the senate at Rome received permission to "serve the shrine as he served it before."[26] Any difficulties faced thereafter by the founders of new cults have left no trace; the decision by the Herakleists of Tyre duly to seek permission to found a shrine has survived (*ID* 1519.12–16). Archaeologically "a good fifteen genuinely oriental sanctuaries,"[27] a figure which seems in fact an underestimate, have been detected; "genuinely oriental" refers here to the form of the sanctuary, at its simplest a modest courtyard containing an altar or altars (occasionally a small temple) and probably surrounded by a terrace. One can list the three Sarapieia; the sanctuary of the Syrian gods; the synagogue; and, on the slopes of Mount Kynthos, the sanctuaries of "Zeus *hypsistos* and the gods to whom (the dedicator) founded altars"(*ID* 2306);[28] that containing altars to several gods of Ascalon in Palestine (*ID* 1719–21); that of the gods of Iamneia, again in Palestine (*ID* 2308); that of the mysterious "First Gods" (*ID* 2310); and some ten further small sanctuaries not associable with particular gods. The sanctuary of the Syrian gods, originally a private foundation with a Syrian priest (*ID* 2226), proved so popular that by 118/7 (*ID* 2227) it was taken over as an Athenian public cult with an Athenian priest; it attracted worshippers of very diverse origins.[29]

Not all these sanctuaries were open to a multinational clientele. The establishment of the Poseidoniasts of Berytus remained a place where Berytians honoured Poseidon and their other "ancestral gods," though they were happy to admit an

25. On the forms of association associated with the Egyptian cults, see Baslez, "Associations à Délos," 236–47. The level of participation by Delians is unknowable: since only nonnatives bear an adjective indicating their origin, and even they only occasionally, any bare name in the relevant documents (primarily *IG* XI.4.1215–72) may belong either to a Delian or to a foreigner.

26. *ID* 1510 (*Syll.*[3] 664). The priest is a Rheneian, i.e., presumably a native Delian who moved to neighbouring Rheneia when the Athenians ethnically cleansed Delos. As for what was at issue, Roussel, *Cultes égyptiens,* 261–63, and Bruneau, *Cultes de Délos,* 461, suppose that what was now the public Sarapieion sought to suppress its private rival; Baslez, *Religions orientales à Délos,* 215, stressing the "as before" of the senate's decision, argues that "c'est un genre de pratiques, c'est une tradition cultuelle qui sont en jeu"; but it is hard to see why, e.g., the ritualised daily opening of the shrine, hinted at in the inscription, should have caused offence.

27. Bruneau, *Cultes de Délos,* 457, with a useful survey, 457–80, from which I take my list.

28. The location and form of the sanctuary, and the four rough altars found in it (mentioned in the inscription), are taken as indications that this is a foreign Zeus (Bruneau, *Cultes de Délos,* 240–41): gods too barbarous to be named? (Plassart, *Mont Cynthe,* 291); on the dedicator as probably oriental, see Baslez, *Religions orientales à Délos,* 90 n. 4.

29. Will, *Déesse syrienne,* 140. The sanctuary seems to be already attested in the first half of the second century by the inscription *BCH* 92 (1968): 359–74 (Bruneau, *Cultes de Délos,* 467). It hosted no non-Syrian god except Asclepius (whom some see as Eshmun): Bonnet, *Enfants de Cadmos,* 512. On the original form of the association, like a Semitic *marzeah,* see Baslez, "Associations à Délos," 231.

Athenian benefactor, a Roman benefactor, and indeed the goddess Rome;[30] other less-known associations may have retained a similar national focus. The Italian societies too seem confined to persons with names on the Roman model; whether the worshippers of *theos hypsistos* at the synagogue included pagan sympathisers (the so-called god-fearers) is not known.[31] But it does not follow that Berytians and Italians were confined to worshipping their national gods, and dedications by them in other sanctuaries show this was not the case.[32] Indeed the Berytian Poseidoniasts resolve in honour of a Roman benefactor: "Let there be led each year for all time at the procession of the Apollonia an ox bearing the inscription 'The society of the Berytian Poseidoniasts on behalf of Marcus Minatius son of Sextus'" (*ID* 1520.49–52). Since the Poseidoniasts are very unlikely to have conducted an elaborate procession of their own in honour of Apollo, it looks as if they participated as a group in the island's greatest festival in honour of its god Apollo.[33] We know independently that the *hieropoioi,* "performers of rites," gymnasiarchs, and ephebes involved in that festival were drawn from the whole greater Greek (and Roman) world;[34] it was doubtless open to all "the Athenians and Romans and other Greeks who reside on Delos or sail to the island as traders and shipmasters." Very likely the same was true of all the public festivals.

By contrast to these sanctuaries with restricted clienteles, a whole series attracted dedications from worshippers of many origins. Some distinctions on chronological and other grounds can be drawn: the sanctuary of the Syrian gods was initially frequented primarily by Syrians from Hierapolis, and retained a strong Syrian

30. Benefactors: *ID* 1520, which mentions (91) the archon of 153/2; *ID* 1780, 1796; Trümper, "Clubhouses in Delos," 117, infers that "Athenian and Roman representatives were invited to the clubhouse on specific occasions or even regularly." Dedications by the Berytians: *ID* 1772–96, 2323–27. The long subscription list *ID* 2611, which has been associated with them through prosopographic coincidences, might, if some other possible coincidences are accepted (see the notes ad loc., and Baslez, *Religions orientales à Délos,* 158–59), attest the inclusion of a Tyrian and two Athenians (naturalised?), but, for instance, no Italians. On the sanctuary, see most recently Trümper, "Clubhouses in Delos," who at 119 (cf. Trümper, *BCH* 126, 2002: 265–330, at 313) follows Bruneau, *Études,* 438–72 and 793–800, in acknowledging just three chapels (the two attested of Rome and Poseidon, and one hypothetically assigned to Astarte/Aphrodite), rejecting the fourth postulated for Herakles/Melkart or Asclepius/Eshmun. Baslez, *Religions orientales à Délos,* 70, notes the similarity between the cult of Rome admitted to the clubhouse and Syro-Phoenician city-cults.

31. As Roussel, *Délos, colonie athénienne,* 94–95, envisages, and Baslez, *Religions orientales à Délos,* 157–58, argues from the non-Jewish names of some dedicators.

32. *ID* 2182 is a dedication by a Berytian to Zeus Kasios from Sarapieion A. Berytian ephebes (*ID* 2598) will have been involved in the Apollonia. Italian dedications are too frequent to need listing.

33. So Bruneau, *Cultes de Délos,* 629; the Posideia also mentioned in the decree (32, 38) probably belong to the association, though as Baslez, *Religions orientales à Délos,* 274, notes, this leaves the "sacrifice to Poseidon" of the Herakleists (*ID* 1519.39) of Tyre mysterious. Those Herakleists hold a meeting "in the sanctuary of Apollo" (*ID* 1519.1–2), presumably that of the island.

34. Bruneau, *Cultes de Délos,* 76–77.

(though not pan-Levantine) emphasis, whereas the Egyptian cults attracted islanders and Italians, later Athenians and Levantines, but never predominantly Alexandrians or Egyptians.[35] But these nuances and preferences do not affect the basic point that religious choices on the island were not decisively determined by ethnic origin. If statues of benefactors dedicated to the triad of Apollo, Artemis, and Leto by easterners may merely reflect a desire for a prestigious location,[36] we have nonetheless already seen that the Apollonia were open to all, and straightforward dedications were made to Zeus Kynthios and Athena Kynthia, "of Mount Kynthos" on the island, by worshippers of many nationalities.[37] The ephebes, drawn as we have seen to a large extent from Italy and the East, are likely to have had a role in almost all the public cults. Just as ethnic non-Greeks honoured Greek gods, so ethnic Greeks honoured non-Greek gods. The two foreign cults taken over by the Athenian state were enthusiastically supported by Athenians as well as by their earlier worshippers: in these two cults above all, barriers of birth disappeared, with the single exception that the Athenians reserved the priesthoods for themselves. Five dedications were offered to the Egyptian gods by the "*demos* of the Athenians" (i.e., notionally that in Athens itself); one in the sanctuary of the Syrian gods was made "in accord with instructions" ([κατὰ προστά]γματα)—instructions presumably received in a dream by one of the officials of the cult and then conveyed to Athens.[38] Dedications by Athenian priests and private individuals were also very common; service by the daughters of the prosperous as "basket-bearers" (*kanephoroi*) in these cults was a source of family pride.[39]

We can also observe the catholic religious choices of individuals or families. The wealthy banker Philostratos son of Philostratos, an Askalonite by origin but later naturalised in Italy as a Neapolitan (*ID* 1724), made dedications "for the city of Askalon and himself and his wife and children" to "Astarte Palaistine Ourania Aphrodite" and to Poseidon of Askalon, while his friend Midas son of Zenon dedicated constructions in his honour in the sanctuary of the Syrian gods (which

35. Baslez, *Religions orientales à Délos*, 43, 148, and 154. Non-Greeks who had not yet got a national sanctuary used the Sarapieion (83, 108). The clientele of the Egyptian sanctuaries before the Athenian period is little known: n. 25 above.

36. E.g., *ID* 2005, 2008; cf. *ID* 1984 by Amisenoi; for Italian dedicators of similar monuments, see *ID* 1965, 1967; for Italian recipients, *ID* 2001–4 and often. For a private dedication by a Hieropolitan (admittedly a rarity), see *ID* 1450 A 113; I am unclear whether the dedications ascribed to a "Mikythos" sometimes described as "Syrian" (Tréheux, *Étrangers*, s.v. Μίκυθος) are all by the same man, and thus about their chronology.

37. Bruneau, *Cultes de Délos*, 230.

38. *ID* 2041–45, 2220. On such dedications "by instruction," see n. 77 below.

39. See Mikalson, *Hellenistic Athens*, 231: 8 of ca. 130 private dedications to the Egyptian gods are made by Athenians; 234: evidence for the "whole-hearted" commitment of Athenian priests and private citizens to the Syrian gods; cf. Bruneau, *Cultes de Délos*, 472, who refers to the discussion of the theatre subscription list *ID* 2628 in Roussel, *Délos, colonie athénienne*, 267–68.

Philostratos too supported financially).[40] But he also dedicated the north portico of the agora of the Italians to "Apollo and the Italians" (*ID* 1717; cf. *ID* 1718, 2454), and the Italians reciprocated by offering a statue of him to Apollo, as did three Roman brothers (*ID* 1722, 1724); his nephew dedicated a statue of him, unctuously described as "saviour and benefactor," to Zeus Kynthios and Athena Kynthia (*ID* 1723). The offerings for his own family are, it is true, made to Levantine gods, but his admirers associated him also with the great gods of the island, Apollo and the Kynthian pair. Among the many cases of members of Athenian families holding office in a variety of cults, the most conspicuous is perhaps that of Gaios son of Gaios (note the Roman name) of the deme Acharnai, who in different years was priest of the (Greek) Theoi Megaloi, (Egyptian) Sarapis, and (Syrian) Aphrodite Hagne (and made dedications in all these sanctuaries); his daughter served under him as *kanēphoros*, "basket-bearer," to Aphrodite Hagne, but also back in Athens as an *ergastinē*, "wool-worker," for Athena Polias.[41]

Obviously interesting in this international environment is the question of how gods were named. Italians follow the normal Greco-Roman convention of choosing the theonym native to the language: thus the Magistreis Mirquri Apollini(s) Neptuni become Hermaists and Apolloniasts and Poseidoniasts in Greek (e.g., *ID* 1753), and the Competaliasts, using Greek, dedicate a statue of Pistis (= Fides, *ID* 1761.19).[42] The goddess Maia who accompanies Hermes/Mercury in several dedications (*ID* 1731–33; cf. 1744) is the same in both languages. Unlike Apollo, a genuinely shared god, she seems to be a conflation between two separate figures, the Greek mother of Hermes, and an Italian goddess of the same name but different origin; in this Greco-Roman amalgam the mother of Hermes (a figure of myth more than of cult) acquired a prominence she had never enjoyed at home.[43] In bilingual texts the popular Delian god Zeus Ourios, "of Fair Wind," is Latinized as Iuppiter Secundanus (*ID* 1754.7), while a freedman and four slaves dedicate a statue of Juppiter Leiber = Zeus Eleutherios, "of Freedom" (*ID* 1771.6, 12), evidently seen as the god of manumission.[44]

Italian names found in Latin aside, the only non-Greek theonyms that occur frequently are those of Hadad and Atargatis (discussed below) and of that cluster

40. *ID* 1719–21; 2253.6 (cf. 2254.5); 2628 a I.29–30.

41. I borrow this case from Mikalson, *Hellenistic Athens*, 238 (who gives references); for others, see 236–41.

42. Cf. Hephaistos/Volcanus in *ID* 2439–40.

43. Not in Delos alone: see, e.g., Link, in *RE* XIV s.v. Maia, 532–33; cf. Radke, *Götter Altitaliens*, 192–93.

44. Which epithet is primary is not clear. In renderings from Latin to Greek, Zeus Eleutherios is used both for Iuppiter Libertas (Mon. Anc. 19) and for Iuppiter Liberator (Dio Cass. 62.26.4, with Tac. *Ann.* 16.35). On the apparent assimilation of Iuppiter Libertas and Iuppiter Liber, see Radke, *Götter Altitaliens*, 179–80. It is interesting that slaves in Plautus dream of *eleutheria* (*Persa* 29, *Stich.* 422).

of major Egyptian or Greco-Egyptian gods whose names became accepted into Greek: Sarapis, Isis, Anubis, Harpocrates, less often Horos and Ammon; also Hydreios, an Egyptian god with a Greek name. A rare intrusion of a less-familiar Egyptian theonym is a dedication by an Egyptian of the region of Mount Kasios to his local gods (but on behalf of a Roman) "Great God and Zeus Kasios and Tachnepsis."[45] Elsewhere, we find that the Minaians from the far south of Arabia give their god Oddos his own name even in the Greek portion of a bilingual (*ID* 2320); a Zeus Ammon–type compound, Zeus Dousares, may appear in a similar text;[46] three men of Iamneia near the coast west of Jerusalem pair "Herakles and Hauronas" an interpreted and an uninterpreted god (*ID* 2308), while two further texts contain non-Greek theonyms of problematic interpretation.[47] Several of these untranslated gods are Arab, just as humans from Arabia may bear non-Greek names contrary to the Delian norm.[48] There are also unnamed clusters of "Gods Above" and "First Gods."[49] But the total of non-Greek theonyms attested on an island largely populated by non-Greeks is modest; we have met already the Herakleists of Tyre, the Poseidoniasts of Berytus, Levantines honouring their native gods under Greek names.

Much the best-documented and most interesting case is that of the goddess—well, what is one to call her?—the goddess who is variously named in inscriptions as Atargatis (eleven times), Atargatis Hagne Theos, that is, "Atargatis Reverend Goddess" (three times), Hagne Aphrodite Atargatis (once),[50] Hagne Theos or Thea Hagne (seventeen times), Hagne Theos Aphrodite (four/five times), Hagne Aphrodite (twenty-three times), Hagne Aphrodite Syria Theos (five times), and finally, plain Aphrodite (once).[51] Amid this plethora of naming possibilities, a development and a bias are visible. Dedications made while the priest was still a Hieropolitan,

45. *ID* 2180 = 2181, Θεῶι Μεγάλωι καὶ Διὶ Κασίωι καὶ Ταχνήψει, Ὧρος Ὥρου Κασιώτης ὑπὲρ Λευκίου Γρανίου τοῦ Ποπλίου Ῥωμαίου. γυναῖκα μὴ προσάγειν μηδὲ ἐν ἐρεοῖς ἄνδρα· κατὰ πρόσταγμα, taken by Baslez, *Religions orientales à Délos*, 36, as addressed to Osiris, Isis, and Horos (Tachnepsis is said to have been the name of Isis in Kasion, *POxy.* 1380.74–75). Boubastis appears (place or goddess?) in *ID* 2110, 2186; a statuette of her: *ID* 1417 B col. I. 5–6. On Hydreios, see p. 91 above.

46. *ID* 2315 as read by Bruneau, *Cultes de Délos*, 244, 9/8 B.C.; cf. Baslez, *Religions orientales à Délos*, 102–3.

47. *ID* 2311 and 2314, where see the notes, and the suggestions cited by Baslez, *Religions orientales à Délos*, 91. Sîn of Alam occurs in the monolingual Hadramitic text *ID* 2319. A self-identified "Arab" dedicates to Helios, *ID* 2321; Temallatos of Gerrha (toward the Persian Gulf) makes dedications to Artemis, *ID* 1444 A a 45 and 51 (the latter decorated with a Helios), in the Kynthion (*ID* 1442 A 82) and the Sarapieion (*ID* 1442 B 58–59).

48. For Temallatos, see the previous note; for Chauan, *ID* 2321 (but with a father Theophilos).

49. *ID* 2309, 2310 (cf. p. 9n35)

50. *ID* 2266, dedicated by a Roman (a freedman?).

51. I borrow the list and numbers from Bruneau, *Cultes de Délos*, 470. For very detailed differentiations, see Baslez, *Religions orientales à Délos*, 83–87.

predominantly deriving from Syrians, speak of Atargatis or Hagne Theos; but once the cult had been taken over by the Athenians, it was to Hagne (Theos) Aphrodite that they dedicated, officially and privately;[52] by contrast, only one dedication of any date by a Syrian demonstrably names the goddess Aphrodite,[53] though they often speak of Hagne Theos (possibly by way of compromise). Again, Syrian dedicators are much more likely to give the goddess a consort Hadad, occasionally extended to Zeus Hadad, and sometimes to add a third figure called "Asclepius"; the Athenian priest was a priest of Hagne Aphrodite, with no mention of Hadad.[54] This is a rare case where choices in naming display a pattern, and one with an ethnic basis; two dedications to Hadad and Atargatis by Syrians from before the Athenian takeover in fact describe them as "ancestral gods" (*ID* 2226, the earliest text from the shrine; and *ID* 2247). Here therefore the assimilation of Atargatis to Aphrodite comes from the Greek side. But there is no sign that anything more was at issue than habit and preference; no Syrian was forced to recognise Aphrodite in Atargatis. The general view is that the "Asclepius" who apparently acquired a regular though subordinate place in the sanctuary (he was included in four dedications to Hadad and Atargatis) is a cover for a Syrian god, probably Eshmun;[55] the prestige of Asclepius on the island will have caused his name to be given preference. The identifiable authors of the Asclepius dedications are Syrian (as are those of dedications speaking of Zeus Hadad); but they were happy to give the god his Greek name.[56]

Zeus Hadad is a double naming of Zeus Ammon type (whereas Aphrodite Atargatis is almost nonexistent). In Sarapieion C there are a small number of fur-

52. *ID* 2258 is a private dedication by an Athenian to Hadad and Atargatis, but predates the takeover of the cult. According to Bruneau, *Cultes de Délos,* 470, no dedication that names Atargatis is securely dated after 113/2 except *ID* 2266 to Hagne Aphrodite Atargatis of ca. 90.

53. *SEG* XXXI 731 (*ID* 2256 with a new fragment), to Zeus Hadad and Hagne Theos Aphrodite. *ID* 2265 to plain Aphrodite is dedicated by a man from Naples.

54. For Hadad in dedications by Athenians I note only *ID* 2237 (by the priest), the early 2258, and possibly 2283 (also early). Asclepius: *ID* 2224, 2248, 2261, 2264. Zeus Hadad: *SEG* XXXI 731 (cf. previous note), *ID* 2262 (these two both dedicated by Syrians), 2291, 2295. The inscribed mosaic *ID* 2244 names Hadran, a figure known from Hierapolis and the Lebanon, and usually taken as a hypostasis of Hadad: Roussel, *Délos, colonie athénienne,* 261 n. 7; Baslez, *Religions orientales à Délos,* 85–86; doubts in Will, *Déesse syrienne,* 146.

55. Asclepius with Hadad: see previous note. Asclepius as *interpretatio:* so S. Reinach, *BCH* 7 (1883): 366–67; Roussel, *Délos, colonie athénienne,* 262–63; H. Seyrig, *Syria* 37 (1960): 374–75; Bruneau, *Cultes de Délos,* 375 and 470–71; Baslez, *Religions orientales à Délos,* 78. But Will, *Déesse syrienne,* 146, allows the possibility that this is Greek Asclepius. The fragmentary Greek-Phoenician bilingual apparently set up by a Tyrian in the Asklepieion (*ID* 2322) attests that "a Phoenician recognised in Asclepius a god of his country" (Bruneau, *Cultes de Délos,* 375); Baslez, 112, points to the absence of dedications in the Asklepieion by Tyrians permanently resident on the island, but that argument proves little in the virtual absence of all dedications from that sanctuary.

56. The claim of a complete lack of rapprochement between the Greek and the Syrian pantheons (Baslez, *Religions orientales à Délos,* 89) appears extreme.

ther instances: in *ID* 2135, for instance—"For Sarapis, Isis, Anoubis Hermes, Apollo Harpocrates, Hermias of Alexandria (dedicates) a thank-offering"—editors are doubtless right to recognise four deities honoured, not six, with Anoubis Hermes and Apollo Harpocrates treated as single entities; at the very least, these two pairs of gods are brought into close association.[57] The Zeus Sarapis combination, later so frequent, occurs just once (*ID* 2152). Two dedications by men of Askalon "to Palestinian Astarte Heavenly Aphrodite" (Ἀστάρτηι Παλαιστινῆι Ἀφροδίτηι Οὐρανίαι) are expanded versions of the same form of juxtaposition (*ID* 1719; 2305). More expanded still is a dedication made by Andromachos for self, wife, and children as a thank-offering "to Isis Saviour Astarte Aphrodite Fair Voyage Listener and Eros Harpocrates Apollo":[58] first the saviours at sea of three cultures are juxtaposed; then the Egypto-Greek juxtaposition of Harpocrates and Apollo (found elsewhere) is expanded, at the start, by the unique addition of a further Greek figure, Eros. Unfortunately the author of this singular dedication failed to reveal his place of origin. Another triple coupling is made by a Sidonian who dedicates "to Isis Mother of the Gods Astarte" (Ἴσιδι μητρὶ θεῶν Ἀστάρτῃ, *ID* 2101): his own goddess Astarte is given an Egyptian and a Greco-Anatolian equivalent. Are the authors of such dedications simply identifying different gods, or are they seeking to create new composite figures?[59] On any view their conception cannot be that the gods of different cultures are wholly discrete and incompatible entities. But such couplings are an occasional practice, irregularly applied. Somewhat different are expressions, numerous on Delos, such as Isis Dikaiosyne or Isis Nemesis; here the second noun seems to indicate a specification or restriction more than an identification.[60]

57. Cf. *ID* 2174, [Γν]α̣ῖος Ἐγν[άτιος] Γναίου Ἡ[λιό]δ̣ω̣ρ̣[ος] Σαράπιδ[ι, Ἴ]σ̣ιδι, Ἁρφοκρ̣[άτει], Ἀ̣νούβιδ[ι] Ἑρμε̣[ῖ, χαρι]σ̣τήριον; probably 2131, Σαρ[άπιδι, Ἴσιδι, Ἀν]ούβιδι, Ἀπόλλ[ωνι Ἁρπο]χράτει, θεοῖς συνάοις κ[α]ὶ συβώμοις; 2185, [Σ]α̣ράπιδι, Εἴσ[ι]δι, Ἀνούβιδι, [Ἀπ]όλλωνι [Ἁρ]φοκράτει; and already *IG* XI.4, Ἰατροκλῆς Δημαγάθου Ἀθυνβριανὸς Πλούτωνι καὶ Κόρει, Δήμητρι, Ἑρμε[ῖ] Ἀνούβι, κατὰ πρόσταγμα. In all these texts commas should probably be removed (as I have given them here): for the two former, cf. *ID* 2184, Σαράπιδι, Ἴσιδι, Ἀνούβιδι, Ἁρφοκράτει; but Baslez, *Religions orientales à Délos,* 48–49, keeps Hermes and Anoubis separate in *ID* 2174. The Anoubis-Hermes link is prefigured by Anoubis Hegemon, *IG* XI.4.1253, Hegemon, "Leader," being a regular epithet of Hermes.

58. *ID* 2132: Ἴσιδι Σωτείραι Ἀστάρτει Ἀφροδίτηι Εὐπλοίαι ἐπ[ηκόωι] καὶ Ἔρωτι Ἁρφοκράτει Ἀπόλλωνι, Ἀνδρόμαχος Φανομάχου [ὑπὲρ ἑαυτοῦ] καὶ γυναικὸς καὶ τέκνων χαριστήριον. *ID* gives no findspot, but it is accepted as coming from the Sarapieion C by Roussel, *Cultes égyptiens,* 197. The dedicator is registered as an Antiochene without a mark of doubt in Tréheux, *Étrangers,* apparently on the insecure basis of other attested Andromachoi from Antioch on Delos. On these dedications (*ID* 1719, 2305, 2132), see now J. Wallensten, "Dedications to Double Deities: Syncretism or Simply Syntax?," *Kernos* 27 (2014): 159–76; she suggests that the link between Eros, Harpocrates, and Apollo might be their shared status as young gods with famous mothers.

59. Cf. Bonnet, as quoted in chapter 2, n. 40, on *ID* 2132.

60. Cf. p. 106n127.

The use of non-Greek theonyms, we have seen, is limited. Not seldom therefore a Greek name is given to a non-Greek god. The Poseidon of the Berytians, the Herakles of the Tyriasts, are clear examples, as is probably the Asclepius worshipped alongside Hadad and Atargatis. The dedication mentioned above to "Sarapis, Isis, Anoubis Hermes, Apollo Harpocrates" makes it likely that the plain "Apollo" of the thank-offering brought by Protos of Cos, when "saved from many great dangers," to "Serapis, Isis, Anoubis, Apollo, gods who share their temple," is an Apollo Harpocrates, whom the Coan dedicator was glad to identify with the great god of the island where he had found safety.[61] But often we cannot proceed beyond a probability. The three dedications to Aphrodite Peistiche (once just Peistiche) have ears attached signifying, in the Egyptian manner, a deity who hearkens;[62] the epithet does not look Greek, but we cannot pin the goddess down precisely, any more than the Artemis Sosikolonos honoured by a man of Gadara[63] or an Artemis Hagia, "Sacred," to whom an Athenian priest of Sarapis made an offering probably in the Sarapieion (*ID* 2068). If a dedication is made to several Egyptian gods and to Zeus Ktesios, "of Property," is the last-named the homely Greek god, or his Alexandrian near equivalent Agathos Daimon, "Good Spirit"?[64] If an altar of Herakles Apallaxikaos, "Averter of Evil," was found in Sarapieion C,[65] does this prove that Horus-Harpocrates underlies him, or that the Egyptian sanctuary (once described by Roussel with some exaggeration as a "Pandemonium") was a hospitable home to visiting gods?

An interesting case is that of Zeus Ourios. He is an old Greek god with an epithet meaning "of Fair Wind," still sufficiently transparent to be rendered in Latin, as we have seen, Secundanus. But four of the six dedications to him derive from the

61. *ID* 2119, Πρῶτος Πυθίωνος Κῶιος, σωθεὶς ἐκ πολλῶν καὶ μεγάλων κινδύνων Σεράπει, Ἴσει, Ἀνούβει, Ἀπόλλωνι, θεοῖς συννάοις, ὑπὲρ αὑτοῦ καὶ τῶν παιδίων Φιλουμένης καὶ Πυθίωνος, χαριστήριον. *Aliter* Baslez, *Religions orientales à Délos,* 49, arguing that at the date of this dedication Harpocrates had not yet been added to the triad Sarapis, Isis, Anoubis.

62. *ID* 2396–98; other dedications to plain Aphrodite with ears, *ID* 2394–95; other private dedications to Aphrodite, *ID* 2390–91, 2393; to Venus v(ictrix?), 2392. Distinct are the official dedications *ID* 1810–11, and those to Hermes and Aphrodite, *ID* 1832–33, 2408. Bruneau, *Cultes de Délos,* 347, thinks of a link with Πειθώ; Baslez, *Religions orientales à Délos,* 81, 90, with πίστις ("faithful Aphrodite"), but what of the suffix? Bruneau points out that the dedications were found near the sanctuary of Apollo, not an oriental environment.

63. *ID* 2377. The epithet is unique and unexplained: protector of the place called Κολωνός on Delos? (a view considered and rejected by Roussel, *Délos, colonie athénienne,* 220 n. 11); of a κολωνός (hill) near Gadara? (cf. W. Kroll in *RE* IIIa, s.v. Sosikolonos, 1159); translinguistically, "of *coloni*," farmers, settlers? Bruneau, *Cultes de Délos,* 206, calls it "oriental" (similarly Roussel, 220 n. 11), but it looks Greek: is he envisaging a translation?

64. *ID* 2186:]δι, Ἀνούβιδι, [——] Βουβάστει, [Δι]ὶ Κτησίωι; contrast Roussel, *Cultes égyptiens,* 198; Bruneau, *Cultes de Délos,* 242.

65. Bruneau, *Cultes de Délos,* 411, citing Roussel, *Cultes égyptiens,* 200, on *ID* 2479; the commentary ad loc. speaks more vaguely of "some Egyptian god." Pandemonium: Roussel, *Délos, colonie athénienne,* 251.

Sarapieion, and two of these put him at the head of a list of recipients which continues with familiar Egyptian gods.[66] Yet there is no obvious Egyptian god to see him as a cover for; it looks rather as if he has become a transethnic provider of that resource so crucial in the Aegean, "fair wind." He enters Sarapieion C in that role (one of the dedications is explicitly "for those who sail"), and is also taken up by the Italians who honour him as Secundanus.[67] The dedication to this Greek god by an Italian from Elea in the sanctuary of the Egyptian gods (*ID* 2415) nicely sums up this international role.

Similarly, Zeus Kynthios and Athena Kynthia, who had their traditional and much-frequented sanctuary on top of Mount Kynthos, are joined with Sarapis and Isis in a dedication from Sarapieion C: two Athenian parents dedicate "to the gods" a statue of their daughter who had served as basket-bearer for the two Greek and the two Egyptian deities (*ID* 2074). (What this refers to precisely is obscure: an office of basket-bearer for Sarapis and Isis is attested, for Zeus and Athena not.) The same group minus Athena are recipients of a later dedication from the same place (*ID* 2104); in another, Zeus Soter is followed by three Egyptian gods (*ID* 2109). Zeus in these texts must be himself, not an alias for Sarapis, who is named separately; so too with Athena and Isis. But it comes naturally to the dedicator to conjoin gods of different origins, and to honour Zeus in a sanctuary in which originally he had no place.[68] Such intrusions of what are apparently Greek gods into the Egyptian sanctuaries are quite common, already and even particularly in their early phases. A man of Athymbra in Caria dedicates to what we know to be his local gods Plouton, Kore, and Demeter, but then adds Hermes-Anoubis, choosing the Egyptian god who continues the underworld theme;[69] a group of "contributors" pay up for Dionysus and Sarapis.[70] Sarapieion C hosted numerous statues of Greek gods, according to the inventories.[71]

66. *ID* 1561 ("for Mithradates Eupator"), 2128 (by a man from Nymphaeum in the Tauric Chersonese), 2179 (by Athenians), 2415 (by a man from Elea). In *ID* 2305, where a man of Askalon links him with Astarte Palaistine Aphrodite Ourania, Baslez, *Religions orientales à Délos,* 82, 89–90, takes him as equivalent to a maritime Baal, comparing *ID* 2182 to Zeus Kasios; such may no doubt have been the Askalonite's understanding.

67. *ID* 1754; on the findspot of this, 2305, and the fragmentary 2416, see Bruneau, *Cultes de Délos,* 246.

68. So too probably the Dioscuri, named after four Egyptian gods in *ID* 2123. Roussel, *Cultes égyptiens,* 195, toys with the idea that Zeus Soter in *ID* 2109 was part assimilated to Agathodaimon; but with a Roman dedicator there is little reason to suspect this.

69. *IG* XI.4.1235 from Sarapieion C, Ἰατροκλῆς Δημαγάθου Ἀθυνβριανὸς Πλούτωνι καὶ Κόρει, Δήμητρι, Ἑρμε[ῖ] Ἀνούβι, κατὰ πρόσταγμα. On the cult of Plouton and Kore at Athymbra, see Cohen, *Hellenistic Settlements in Europe, the Islands, and Asia Minor,* 258.

70. *IG* XI.4.1224, from Sarapieion C. Note too, e.g., Eros Nikephoros from Sarapieion A (*IG* XI.4.1304), and Artemis Phosphoros from Sarapieion B (*IG* XI.4.1275; cf. 1276). Remarkable is the plaque from Sarapieion C (*ID* 2475) inscribed "of Demeter Eleusinia and Kore and (my) wife," apparently attesting eschatological hopes and also (?) the Isis-Demeter association: cf. Roussel, *Cultes égyptiens,* 199–200. Why Aphrodite joined in ordering works conducted in the Sarapieion, κατὰ πρόσταγμα Σαράπιος, Ἴσιος, Ἀνούβιος, Ἀφροδίτης (*ID* 2098), is not obvious.

71. Baslez, *Religions orientales à Délos,* 47 n. 5.

The main Egyptian gods, Isis, Serapis, and Anoubis, usually, we have seen, retain their names. But there may be exceptions. Two dedications perhaps attest full *interpretatio* applied even to the major Egyptian deities. Aristokydes son of Demaretos was joined by Artemon son of Pytheas in a dedication in Sarapieion C made "on the orders of Osiris" "to Zeus who rules everything and Great Mother who rules everything"; attached to it was a *lex sacra,* "Do not approach from wine or in flowered garments."[72] The same man also made a dedication, again "on instructions," to Osiris and Isis in Sarapieion B (*IG* XI.4.1233), and, though a native Delian, was apparently able to stay on under the Athenian occupation and work as a dream-interpreter in Sarapieion C, where he dedicated to the regular triad Sarapis, Isis, Anoubis (*ID* 2071). For this committed devotee of the Egyptian gods, the figures who "ruled everything," it is argued,[73] should have been Osiris/Sarapis and Isis. Yet one may wonder why, in this one case, he avoided their familar names: was he influenced by the dream instructions from Osiris, and, if so, was the instruction to honour the Egyptian gods under unusually magniloquent Greek titles? Or did Osiris tell him to turn for a change to some of his own native gods? Also problematic is *ID* 2387, a dedication "to Asclepius and Hygieia and Apollo and Leto and Artemis Agrotera and all the gods and goddesses who share their altars and shrines," found near Sarapieion C.[74] The findspot and the typically Egyptian reference to altar-sharing gods suggest that the first three recipients may be what a Roman senator once called "Egyptian Hygieia and Asclepius and Apollo,"[75] that is, Isis, Sarapis, Horos. Perhaps in this case, as in the sanctuary of the Syrian gods, it was the healing prestige of Asclepius that caused the Greek name to be used, contrary to the norm, in preference to the Egyptian; two other dedications from the Sarapieion also name Asclepius.[76] What is interesting if so is that in *ID* 2387 Apollo, even if covering for Horos, drew in his familiar Delian associates Leto and Artemis Agrotera (for whom in this context no Egyptian equivalents can be named).

It may seem that in worrying whether in a given case Apollo, say, is a pure Greek Apollo, or rather an Apollo Horos/Harpocrates, one is insisting on a distinction that his worshippers declined to recognise. Certainly, many crossovers can be observed between originally distinct religious traditions. Dedications made "in accordance with an instruction" (from a god) or "in accordance with a vision" are

72. *IG* XI.4.1234, κατὰ πρόσταγμα Ὀσείριδος, Διὶ τῶι πάντων κρατοῦντι καὶ Μητρὶ Μεγάληι τῆι πάντων κρατούσηι Ἀριστοκύδης Δημαρήτου καὶ Ἀρτέμων Πυθέου; the *lex sacra, IG* XI.4.1300.

73. Roussel, *Cultes égyptiens,* 112; Y. Grandjean, *Une nouvelle arétalogie d'Isis à Maronée* (Leiden, 1975), 69 n. 157.

74. *ID* 2387, Ἀ[σ]κληπιῶι καὶ Ὑγιείαι καὶ Ἀπ[ό]λλωνι καὶ Λητοῖ καὶ Ἀρτέμιδι Ἀγροτέραι καὶ θεοῖς συμβώμοις καὶ συννάοις πᾶσι καὶ πά[σ]αις, Δάμων Πάτρωνος Ἀθηναῖος ὑπὲρ ἑαυτοῦ καὶ τῶν τέκνων καὶ ὑπαρχόντων, χαριστήρια.

75. Paus. 2.27.6; cf. Roussel, *Cultes égyptiens,* 150; Bruneau, *Cultes de Délos,* 375.

76. *ID* 2384 and 2386, where Bruneau, *Cultes de Délos,* 375, suggests detecting an *aretalogos* of Asclepius and Hygieia.

extremely numerous in the cult of the Egyptian gods, where there were special "dream-interpreters" to give the instructions clear form; but in the Athenian period any god can receive such a dedication.[77] The plaques with ears that by old Egyptian tradition indicate a "hearing god" similarly move from the Egyptian gods not just to the mysterious Aphrodite Peistiche but once even to Apollo in his main sanctuary (*ID* 2365).[78] The early dedications in the Egyptian sanctuaries tend to be addressed to Sarapis, Isis, and Anoubis, this on an island the religious life of which was famously dominated by the triad of Apollo, Artemis, and Leto; a three-sided altar which shows one each of the Delian triad on its three sides but bears a dedication to the three Egyptians underlines the possibility of influence.[79] The "Herakles" worshipped by three men of Iamneia in Palestine, paired as he is with Hauron as, is manifestly no Greek by origin; but the outline of the Greek Herakles' club is traced above the text of the inscription. On Delos as elsewhere, Herakles protected the doors of people of all nationalities; the shared Delian religion, it has been said, is revealed above all at the domestic level.[80] Atargatis was not the mother of Eros, but an Athenian felt able to dedicate Erotes to "Hagne Aphrodite Syrian goddess" (*ID* 2251), and statuettes of Aphrodite and Erotes have been found in her sanctuary.[81]

Mount Kynthos has often been noted as the place where, above all, interchange is visible.[82] Here long-established precincts of old Greek gods acquired as neighbours many small open-air sanctuaries of oriental newcomers; perhaps more importantly, one of the old gods worshipped here was Zeus Kynthios, that is, a mountain Zeus, a figure with an equivalent, *mutato nomine,* for every worshipper from Anatolia and Syria. Many of the oriental sanctuaries on the slopes of Kynthos in fact hosted male deities of much that type.[83] Zeus Kynthios was associated with a goddess, Athena

77. See Baslez, *Religions orientales à Délos,* 289 nn. 1–3; also *ID* 1428 II 49–50. On the specifically Egyptian origins of "on instruction" dedications, see Moyer, *Limits of Hellenism,* 165–71.

78. *Guide de Délos*⁴, fig. 90, shows the ears dedicated to Isis *epēkoos, ID* 2173; for a full list of votive ears from Delos, see Baslez, *Religions orientales à Délos,* 295 n. 8.

79. *IG* XI.4.1264, noted by Baslez, *Religions orientales à Délos,* 43–44. She notes that up to 116 the worshippers of Sarapis-Isis-Anoubis were almost all Italians, Athenians, and Delians: the triad was not an Alexandrian product.

80. Baslez, *Religions orientales à Délos,* 117, 124. Hercules of Iamneia: *ID* 2308, with the note; and Bruneau, *Cultes de Délos,* 404–7, 408–9. Cf. above, p. 74, on the appropriation of Herakles the benefactor of mankind by the Herakleists of Tyre. "At the same time Greek, Italian, Phoenician, the Herakles of Delos seems to me an interesting example of a kind of Delian religious *koine*": Bruneau, *Cultes de Délos,* 412. *ID* 2309, from the same findspot as 2308, pairs Herakles mysteriously with "the gods above": the gods of Mount Kynthos? (Plassart, *Mont Cynthe,* 279); scarcely the Dioscuri (Baslez, *Religions orientales à Délos,* 117).

81. A. Laumonier, *Les figurines de terre cuite,* Exploration Archéologique de Délos 23 (Paris, 1956), 15. Note, however, Will, *Déesse syrienne,* 149: a goddess with accessory figures (usually animals) is an oriental type.

82. See Plassart, *Mont Cynthe,* passim.

83. The dedications from sanctuaries B and C (which cannot be divided between the two) are listed by Bruneau, *Cultes de Délos,* 476: recipients include inter alios Zeus Megistos, Megistos Ares (or Zeus)

Kynthia, in a kind of pairing that was again widely familiar; Athena's somewhat faded cult recovered lustre in the Hellenistic period, it has been suggested, because "foreigners saw in her a suitable equivalent to their mighty Baalat."[84] We have already noted the multiethnic clientele of Zeus Kynthios and Athena Kynthia, and their association with Egyptian gods in dedications in Sarapieion C. An unknown dedicator may have offered them a statue of the ancient oriental type of a god mounted on the back of an animal, in this case a lion.[85] We cannot be sure whether the man of Gaza who dedicated *bathra*, "benches," to them (*ID* 1896) was thinking of the "Zeus," there known as Marnas, who according to Stephanus founded his home city,[86] but the worshipper (ethnic unfortunately lost) who dedicated to Zeus Helios (*ID* 2414) was certainly extending the scope of Zeus Kynthios (whether by actually addressing him as Zeus Helios, or at the least by introducing Zeus Helios into his sanctuary). Another worshipper spoke rather charmingly of "the god of the air on Kynthos":[87] again, it is not clear whether Zeus Kynthios himself was meant, or a related figure. Remarkable above all is the fact that one Leukios Pinarios made a dedication to Zeus Kynthios in a private house, "the house of jewels":[88] for this Italian (Roman?), the civic god has become domestic.

The issues in relation to iconography, sanctuary design, and architecture are too complicated to discuss adequately here, but transfer there certainly was. Sanctuaries of Eastern gods retained their oriental shape based on enclosed open space; the Sarapieion displayed distinctive Egyptian features, a "track" (*dromos*), monumental gateways (*pylōnes*), sphinxes; the sanctuary of the Syrian gods had a cult theatre (a thing rare in Greece), an empty throne of the goddess, as well as two further features designated by non-Greek words.[89] These dispositions were necessary, because these sanctuaries had ceremonial and also day-to-day uses different from those of normal Greek cult. But both also contained small temples of broadly Greek aspect,

Ouranios (?), Zeus Dousares, Sîn of Alam (in non-Greek letters), Zeus Ido[genes]; Zeus Hypistos and Zeus Helios were nearby (*ID* 2306, 2414).

84. Bonnet, *Enfants de Cadmos*, 515.

85. *ID* 2428, with the note (but note the doubt of Marcadé, *Musée de Délos*, 398 n. 1).

86. As suggested by Plasssart, *Mont Cynthe*, 112, citing Steph. Byz. s.v. Gaza; on Marnas, see Cook, *Zeus*, 3:549–58.

87. *ID* 2431, [τῶι] ἀερίωι [ἐν Κύν]θῳ θεῶι: from sanctuary F, not the Kynthion itself.

88. Published by Bruneau, *Cultes de Délos*, 223, stressed by Baslez, *Religions orientales à Délos*, 119. *The lex sacra ID* 2529 = *LSS* 59 has also been seen as marking a crossover, but its prescriptions do not show any distinctively Eastern feature unless an adjective at the lost beginning of line 18 specified a particular species of meat to be avoided; as it stands, the requirement of purity "from meat" (ἀπὸ κρέως) is unique.

89. ΕΙΓΑΝ, *ID* 2234; and ΝΑΜΑΡΑΝ, *ID* 2240–41; cf. Will, *Déesse syrienne*, 154–55 (111–14, theatre; 152–54, throne); on the physical remains of Sarapieion C, see H. Siard, *RA*, 2009, 155–61; in general, the summary in Baslez, *Religions orientales à Délos*, 250–72; *Guide de Délos*⁴, nos. 91, 98, 100, 106–9.

even if the internal arrangements might diverge from Greek norms.[90] As for images of non-Greek gods, now of marble and (a few Egyptian imports aside) broadly Greek in aspect,[91] the challenge is to detect some slight remaining distinctive attributes. Isis retains her knot and distinctive coiffure; Hadad and Atargatis can probably still be flanked by lions and bulls; a new clothed type was perhaps developed for Harpocrates, to separate him from that other (Greek) divine child Eros. But the theriomorphism of what appears to be a dog-headed Anoubis is very unusual.[92] The reverse process, of Greek gods assuming oriental iconographic traits, is harder to detect, but a possible case of a "god on animal" statue dedicated to Zeus Kynthios was mentioned above; an inventory mentions a statue of Apollo holding the falcon associated with his Egyptian equivalent Horos (*ID* 1416 A col. 1.24–25).

At the level of organisation above all we find conflation and compromise. There are more cult functionaries attached to Sarapieion C and the Syrian precinct than to a normal Greek sanctuary, but many fewer than in Memphis or Hierapolis; the posts of priest, "key-bearer" (κλεϊδοῦχος), and basket-bearer are reserved for Athenians, among whom they rotate annually in the Athenian manner. In these two cults there emerges the new concept of a community of worshippers, the *therapeutai;* but the *eranoi* and *thiasoi* and *koina* associated with these and other cults are in varying measure cut from Greek cloth.[93] Some of the specific details that would make the picture vivid escape us: true oriental priests of Isis and Atargatis looked exotic to a Greek eye, and we would like to know how their Athenian replacements dressed and did their hair.[94]

To what extent does the evidence assembled in this chapter justify the claims about a "universal polytheism" made in chapter 2? The situation of Hellenistic Delos was unique, an island from whom most of the native Delians had been expelled but which played host to traders from both Italy and the East on a large scale; it was governed from Athens, and on the island the only decision-making body was the mixed one of "the Athenians and Romans and other Greeks who reside on Delos or sail to the island as traders and shipmasters." Polis religion

90. Will, *Déesse syrienne,* 109–10.

91. Marcadé, *Musée de Délos,* 408–11, and (references in inventories) 429; cf. Moyer, *Limits of Hellenism,* 198.

92. Isis: Marcadé, *Musée de Délos,* 430–31; Hadad and Atargatis: 382; Will, *Déesse syrienne,* 148–49; Harpocrates: Marcadé, *Musée de Délos,* 434–35; Anoubis: *BCH* 76 (1952): 122–23, fig. 22, museum inventory A 5280; cf. Marcadé, *Musée de Délos,* 413, 433 (envisaging other examples). On Sarapis, see Marcadé 427–28.

93. See Baslez, *Religions orientales à Délos,* 192–95 (*therapeutai;* cf. Baslez, "Associations à Délos," 244–47); 197–203 (*thiasoi* etc.; cf. Baslez, 233, 281, on accountability and discipline).

94. For tantalising hints, see Bruneau, *Cultes de Délos,* 473; Baslez, *Religions orientales à Délos,* 234 n. 44; and cf. E. Walters, *Attic Grave Reliefs That Represent Women in the Dress of Isis,* Hesperia Suppl. 22 (Princeton, 1988).

therefore was in abeyance. Even in this melting pot, not everything melted; on the contrary, the proliferation on the island of little sacred laws requiring worshippers, for instance, "not to sacrifice pig or goat" (in this sanctuary), or to observe certain purity rules, shows how important it was not to allow a total fusion of conventions.[95] One could worship in many different sanctuaries whatever one's origin, but when there one had to observe the specific traditions. But this was to be predicted: even if the gods of every nation were in some sense the same, no one ever supposed that the local traditions for worshipping them, sanctified by time, should be neglected. In that sense, *interpretatio* is revealed as ineffective: your god may be my god, but their cults are not combined: Delos is a place where individuals honour gods originating from many different cultures,[96] not one where, so to speak, Aphrodite and Astarte and Atargatis all blend into one. Delos is an extreme case, which illustrates possibilities less evident in less unusual situations, but also boundaries. But the island's special status in this sense lasted only a few decades, from somewhere in the second half of the second century to the sack by Mithradates in 88 from which it never recovered.[97] How things would have developed if the "Delian moment" had been prolonged we can only speculate.

95. *LSCG* 94–95 (*IG* XI.4.1300 and *ID* 2367), *LSS* 54–59 (*ID* 2530, 2305, 2180, 2308, *BCH* 73, 1949: 154; *ID* 2529). A point stressed by Bonnet, *Enfants de Cadmos*, 5–9: names are porous, rituals not.

96. But particular preferences persisted: Baslez, *Religions orientales à Délos*, 79 n. 40.

97. Only from 120 to 88 for Baslez, *Religions orientales à Délos*, 12.

Conclusion

Early in its first chapter, this book evoked the *Griechische Götternamen* of Hermann Usener. The great Wilamowitz, thanking Usener, his quondam teacher, for the gift of the book, spoke out, after some emollient paragraphs, with a blend of hyper-protestantism and brutal frankness: "Gefühl ist alles," he wrote, "Name ist Schall und Rauch," "Feeling is everything, name is noise and smoke" (a common German expression for a thing of no substance).[1] Undoubtedly the book was deeply flawed; it relied heavily on etymological speculation, which was outdated even in its own day; and it is doubtful whether etymological speculation has ever been of much use in the study of religion. To that extent one can agree with Wilamowitz. But the sharp polarisation that he introduces between "name" and "feeling" cannot be accepted. Attention not so much to single names as more broadly to naming—so all modes of address, including epithets—is essential to a study of ancient and perhaps modern religious psychology.

In naming, both the likeness and the unlikeness of gods to men are reflected. Gods, like powerful mortals, attract respectful and indirect forms of address (title not name); they also attract euphemism, something again familiar in the mortal sphere even if more commonly applied to forces and conditions than to people.

1. Letter of 7/12/1895 (cf. more fully Wilamowitz, *Glaube*, 11), cited by Mette, "Usener und seine Schule," 79–81, with 81–90, Usener's response and letters from other scholars; the reaction of H. Diels was similar (for Rohde, see p. 18n69). Wilamowitz also opposed Usener's "From the many to the one" approach, writing: "God is not younger than the gods" ("Gott ist nicht jünger als die Götter"). Cf. M. M. Sassi, in *Hermann Usener filologo della religione*, 80; A. Wessels, *Ursprungszauber: Zur Rezeption von Hermann Useners Lehre von der religiösen Begriffsbildung* (Berlin, 2003), esp. chap. 1; Konaris, *Greek Gods in Modern Scholarship*, 192–93.

But through the cultic epithet gods are fragmented into multiple aspects in a way that monarchs, for instance, are not. This fragmentation can be explained "theologically" by the argument that gods have multiple powers which are accessible everywhere: Zeus Meilichios is an aspect of Zeus, the precinct of Artemis Brauronia is the sanctuary at Brauron which gives local access to the universal Artemis. But these fragmented aspects acquire a kind of autonomy: Zeus Meilichios is shaped by the particular expectations that worshippers have of him, the particular offerings they bring to him, the particular form in which they depict him; Artemis Brauronia is powerful because plain Artemis, like all deities, is invisible, and a sense of her reality is achieved through the symbols and rites seen and performed in a particular place. So strong is this fragmentation that there is a tendency to give preference to epithets, whether local or functional, over actual proper names of gods, as if these local connections or specific powers were more important or more reliable than the name; the mythological persona gives way to the perception of presence, or power. But what power? How many powers? Gods can coalesce into forms such as Artemis Hekate.

Different issues arise when cultures meet. *Interpretatio* was one solution, a solution that came very naturally to the ancients but which confronts us with many questions. A proposition such as "Neith is Egyptian for Athena" never meant that their rituals were identical (though similarities might underlie it), still less that Neith's traditional rituals could be replaced by those of Athena, or vice versa. To that extent it can be parsed as no more than "Neith is the nearest Egyptian equivalent to Athena." The inconsistency and imprecision with which identifications were made might seem to suggest that what was at issue was merely the need, if writing in Greek, to use a Greek theonym (and similarly mutatis mutandis with other languages), not to make any stronger claims about essences. Yet the ancients again and again express themselves without qualification in the form "Athena is Neith," and much less often in the form "Athena is the closest equivalent to Neith" or something similar. And most philosophers in the Greco-Roman tradition were clear that the gods were the same everywhere, only names differing. A remarkable feature of *interpretatio* is that it was pervasively practised, never as it seems discussed. Multiple understandings could coexist; or perhaps one should rather say that the need was not felt to achieve clarity, to dispel the fog that so often surrounds the big concepts that we apply in everyday life. Perhaps some worshippers of Hermes Anoubis or Isis Soteira Astarte Aphrodite Euploia hoped that they were bundling up the powers of the gods and goddesses of different nations, not addressing a single essence. But the tolerated vagueness is itself a historical datum. The attempt by Catholic missionaries to find Chinese equivalents for Christian theonyms, by contrast, formed an important part of the "Chinese rites" controversy of the seventeenth and eighteenth centuries, and ended in condemnation; the relation of God to Allah is also problematic.[2] It is in this sense that one can speak of a "universal polytheism" of the ancient world: not that the gods of one people blended into those of another, not that differences in ritual or in the fundamental shape and structure of a pantheon were effaced, but that compatibilities were taken for granted, not resisted, bridges therefore existed without the need to be consciously constructed, and were crossed.

This potential for easy interchange is revealed, above all, on the entrepôt of Delos in the period of Athenian occupation after 166 B.C. Greeks, Italians, and Syrians mingle and, with some limitations, pay honour to one another's gods; Egyptian gods too enjoy exceptional popularity with non-Egyptian worshippers. Different cults, it is true, retain different traditions; sacred laws remind worshippers what they may and may not wear, carry, eat before or during entry to a given sanctuary. But many crossovers and combinations also occur. The main cults of the Syrian gods and of Sarapis and his team acquire Athenian priests. As is usual outside literature, *interpretatio* is applied inconsistently. The goddess commonly called Atargatis by her Syrian worshippers becomes Aphrodite Hagne in most Athenian parlance, and her consort Hadad drops out of view; on the other hand, the Egyptian gods Isis, Sarapis, Osiris, and others retain their own names most of the time. We also encounter spectacular combinations such as the Isis Soteira Astarte Aphrodite Euploia mentioned above. The persistence of foreign names might seem to show that *interpretatio* cannot be based on the implicit assumption that the gods of Greece and those of other countries are the same, differently named. If Isis is Demeter, why, as a Greek, worship Isis? Two responses would be possible: (i) the gods are the same everywhere, but different countries have discovered the powers of a different selection from among them: Isis therefore is not Demeter (though other identifications may be correct), but a distinct figure with valuable attributes whom Greeks would be wise to cultivate; (ii) they are indeed at bottom the same, but the different traditions of naming and worship that have developed are both pleasing to the goddess and should be preserved. But, again, to reason so may be to reason too nicely. One need not look for theological consistency in this lived religion. On any view, the Egyptian goddess did not represent a threatening or unapproachable "other" for the Greek, any more than Apollo did for Lydian Croesus. The ancient polytheisms were universal in the sense that no god's power was available only to members of a particular nation.

In the regions outside Greece where Greek became the standard written language, *interpretatio* was not the only solution to rendering a non-Greek god in Greek. The original name could be preserved—a rarity in literature, except for a few powers such as Isis whose cult spread into the Greek world, but commoner in

2. Bettini, *Elogio del politeismo,* 60–61, 63 n. 8.

inscriptions. Gods could be addressed by periphrasis or title or location, not name: "Maiden," "Demi-Goddess," "the God at Doliche." Delicate and often unanswerable questions arise about these choices: some may be simple transferences into Greek of an indigenous naming convention whereby proper names were avoided; others may represent a preference for periphrasis over an indigenous theonym deemed barbarous. Another phenomenon was the emergence of gods with Greek names unknown to Greece: the Arab god Theandrios probably also had an Aramaic name alongside the Greek Theandrios, though Theandrios is not a direct translation of it, but Sozon and Hosios kai Dikaios seem to have emerged without antecedents in a western Anatolia which was already Greek-speaking.[3] Such cases reveal how complex was the interplay between language and local religious culture. We see this complexity also in the flourishing of the "cultic double name" outside old Greece in (particularly) the Hellenistic and Roman periods. New epithets are created, occasionally by transcription or translation from another language, more commonly by fresh coining: thousands of new toponymic epithets use the existing Greek form of expression (one in fact thoroughly compatible with indigenous traditions) to identify a relation between god and place often very different from that characteristic of Greece itself; new functional epithets show how, for example, Phrygian religious sentiment could be conveyed within the transplanted idiom.

Forms of expression emerge that present a relation between men and gods unknown in old Greece. There was no "Athena of Pericles." But such "god of X" (where X is a mortal) combinations are found in both Anatolia and eastern Syria in the Roman period. In Syria they directly translate an idiom found also in Nabataean texts, vaguely reminiscent of the Old Testament "God of Abraham," but not yet clearly understood: X is doubtless a forebear, not necessarily distant, of a worshipping group, but gods so named could spring such confines and be attested over a substantial area. In Anatolia the assumption has been that "Men of Artemidoros," for instance, is a cult of Men founded by Artemidoros that acquired sufficient reputation to be named from its founder and spread to new sites. But that X was founder of the cult is never demonstrable, and a recent discovery suggests that he might rather have been someone to whom the god manifested himself;[4] these two explanations are obviously not incompatible. On any view this strong linkage of a cult with an individual (which may, admittedly, have become nominal and disregarded in time, like the origin of any name) is alien to the old Greek world of polis religion. Similarly alien is the language of "my/your/his God" familiar from the Old Testament and characteristic in fact of all the Semitic languages. By contrast with the "god of X" form, this idiom crosses over into the Greek of the rele-

3. See pp. 89–92.
4. See pp. 115–16.

vant regions much less than might have been predicted. The closest equivalent to "my god" for a Greek was "(my) ancestral god," and that expression became very popular in both Anatolia and Syria, in the latter probably substituting in part for the more intimate "my god" of the vernacular. Doubtless the relation between language and sentiment is complex. To suppose that all speakers of languages that allowed speaking of "my God" felt closer to the divine than all speakers of Greek would be rash. But the idiom must have encouraged a sense of a more personal relation to a particular god picked out in this way from a broader pantheon. Again we see, from all the cases mentioned in this paragraph, how naming cannot be separated from feeling.

The emphasis in this book has lain on the language used in inscribed prose dedications, rather than on that of hymns or works of literature or verse inscriptions that borrow literary epithets. The approach may appear reductionist: these spheres are in constant interaction. But the inscribed epithet represents the encapsulation of what is essential for the dedicator. The lexicon of epithets increased noticeably from the Hellenistic period onward, partly through a new tendency to include praise and glorification within the dedication, partly through interaction with other religious cultures and the calquing of titles of respect traditional in them. These are concrete manifestations of two often noted but elusive features of Hellenistic and later religion, an intensification of the relation between man and god, and an intermingling of religious traditions. A controversial but crucial development is the growth of dedications to a figure known as Highest God or something similar: whether Highest God is seen as a distinct figure with a cult of his own, or as is more plausible a title of respect applicable to existing figures who remain separate, this anonymisation seems to crystallise and express in intensified form the uncertainty always felt by pagans about the nature of the figures to whom they prayed—combined with a certainty that a highest god did indeed exist.

Highest God is unusual among subjects discussed in this book in that his cult has been the subject of abundant recent debate. With a few other notable exceptions, *Griechische Götternamen* have tended to be discussed only en passant and in fragments since the publication of Usener's flawed and mighty book of that name. The subject is indeed impossible. Classifying epithets, it has rightly been said, is like herding cats,[5] while interpreting their meaning is often shooting in the dark. Dedicators were seeking to communicate with gods, not with us; if they addressed a figure as "King," or "Great God," they knew, but we often do not, whether he was for them just that and that only, or also had a name. One can often feel that one is trying to catch "noise and smoke." Almost all generalisations can be met by countercases; despite what was said in the previous paragraph, many dedications from late antiquity are addressed to named gods without titles of respect and, spelling

5. Hornblower, "Cult Epithets in the *Alexandra*," 96.

and letterforms aside, could be supposed to have been made half a millennium earlier; some geographical distinctions were touched on in chapter 5. The bulk of relevant material is far beyond the control of one individual. There is no definitive book to be written on this topic. What this book has tried to do is to sensitize the reader to these issues, to stimulate curiosity, to show how attention to them brings us close to central aspects of pagan religious feeling.

APPENDIX A

Postclassical Use of the Epithet Οὐράνιος

Οὐράνιος in the classical period is applied to Zeus in Sparta (Hdt. 6.56) and very frequently to Aphrodite.[1] Both gods retain the epithet later, Zeus receiving it more widely than before (sometimes in the form ἐπουράνιος, as in *INikaia* 1114–15, or as a supplementary epithet, Διὸς Σω[τ]ῆρος Οὐρανίου, *IAnazarbos* 50), and it becomes a regular epithet of Men. Other male gods are given it seldom and mostly in outlying regions.[2] For its special application to Syrian supreme gods, see pp. 122–24. There are some references to a single οὐράνιος θεὸς: *IG* XII Suppl. 30 (Lesbos), Ζωσίμη οὐρανίοιο θεοῦ σωθεῖσα προνοίαις (referring back to a Zeus mentioned earlier?), particularly where he is called on to protect a grave, as in *ÖJhBeibl* 18 (1915): 45, ἐξορκίζομεν ὑμᾶς τὸν ἐπουράνιον θεὸν καὶ Ἥλιον καὶ Σελήνην (distinguished from underworld gods mentioned later); *ISultanDaği* 509; and other examples collected by Robert, *Hellenica* XIII, 100–103;[3] in these cases Robert raises the question whether the reference is to an anonymous god of the sky or to Men Ouranios.

Goddesses receive the epithet more often, but apparently, Aphrodite aside, not before the imperial period:[4] Nemesis (*IG* II² 5070), Muses (*IG* V.2.287), Tyche? (*Forschungen in Salona*, Vienna 1926, 2:30), Nike (*IGB* IV 2266), Meter Theon (*IGB* II 805), Hera (*IG* XII.4.2.521–25, Cos), Artemis (*SEG* LVII 527, Mopseion in Thessaly), Artemis Ephesia (*TAM* III.1.390), Kyria Artemis (*SEG* XXXVIII 1652, Gerasa). The identity of an unnamed "heavenly goddess" can therefore be obscure: *IGerasa* 24 (cf. 26, Διὶ Κρόνῳ καὶ [θ]εᾷ Οὐρανίᾳ);

1. See V. Pirenne-Delforge, in *Nommer les dieux*, 271–90; pp. 101 and 103 above.

2. Hermes, *IG* V.1.559; Bes, Perdrizet, *Memnonion*, nos. 489, 500 (amid a torrent of epithets); Apis, at Abila in Syria, *RICIS* 402/1004; Eros, from his mother, *IG* II² 5017; the "Messenger Divine" (*IStrat* 1307). Ares, *ID* 2312, is very questionable.

3. Cf. *BÉ* 1966, no. 213; and probably *AMS* 53 (2004): 245, no. 52.3, where θεῷ Οὐρανῷ is printed.

4. The goddess so identified on a problematic relief from Isthmia is debated: *SEG* XXXVI 327.

SEG XXXI 999 (= *SEG* IV 646, from Kula); several further Syrian examples in Aliquot, *Vie religieuse au Liban*, 150 n. 139.[5]

Heavenly gods are often distinguished from chthonian in curses (e.g., *TAM* V.2.1371), and Hekate acquires both epithets in such a context (e.g., *SEG* XXX 326, σὺ δὲ, δέσποινα Ἑκάτη οὐρανία Ἑκάτη καταχθονία).[6] Οὐράνιος is occasionally used as a dignifying title (*CIRB* 36 A 5–6, θεοῖς ἐπουραν[ί]οις Διὶ Σωτῆρι καὶ Ἥρᾳ Σωτείρᾳ; *IG* XII.5.655, είέρειά τε κατασταθεῖσα τῶν οὐρανίων θεῶν Δήμητρος καὶ Κόρης τῶν σεμνοτάτων), and the gods worshipped collectively can be so designated: *IG* V.1.40, ἱερεὺς Οὐρανίων; *IG* XII Suppl. 165, ἱερεὺς Διὸς Κεραυνίου καὶ θεῶν Οὐρανίων . . . Διὶ καὶ θεοῖς Οὐρανίοις; *ABSA* 42 (1946): 206, no. 2 (Amathus, on Cyprus), κνεισευτ[ὴρ] Θεῶν Οὐρανίων καὶ Διὸς Λαβρα[νίου.

5. Whether there is influence here from the "Οὐρανία of the Carthaginians" (as Dio Cass. 80.12 [Loeb, ix, 460] calls the great African goddess Caelestis) is not clear; Herodian 5.6.4–5 gives her Phoenician name as Ἀστροάρχη. In *Hymn. Orph.* proem 41, paired with Adonis, Οὐρανία is probably Aphrodite.

6. Cf. Plut. *De def. or.* 13, 416 E; similarly with Gaia, A. Audollent, *Defixionum tabellae* (Paris, 1904), no. 79.

APPENDIX B

Translated Theophoric Names

In bilingual texts a theophoric name[1] is sometimes[2] "translated" by substitution of an appropriate theonym according to familiar principles of *interpretatio.* In the trilingual from Xanthos, Lycian Nattrbbiyẽmi is Greek Apollodotos.[3] On a Phoenician-Greek tombstone from Attica (*KAI* 53; Bonnet, *Enfants de Cadmos,* 452–53: ca. 400 B.C.), the Sidonian Abdtanit, "servant of Tanit," son of Abdshamash, "servant of Shamash," becomes Artemidoros, son of Heliodoros, in Greek. On another from Rhodes (P. M. Fraser, *ABSA* 65, 1970: 31: "c. 200 B.C. or slightly later"), Abdelmelqart, "servant of Melqart," is Herakleides (of Kition) in Greek. In *KAI* 54 (Bonnet, *Enfants de Cadmos,* 462–72: Athens, 4th/3rd c.), Abdashtart, "servant of Astarte," is Aphrodisios; at Demetrias in Thessaly a man bearing a name formed from Eshmun becomes Asklepiadas (O. Masson, *BCH* 93, 1969: 694–96, no. 4: 3rd c.) A dedication from Malta by two Tyrians to Herakles/Melqart is made by Abdosir, "servant of Osiris," and his brother Osirshamor, sons of Osirshamor, who in Greek are Dionysios and Sarapion, sons of Sarapion (*KAI* 47, 3rd/2nd c. B.C.; cf. p. 199); here we find both the old Herodotean equivalence of Osiris/Dionysus and the Hellenistic modernisation of Osiris as Sarapis.[4] As these examples show, the Greek suffix or second element in such renderings can vary; in *KAI* 60 (Piraeus, late 4th c.; Bonnet, *Enfants de Cadmos,* 446–51), the rendering of ŠM'B'L,

1. The pioneer work was R. Herzog, "Namensübersetzungen und Verwandtes," *Philologus* 56 (1897): 33–70, building on ideas of, e.g., the Semiticist C. Clermont-Ganneau; on Phoenician and Greek in brief, F. Briquel Chatonnet, *CRAI,* 2012, 626.

2. Not invariably: e.g., in *KAI* 42 an individual who bears a Ba'al name in Phoenician is Praxidemos in Greek.

3. H. Metzger et al., *Fouilles de Xanthos VI: La stèle trilingue du Létôon* (Paris, 1979), 53, 58. This individual is absent from the Aramaic.

4. Fraser, *Ptolemaic Alexandria,* 2:404–5 n. 514, claims that the Greek names are primary; this is not clear to me.

'Ba'al/god has heard," as Diopeithes, "obedient to Zeus" on the most obvious understanding, somewhat changes the sense. A standard Greek name is sought even if this does not give full semantic equivalence with the Phoenician.[5] Such translation of theophoric names does not represent a unique case, but a special application of a more general tendency to translate names where an equivalent can be found: thus for instance Phoenician Benhodesh, "son of the new moon," is rendered Noumenios (*KAI* 41 and 55; Bonnet, *Enfants de Cadmos,* 454);[6] Egyptians too when taking Greek names often sought semantic equivalents of the Egyptian original.[7]

Bilinguals provide unambiguous evidence.[8] The case can never be so clear with Greek names borne by non-Greeks in monolingual texts, but the predominance of certain names in certain contexts can make it virtually certain that they are products of translation, or at all events of popularity that a name has gained because of a similarity in sense to an indigenous name. It is generally accepted that the occurrence of, for instance, Pan- and Kron-names in Egypt, much more frequent than in the rest of the Greek world, rests on identification of Pan and Kronos with Min and Geb respectively.[9] (In Egypt, the second elements in compound theophoric names could apparently be chosen to reflect different nuances of indigenous theophorics, with, e.g., the endings -onios, -ides, -odoros, -okrates, -ophanes, -ophilos when added to Apoll- reflecting different names based on Horos.[10]) The great frequency of names such as Herakleides, Dio- and Apoll- compounds, Zenon, Heliodoros, Artemidoros, and the like among Phoenicians is doubtless to be similarly explained,[11] especially since, as we have seen, some of these Greek names are actually attested with Phoenician equivalents in bilinguals. Robert held that the popularity of Athenaios/Athenaïs in Cappadocia rested on the *interpretatio* of the indigenous Ma as Athena.[12]

The decision whether to translate could be determined by context. In a Palmyrene bilingual from Rome (*CIS* II.iii.3902), Iarhai, a form shortened from Iarhabol, is translated as

5. See Herzog, "Namensübersetzungen," 39–40. In a neo-Punic bilingual (Jongeling, *Neo-Punic Inscriptions,* 30 Labdah N 23), Diodorus in Latin translates Mutunbal.

6. In *KAI* 57 (bilingual, though the Greek text *IG* II^2 9035 is missing from *KAI*), Mahdash, "from the New Moon" (so *KAI*), is again Noumenios in Greek.

7. On translation, including translated theophorics, and phonetic similarities in the "double names" of Ptolemaic Egypt, see S. Coussement, *"Because I am Greek": Polyonymy as an Expression of Ethnicity in Ptolemaic Egypt,* Studia Hellenistica 55 (Leuven, 2016), 87–92. See too W. Peremans, "Ethnies et classes dans l'Égypte ptolémaïque," in C. Nicolet, ed., *Recherches sur les structures sociales dans l'antiquité classique* (Paris, 1970), 213–23, at 221; Y. Broux, *ChrÉg* 88 (2013): 319 (on *PAmh.* 1.72, probably ca. 202 A.D.).

8. Double names joined by the "also known as" formula can work similarly: P. Bernard, *JSav,* 1990, 58, draws attention to *IGLS* I.166, Βαρνεβουν (i.e., "son of Nebo," often rendered as Apollo) ṭὸν καὶ Ἀπολλινάριον.

9. On Geb/Kronos, see C. E. Holm, *Griechisch-ägyptische Namenstudien* (Uppsala, 1936), passim; on Pan/Min, 91 n. 3.

10. So Peremans, "Ethnies et classes," 221.

11. So Herzog, "Namensübersetzungen," 41–46; cf. F. Briquel Chatonnet, "Quelques remarques sur l'onomastique des Phéniciens d'après les inscriptions grecques," in *Actes du IIIe Congrès International des Études Phéniciennes et Puniques* (Tunis, 1995), 1:203–10, at 205–6; chapter 6, n. 15.

12. Robert, *Noms indigènes,* 494; cf. 508 for Apollonios and Heliodoros in Cappadocia as deriving from "some native god assimilated to Apollo."

Heliodoros. But Iarhai appears transcribed in Palmyrene texts written in Greek;[13] the pressure to translate was apparently felt more strongly abroad. Greek names were not normally translated into Latin or vice versa, presumably because the cultures were so intermingled that there was no chasm of unfamiliarity to be bridged. Translation of Greek theophoric names into Latin would have been highly artificial anyway, since there was no Latin tradition of such names. A historian who supposedly rendered Saturninus as Kronios incurs the ridicule of Lucian (*Hist. conscr.* 21).

A different approach is seen in the name of Abdashtart/Straton, king of Sidon in the fourth century: he and other Phoenician bearers of similar names seem to have sought a correspondence not of sense but of sound, with the common Greek element -strat— echoing the "Astarte" of the original theophoric name.[14] That rendering changes the name from theophoric to secular; conversely, Antiochus I's wife Stratonice becomes theophoric when rendered Astartanikku in Akkadian in the Borsippa cylinder.[15]

The popularity of certain Greek names in Caria, including several that are theophoric (Artemi-, Herm-, Hekat-), has in all probability been influenced by Carian; translation and phonetic assimilation may both have operated, in different cases.[16]

13. See the commentary ad loc. in *CIS* II.iii.3902. The same text translates this man's father's name Ḥalîphî, explained by the editor as deriving from a root meaning "change," as Antiochos (perhaps playing on *anti-*, "in place of").

14. On him, see Bonnet, *Enfants de Cadmos,* 206–7; on -strat- names, J. Wallensten, *Kernos* 27 (2014): 172, with bibliography. Choice of name through sound similarity is again not confined to theophoric names: see, e.g., M. H. Williams, "Semitic Name-Use by Jews in Roman Asia Minor and the Dating of the Aphrodisias *Stele* Inscriptions," in E. Matthews, ed., *Old and New Worlds in Greek Onomastics* (Oxford, 2007), 173–97, at 180–83 (or *BÉ* 1977, no. 542 on Silas/Zoilos) That article also discusses names that translate (Eirenaios, from Shalom) or express (Theophilos) Jewish religious values. For types of name change ("noms d'assonance," translated names, etc.), see A. Mullen, *Southern Gaul and the Mediterranean* (Cambridge, 2013), 134–35.

15. Kosmin, *Elephant Kings,* 114.

16. See I. J. Adiego, in *RÉA* 107 (2005): 614, for Hermias as a possible translation of Imbrassis; Adiego, in O. Henry, ed., *4th-Century Karia: Defining a Karian Identity under the Hekatomnids* (Istanbul, 2013), 17–18, for phonetic assimilation.

APPENDIX C

Interpretatio in India

India represents in some ways a special case. The simple process of *interpretatio* was complicated by the traditions that Dionysus and Herakles had led expeditions to India (Polyaenus, *Strat.* 1.1.1–3 can even record a stratagem used by Dionysus on campaign). Whereas the identification of Zeus with Ammon rests simply on the proposition that Zeus in some undefined sense "is" Ammon, Alexander and his courtiers and troops supposed that they were following in the footsteps of two previous sons of Zeus whose imprint on the culture of the lands they traversed was still visible. Eratosthenes was to argue that the Indian campaigns of Dionysus and Herakles were invented by the Macedonians to glorify their leader who was following, and surpassing, such great gods (Arr. *Anab.* 5.3.1–4). It is certainly true that in earlier sources Dionysus penetrates no further east than Bactria, Herakles much less far;[1] but some Macedonians no doubt convinced themselves in good faith that the two gods had preceded them, on the basis of the signs of their passing that they thought they could detect.

Sources sometimes say that these traditions were accepted by the Indians themselves. According to Arrian,[2] Akouphis, an elder of Nysa in the region between the Kophen and the Indus, claimed that his city had been founded by Dionysus himself for his veteran soldiers/*bacchoi* no longer fit for service, and was named by Dionysus after his nurse Nyse. The name Meros, "Thigh," of the nearby mountain commemorated the god's birth from Zeus's thigh; a proof of the tradition was that ivy grew there alone in India. The case is slightly different with the Sibai: in Arrian it is Greek observers who identify them as remnants of Herakles' army because they dress in skins, bear clubs, and brand their cattle with a club mark, whereas in the vulgate they make the same claim about themselves in negotiation with

1. Bosworth, *Commentary,* 2:200.

2. *Anab.* 5.1.5–6; the same is stated as fact in Arr. *Ind.* 1.4–6, which will be the source for the *Anab.* passage acccording to Bosworth, *Commentary,* 2:204.

Alexander, and duly secure their "freedom."[3] And there are three more instances (one in Arrian, two in the vulgate) of tribes or groups of tribes who supposedly accepted the legends.[4] Perhaps these claims are reliable: one can see that if through diplomatic contacts native rulers and politicians heard the stories circulating in the Macedonian camp, they might have gratefully adopted traditions that so well suited their appeals for friendly treatment; it has even been suggested[5] that they may have contributed to their elaboration. But some doubt is possible. Greek sources may not have asked themselves what Greek stories Indian diplomats would have known; they may simply have assumed that they would exploit traditions that so obviously served their interests.

It is widely believed that in these traditions, whatever side they emanate from, "Herakles" and "Dionysus" are *interpretationes* of indigenous deities; efforts are accordingly made to identify a prominent deity of the region in question who should underlie the Greek god. Because chief gods varied regionally, the identifications are not the same in all cases: the Dionysus of Nysa will be (perhaps) Indra, of the Oxydracae Śiva, but the Herakles of the Sibai will also be Śiva.[6] But the dominant impulse behind these traditions was not the desire to make sense of the Indian religious universe in Greek terms; the point was to identify traces of the wandering gods. It was given ex hypothesi that the only gods to be looked for were Dionysus and Herakles: contrast Strabo's statement (15.1.69, C 718) that the gods worshipped by the Indians were Zeus Ombrios, the Ganges, and "local deities" (*enchōrioi daimones*), or Ctesias's claim of a desert sanctuary of Sun and Moon.[7] The context in which Dionysus and Herakles are spoken of is never that of sanctuaries, and the evidence for their presence that is appealed to is that of the landscape (ivy, vines), toponyms (Nysa, Meros), and customs

3. Arr. *Ind.* 5.12–13; cf. *Anab.* 5.3.4; Quintus Curtius 9.4.1; Diod. Sic. 17.96.1–3.

4. Arr. *Anab.* 6.14.2: the Oxydracae (Sydracae) supposedly claimed to have been free since the time of Dionysus's campaign; elsewhere it was noted that they cultivated the vine, fought to the music of drums and cymbals, and wore dappled costume (Eratosthenes, in Strabo 15.1.8, C 687–88; cf. Arr. *Ind.* 5.9), and they may themselves have claimed descent from Dionysus on the basis of these Dionysiac features of their culture. Quintus Curtius 8.10.1, *Metz Epit.* 34: on his entry to India Alexander was met by "the kinglets of the neighbouring tribes" who declared that he was the third son of Zeus (after Dionysus and Herakles) to have come to them. Quintus Curtius 9.8.5: the Sambastai or Sambagrai, a people apparently of the southern Punjab unknown to Arrian (Goukowsky, *Alexandre et Dionysos,* 44), feared Alexander as a "second Dionysus, a name well-known among those peoples" (cf. Diod. Sic. 17.102.4, who speaks of their granting him "heroic honours").

5. Bosworth, *Commentary,* 2:200. For doubt, see Karttunen, *India in Early Greek Literature,* 56 n. 403. Bosworth himself notes that Akouphis's speech (n. 2 above) is probably based on what Nearchus believed, not native tradition.

6. So Goukowsky, *Alexandre et Dionysos,* 27–28 (Nysa; commoner has been the identification with Śiva-Rudra; the possibility of an identification is doubted by Bosworth, *Commentary,* 2:200), 38 (Oxydracae), 38–39 (Sibai). The popular Śiva-Dionysus equation is rejected by A. Dahlquist, *Megasthenes and Indian Religion* (Uppsala, 1962), 184–89, but carefully restated by Karttunen, *India in Early Greek Literature,* 214–19. Whether one can speak at all at this date of Śiva, as opposed to an "antecedent of Śiva," is unclear: cf. Doniger O'Flaherty, *Śiva,* chap. 3 "The Vedic Antecedents of Śiva."

7. *FGrH* 688 F 45 (17). Clem. Al. *Protr.* 2.26.1 and other late sources (Karttunen, *India in Early Greek Literature,* 219–23; cf. n. 22 below on Philostratus) speak of Indian worship of the Sun (also known from the *Rigveda*), but a reality to which they are certainly alluding has not been identified (Karttunen, 219–23; cf. D. Lenfant, *Ctésias de Cnide,* Paris 2004, 303–5).

(dress, military use of drums and cymbals, and so on),[8] not that of cult. In an often quoted fragment a little later[9] Megasthenes distinguished between the "philosophers of the mountains" as "singers of Dionysus" and those of the plains who worshipped Herakles; but the proofs that even Megasthenes adduces for the former claim (those for the latter are not preserved) are that "wild vines grow among them alone, and ivy and laurel and myrtle and box and other evergreen plants, of which none otherwise grows beyond the Euphrates. Dionysiac too is dressing in muslin and wearing headdresses and using perfume and wearing brightly dyed clothes and the kings being accompanied by music of gongs and drums when they go out."[10] So to ask in each case what native god Herakles or Dionysus is standing in for may be misguided; they are being superimposed, one might say, on landscapes and customs, not on gods. It has been thought that some stories about Herakles (that he released Prometheus from a Caucasus located in the Hindu Kush; that "not even Herakles" could capture Mount Aornos) are based on (unattested) local legends about other heroes.[11] But motif transfer from one figure to another, if it occurred, is not the same as identification of the two.

A few texts may attach more specifically to Indian religious traditions. Quintus Curtius states that a statue of Herakles was carried in front of the infantry of Poros (8.14.11); if the notice is reliable, this statue must in fact have represented a native god, who is generally taken to be Krishna.[12] Megasthenes, within an account of Dionysus that presents him in Greek terms as a culture bringer,[13] states that he taught the natives "to worship other gods and particularly himself, using cymbals and drums," and "to wear their hair long for the god."[14] So here for once it seems that Megasthenes has a particular cult in view. Megasthenes

8. See especially Arr. *Anab.* 5.1.5–6, 6.3.5; *Ind.* 5.9, 12–13 (cf. *Anab.* 5.3.4); Megasthenes *FGrH* 715 F 12 = Arr. *Ind.* 7.9. Cf. Megasthenes F 32 = Strabo 15.1.55, C 710 (which describes the "bacchic" exodos of the king to hunts, accompanied by armed women). An actual sanctuary of Dionysus on Mount Nysa is later described by Philostr. *VA* 2.8: how reliably? (cf. n. 22).

9. On the date of Megasthenes' *Indica,* see Kosmin, *Elephant Kings,* 261–72, restating the case for a date in the last years of the fourth century against Bosworth's 319/8.

10. *FGrH* 715 F 33 ap. Strabo 15.1.58, C 711–12. For the last detail, cf. F 12 = Arr. *Ind.* 7.9; F 31 = Strabo 15.1.52, C 708; F 32 = Strabo 15.1.55, C 710. Dionysus is again associated with the mountains in Diod. 2.38.4 (in an incident derived from the *Mahābhārata* according to Dumezil, *La courtisane,* 48–49), and Herakles with (a region in) the plains by Megasthenes, *FGrH* 715 F 13. But note that in F 33 Megasthenes is speaking only of philosophers, and elsewhere (F 14 ap. Arr. *Ind.* 9.9) he made Dionysus ancestor of Sandrokottos, a king of all India.

11. Caucasus: so already (as a possibility) Eratosthenes ap. Arr. *Anab.* 5.3.2; Aornos: Arr. *Anab.* 4.28.1 with Bosworth's note (*Commentary,* 2:180–81), reviewing possibilities for an underlying local god.

12. Principally because of the stronger argument for that equation provided by Megasthenes in regard to the Herakles of the Sourasenoi (see below), a Herakles "of the plains" like that of Poros: Goukowsky, *Alexandre et Dionysos,* 38.

13. The argument of Dahlquist, *Megasthenes and Indian Religion,* 190–278, is flawed by neglect of this background; the Greek context of Megasthenes is stressed above all by A. Zambrini, *AnnPisa* 12 (1982): 71–149; and *AnnPisa* 15 (1985): 781–853.

14. *FGrH* 715 F 12 = Arr. *Ind.* 7.8. The same fragment continues (Arr. *Ind.* 8.1–3) with a genealogy of Indian kings Dionysus-Spatembas-Boudeuas-Kradeuas-Herakles, where the Greek gods may replace Indian originals (distinct from those represented by the Greek theonyms elsewhere in Megasthenes): see G. Ducoeur, "Interpretatio, relectures et confusions chez les auteurs gréco-romains: Le cas du

also reported at some length, supposedly on the basis of Indian informants, about a Herakles honoured by the Sourasenoi in the region of Methora and Kleisoboura. This Herakles dressed like the Theban; he had many sons but just one daughter, Pandaie, to whom he assigned a kingdom named Pandaie after her; having discovered the pearl on his travels he brought specimens from all the world to India, as an adornment for Pandaie;[15] close to death and unable to find a husband for Pandaie, still only seven, he magically brought her to puberty and slept with her himself, whence all girls of her race mature at that age.[16]

There is nothing Greek in this except the reference to Herakles' travels, whereas some details find parallels in Tamil traditions from the region of Pāṇḏya in the extreme southeast of India, far distant indeed from Methora, of which Megasthenes speaks, but variously associated with it;[17] the precise localisation by Megasthenes of Herakles in the region of Methora, a centre of the cult of Krishna, suggests that Herakles here again overlays that god.[18] But we are not dealing here with simple *interpretatio,* because this Herakles, though named and dressed like the Theban hero, is nonetheless distinct from him; Megasthenes reports that in the native account he was born among the Indians. Here the "quest for Herakles" led to assimilation, probably aided by physical resemblance, which was followed in turn by a separation that left the two figures still strangely similar.[19] One would like to know the position of the two parties in the conversation. Was the identification of Krishna as Herakles one proposed by Greeks and resisted by the Indian devotees of Krishna, or is the relation more complicated? A similar reclamation of Dionysus for India is found in Diodorus and Philostratus.[20]

Dionysos indien," in *Dieux des autres,* 143–58; on possible originals for Spatembas et al., P. Chantraine, *Arrien, L'Inde* (Paris, 1927), 33. In Diod. Sic. 2.38.3–6 the conception of Dionysus as a culture hero has become an account given by Indian λογιώτατοι, and Dionysus has become a mortal granted divine honours, not a god.

15. A. B. Bosworth, "Arrian, Megasthenes, and the Making of Myth," in J. A. López Férez, ed., *Mitos en la literatura griega helenística e imperial* (Madrid, 2002), 299–320, at 309, postulates an underlying myth of the pearl as a gift of Krishna.

16. *FGrH* 715 F 13; cf. Diod. 2.39.1–4.

17. See J. Filliozat, *JSav,* 1969, 76; and in the Budé of Pliny, *HN,* book 6, part 2 (1980), 156–58. The location of Pandaie's realm in the south is known to Polyaenus, *Strat.* 1.3.4. But Dumezil, *Courtisane,* 46–65, rejects the connection with Pāṇḏya for one with the Pāṇḏava of the *Mahābhārata.* Hesych. δ 2248 s.v. Dorsanes claims that Dorsanes is an Indian name of Herakles: cf. K. Latte's note ad loc. (Daśārha, a title of Krishna?); Dahlquist, *Megasthenes and Indian Religion,* 172–73. Richard Stoneman points out to me that father-daughter incest is a recurrent motif of Indian divine mythology: Doniger O'Flaherty, *Śiva,* 380, motif 39e.

18. See R. Audouin and P. Bernard, *RN* 16[6] (1974): 17–19, against the counterarguments of Dahlquist, *Megasthenes and Indian Religion,* 73–88. Audouin and Bernard note (19 n. 6) that the normal bearer of a club is Śiva; but Krishna's brother Balarāma, with whom he might have been confounded, bears a similar object (see, e.g., Audouin and Bernard, 9, fig. 1).

19. On this process, see p. 38 above.

20. Diod. Sic. 3.63.2–5 (with explanation of his titles καταπώγων and Ληναῖος)—not included as a fragment of Megasthenes by Jacoby; Philostr. *VA* 2.9, where two locally differentiated Indian versions are offered: that of those living "around the Caucasus and river Kophen" (i.e., further west) makes him (mysteriously) an Assyrian acquainted with the "things" (rites?) of the Theban; those living between the Indus and Hydraotes (i.e., further east) made him a son of Ganges from whom apparently the Theban stole his rites.

To conclude this section, Alexander and his followers saw evidences of Dionysus's and Herakles' passage on all sides, but it is not clear to what extent they detected them in the guise of particular Indian gods. Megasthenes took that step, but in the case of Dionysus continued to be chiefly influenced by the Dionysiac traces that he observed in Indian culture; in relation to Herakles he had to allow that the Indian Herakles differed from the Greek. We do not know when, or exactly why, the city known to Ptolemy as "Nagara, also called Dionysopolis" (Nagarahāra) received its Greek second name (perhaps not until the later period of Greek presence in India).[21] Neither now nor later do we encounter anything resembling a close engagement with the complex polytheism of India; it is notable, for instance, how lacking in detail in this regard are Philostratus's two books recounting the supposed travels of Apollonius in India.[22]

The story is interrupted in 305/4 or thereabouts when Seleucus cedes India to Chandragupta. It resumes when Greek kings of Bactria reinvade northwest India early in the second century and establish a control which lasts for almost 150 years. But almost the only relevant evidence to survive from this period is that of coins, often bilingual in Greek and an Indian language, some monolingual in one or the other.[23] For one brief period—the reigns of the closely associated kings, perhaps brothers, Agathocles and Pantaleon, ca. 190–180—what are manifestly Indian gods, some of them clearly identifiable by attributes, appear on some issues.[24] These remarkable coins break with Greek tradition in various ways: they are square, in the Indian manner, struck to the Indian standard, and (for the first time) bilingual. The

21. Ptol. *Geog.* 7.1.43. There is to my knowledge no special reason to associate the Dionysiac bronze coins of Agathocles and Pantaleon (below, n. 26) with this town.

22. A temple at Taxila is described in some detail, Philostr. *VA* 2.20.2, but no god is named (on the excavated temples at Taxila, some showing strong elements of Greek style, see S. R. Dar, *Taxila and the Western World,* Lahore 1984, 45–71; Cohen, *Hellenistic Settlements in the East,* 327). Rites are described in *VA* 3.17.1–2, but not the god they honoured. Philostr. *VA* 2.8 treats a shrine of Dionysus on Mount Nysa, established by the god himself; a concern with the Dionysus of Nysa was given by Greek tradition (so too with Aornos, 2.10), and one wonders how much autopsy has gone into the account; there is no trace in earlier sources of the vines Philostratus describes on Nysa. Interestingly, the statue is said to resemble "an Indian ephebe," but this might be an invention in accord with probability. Herakles and Dionysus, treated as a pair of invaders, remain prominent in Philostratus (*VA* 2.33.2, 3.13). The only gods Philostratus adds to the Greek canon of Indian gods are Aphrodite (*VA* 3.3: bicoloured women are considered sacred to Aphrodite) and Helios (2.24: shrine in Taxila; cf. 2.32.2; 2.43: among the altars founded by Alexander at the Hyphasis, one to "Indos Helios"; 3.28.2: claim of an Indian king to be "the same as the sun"); he reinforces (cf. Strabo 15.1.69, C 718) the importance of rivers (*VA* 2.19.2, 2.21.1), and of the hero Ganges (3.20.2, 21). Purely fantastic are the supposed ancient images of Greek gods, Athena Polias, Apollo Delios, Dionysus Limnaios, the Amyclaian, along with images of native gods, among the Indian sages (*VA* 3.14.3). J. Charpentier, *The Indian Travels of Apollonius of Tyana* (Uppsala, 1934), 65–66, supposes an authentic travel record to underlie book 2, while book 3 is travellers' tales.

23. See Allouche-Le Page, *Art monétaire;* Bopearachchi, *Monnaies . . . indo-grecques;* Bopearachchi, *Indo-Greek, Indo-Scythian, and Indo-Parthian Coins in the Smithsonian Institution* (Washington, D.C., 1993); Hoover, *Coins of Baktria and Ancient India;* for a useful introduction, *Crossroads of Asia.*

24. See especially R. Audouin and P. Bernard, *RN* 16[6] (1974): 7–41, on six silver coins from Ai Khanum (cf. *Crossroads of Asia,* 61, no. 21; and 62); earlier Allouche-Le Page, *Art monétaire,* 20–21.

iconographic aspect of this experiment was not continued, and thereafter Greek gods appear, mostly in the same guises that they have three thousand miles to the west. What diversity of interpretation such images may have been open to we can only guess. There is nothing unusual about a Poseidon bearing a tripod,[25] but a tripod was also a familiar emblem of Śiva. The most interesting aspect of this material is perhaps a negative, the neglect of Dionysus by all kings after Agathocles and Pantaleon despite his central role in the Greek image of Indian religion.[26]

Alongside clearly Indian or clearly Greek images, a third class, for the present argument the most important, is that of Greek gods depicted in somewhat unfamiliar guise. On coins of Agathocles and Pantaleon a seated or standing Zeus is seen holding an image of Hecate.[27] On coins of Antialcidas, ca. 115–95, Zeus is often shown holding a novel mixed "trident-sceptre" and a small Nike, and is approached by an elephant; different types set Zeus, the Nike, and the elephant in varying relations.[28] On silver tetradrachms of Hermaeus, ca. 90–70, a throned Zeus appears in a Phrygian cap with rayed crown, making a gesture of blessing. This last image has been plausibly identified as a Zeus as Mithras,[29] but others defy precise interpretation. It is doubtful whether the humped ox seen on several coinages from Apollodotos onward represents the god Śiva, or merely a typical animal of India;[30] a similar

25. See, e.g., *Crossroads of Asia,* 85, no. 85, Poseidon with trident on a coin of Antimachus *theos,* ca. 185–170. "Antimachus cannot have been unaware that the trident was the emblem of Shiva," G. Woodcock, *The Greeks in India* (London, 1966), 86. For the later Kushan identification of Poseidon with Śiva master of triśūla, see Allouche-Le Page, *Art monétaire,* 104.

26. For their Dionysiac coins, see *Crossroads of Asia,* 82, nos. 79–80; Hoover, *Coins of Baktria and Ancient India,* 29–30, nos. 94–96; 33, nos. 103–4 (coins of Agathocles and Pantaleon). Of them Tarn, *Greeks in Bactria and India,* 158, writes: "As they are not bilingual, they were intended for Greeks, and Dionysus is simply Dionysus and has nothing to do with Siva." But a bilingual square bronze coin of Agathocles (Hoover, *Coins of Baktria and Ancient India,* 30, no. 98) has a Dionysiac panther on the obverse and an Indian "dancing girl" holding a flower on the reverse: Tarn sees here a contrast between Greek and indigenous cults, but Allouche-Le Page, *Art monétaire,* 134, suggests that a "Yakṣiṇi champêtre," a nature spirit, has here been assimilated to a maenad. Tarn also suggests, without strong reason, that the coins were issued from or for Nagarahāra/Dionysopolis near Jalalabad.

27. Hoover, *Coins of Baktria and Ancient India,* 26, nos. 81–82; 32–33, nos. 100–102; Allouche-Le Page, *Art monétaire,* 115; on p. 116 Allouche-Le Page mentions A. Foucher's suggestion that Hecate stands for Čista, a local goddess of successful journeys; but Tarn, *Greeks in Bactria and India,* 157–58, 460–61, took the reference to be merely topographic, as do apparently R. Audouin and P. Bernard, *RN* 16[6] (1974): 31 n. 2.

28. Hoover, *Coins of Baktria and Ancient India,* 90–93, nos. 247–59 (some of these bilingual). On the trident-sceptre, see *Crossroads of Asia,* 76.

29. Hoover, *Coins of Baktria and Ancient India,* 105–6, nos. 289–96 (*Crossroads of Asia,* 61, no. 24; 75, no. 57). So too on coins of Amyntas; he already acquired rays on coins of Heliocles I (e.g., Hoover, *Coins of Baktria and Ancient India,* 62, nos. 169–72): cf Boyce and Grenet, *Zoroastrianism,* 3:163.

30. R. Audouin and P. Bernard, *RN* 16[6] (1974): 38 n. 1, against, e.g., Tarn, *Greeks in Bactria and India,* 213; or Allouche-Le Page, *Art monétaire,* 135–36. See Audouin and Bernard, 38 n. 2, for the wheel (a Buddhist symbol?) on a rare bronze of Menander.

uncertainty attaches to elephants.[31] The religious culture of Greek India was inevitably a mixed one. A high-ranking Greek diplomat serving Antialcidas, Heliodoros son of Dion of Taxila, made a dedication in Prākrit to Vāsudeva and declared a religious allegiance by describing himself as a *bhāgavata.* (This is to say nothing of the Indian tradition that the great king Menander was himself a convert to Buddhism.[32]) A bilingual dedication by another official from roughly the same period is addressed to an enigmatic "Chaosei" in Greek, to "Boa" (taken to be the Indian Bhava) in Gandhari Prakrit; some kind of equivalence is being sought, even if "Chaosei" defies interpretation for the moment.[33] Even after the fall of the Greek dynasties the forms, at least, of the Greek gods lived on on coins and in other contexts; a jumping-weight of the second century A.D. (?) depicts an unmistakable fine Herakles.[34] But, without texts, the detailed processes of convergence, adaptation, and eventually divergence can be inferred but not described.

31. Tarn, *Greeks in Bactria and India,* 212–13, identified the figure described (in Kharosthi) as "the god of the city of Kapisi" on (as is now supposed posthumous) bronzes of Eucratidas I as Zeus; the coins also show the forepart of an elephant and a mountain, and Tarn therefore took Zeus to be identified with an elephant-god of Mount Pilusāra attested by a Chinese traveller of the sixth century A.D. (cf. Allouche-Le Page, *Art monétaire,* 130, who specifies that the elephant would have been a vāhana of Indra). But the supposed Zeus is in fact a city goddess: A. K. Narain, *The Indo-Greeks* (Delhi, 1980), 62–64; Bopearachchi, *Monnaies . . . indo-grecques,* 216 series 24; Hoover, *Coins of Baktria and Ancient India,* 56, no. 160. One may wonder nonetheless about the juxtaposition of symbolically rendered elephant and mountain.

32. See R. Mairs, *The Hellenistic Far East* (Oakland, Calif., 2014), 117–28; for the Menander legend, Tarn, *Greeks in Bactria and India,* 265–69.

33. See G. Rougemont, *Corpus inscriptionum iranicarum* II.I, *Inscriptions grecques d' Iran et d' Asie centrale* (London, 2012), 268, no. 88bis.

34. P. Bernard and C. P. Jullien, *Studia Iranica* 11 (1982): 33–47; coins: *Crossroads of Asia,* 77–88. Contrast N. Sims-Williams and J. Cribb, *Silk Road Art and Archaeology* 4 (1996): 110–11, for the Kushan king Kanishka's policy (2nd c. A.D.) of replacing Greek in edicts and Greek theonyms on coins.

APPENDIX D

Some Non-Greek Theonyms in Anatolia

I list the non-Greek theonyms known to me from Greek texts (other than familiar cases such as Agdistis,[1] Anaitis, Ma, Men, Sabazios) under the places where they are attested, in the order of places adopted in *SEG*. The proper ending of the theonym is often unclear. My list is doubtless not complete, and it should be remembered that the place where a god is attested is not certain to be that god's place of origin.[2] In a tiny number of cases an original in an epichoric language can be identified.[3] Included below are the figures who occur, sometimes or always, independently, not attached as an epithet to another god. But it is very likely that an indeterminable number of words attested only as epithets are by origin, or derive from, theonyms: this is doubtless the case with, for instance, some of the Zeuses Hyllos, Kannokos, Komyros, Oaloinos, Osogollis, Spaloxios of Caria (for all these, see Schwabl, "Epiklesen"; on Osogollis, also pp. 94–95), the Zeuses Driktes, Glaukas, Timaios of Lydia (Paz de Hoz, *Die lydischen Kulte*, 49), the Zeuses Aseis and Dagoustes of Phrygia (see Schwabl, "Epiklesen"), Zeus Kotanes of Pisidia (P. Weiß, *AMS* 6, 1992: 151), the Mothers Atimis, Hipta,[4] Phileis of Lydia (Paz de Hoz, *Die lydischen Kulte*, 31–32), Zeus Sarsos from Amastris (*IGRR* 3.1433; Zgusta, *Ortsnamen*, ¶ 1175), Zeus Kodopas from the Corycian cave in Cilicia (*SEG* LVIII 1656).

1. On which, see Drew-Bear and Naour, "Divinités de Phrygie," 2020–21 n. 436.
2. For the warning that a god attested, e.g., in Caria is not necessarily Carian by origin, see *RÉA* 117 (2015): 392.
3. See under Kakasbos (Caria), Arkesimas, and Trosobios (Lycia), and the following note.
4. Cf. p. 80n12 for the plausible Luvian origin of this name.

CARIA

Gerga (?), at Gergakome: W. Held, *Gergakome* (Würzburg, 2008), 146–48. For a different view, see S. Szidat, *IstMitt* 63 (2013): 379–420, who associates the name with the Gorgon.

Kanebos: *SEG* XLVII 1599 = *IStrat* III 1419; *BÉ* 1944, no. 170a; P. Debord and E. Varinlioğlu, *RÉA* 107 (2005): 630 (*SEG* LV 1113B.11).

Sinuri: L. Robert, *Le sanctuaire de Sinuri près de Mylasa* (Paris, 1945).

Toubassis, at Pidasa: display of a document in his sanctuary, *RÉA* 117 (2015): p. 379, line 15 (302 B.C.).

IONIA

Baite: ὀφθαλμοὺς ὑπὸ θεῶν εὐχαριστῶ τῇ Βαιτῃ, *SEG* LXI 969 (where is added "Baite also occurs in an inscription in Bay 8"), among graffiti in thanks for healing from the basilica in Smyrna (1st/2nd c. A.D.).

LYDIA

Bruzi: dedications θεᾷ ἁγνῇ Βρυζι Ἀδυτηνῇ, Paz de Hoz, *Die lydischen Kulte*, 267–68, nos. 55.1–4.

Papias: οἱ ἐγ Λασνέδδων Παπίαι εὐχήν, *TAM* V.ii.1321; Malay, *Researches in Lydia*, nos. 37–43; cf. s.v. Phrygia.

Tyrimnos/Apollo Tyrimnos: Paz de Hoz, *Die lydischen Kulte*, 280–86.

BITHYNIA

Pr(e)ietos: dedications, *TAM* IV.1.74–78 (Nicomedia); in the form Priettos, *SEG* XXXVI 1155, with a rite (κατασπείσας τὰ Πριέτηα). Lucian, *Salt.* 21 probably speaks of him (but the MSS give Priapos) as a Bithynian δαίμων πολεμιστής; on the Bithynian month Prietios, see L. Robert, *ArchEph*, 1979, 231–36 = *OMS*, 7:789–94.

Proustene (?): dedication of (a statue of) Proustene (or Groustene), *IPrusa* 1022; but P/Groustene might be a toponymic epithet agreeing with an unexpressed theonym (so T. Corsten ad loc. in his note on *IPrusa* 1022).

Tataulas: a statue of a naked Apollo with cithara dedicated Ταταυλᾳ, *IPrusa* 42; for a village of Tataulenoi, see *IPrusa* 43.

Xyras: θεῷ Ξυρεῳ, *IHadrian* 19; θεῶι Ξυραι, *IHadrian* 20; possibly based on a toponymic epithet, according to Zgusta, *Ortsnamen*, ¶ 908.

PONTOS

Asbames: Θεῷ ἀνεικήτῳ Ἀσβαμει: Marek, *Stadt, Ära und Territorium*, 165, no. 27 (Amastris); cf. the Zeus Asbameios of Tyana in Cappadocia. "A Persian god associated with horses" according to Mitchell, "Iranian Names," 167–68, with references.

GALATIA

Olybris: Δὶ Ὀλυβρι κυρίῳ Χίλω<ν> ἱερεύς, Robert, *Hellenica* IX, 67, no. 1 (Ankyra). Cf. s.v. Cilicia.

PHRYGIA

Lola: Zgusta, *Ortsnamen,* ¶ 717, considers this possibility for ΛΟΛΑ in *MAMA* 5, R 18 (Nakoleia): Ἀπολλώνιος υἱὸς Συνέσιδος Κοκκειανοῦ δοῦλος Παπα καὶ Ἡρακλέα ΛΟΛΑ εὐχήν.[5]

Papas, Papias, sometimes expanded to Pap(i)as Zeus (Soter): Drew-Bear and Naour, "Divinités de Phrygie," 2018–22.

PISIDIA (AND NORTHERN LYCIA)

Alandros: a dedication to Theos Alandros, near Balboura, *SEG* XXXIII 1173 (Delemen, *Rider-Gods,* no. 391; J. J. Coulton, *The Balboura Survey,* London 2012, 2:411, IDr 4).

Kakasbos and Maseis: Delemen, *Rider-Gods,* 5–38. Kakasbos has recently acquired an alternative name/epithet Kursas near Elmalı: *SEG* LIII 1653–68; near Termessos he also now has the epithet Astron, *SEG* LVII 1483: cf. T. Corsten, *Anatolian Archaeology* 12 (2006): 25 (a depiction of Heracles, Kakasbos's near double in this region, with a star). The original of Kakasbos is apparently attested in Lycian XaXakba: see Frei, "Götterkulte Lykiens," 1808; and R. Lebrun, in *Syncrétismes religieux,* 186.

Mida: DEAE MIDAE on coins of the Roman colony at Cremna, H. von Aulock, *Münzen und Städte Pisidiens* II (Tübingen, 1979), nos. 1112–17, 1180, 1249–52, 1263–66, 1446–53; P. Weiß, *AMS* 6 (1992): 151.

Tisanis:]α Τρωΐλου Τισανει ἱερασα[. . ., *Burdur Museum,* no. 185.

Turiose: dedications, *ICentral Pisidia* 79–82.

Ximei: Μαισωλος Βλιδδεως ἱερεὺς Ξιμει ἐποίησεν (ἐ)ξέδριον κτλ: *Gephyra* 7 (2010): 42 (*SEG* LX 1453), between Pednelissos and Adada (cf. *BÉ* 2102, no. 434).[6]

5. But Drew-Bear and Naour, "Divinités de Phrygie," 2019 n. 430, seem to take it as a toponym.

6. The supposed Termessian goddess Bessis should now be rejected in favour of Artemis Kelbes(s)is: *SEG* LIII 1616.

PAMPHYLIA

Klea: altars inscribed Κλεας, *ISide* 10 and 18.

LYCIA

Arkesimas: Greek rendering of Arββazuma in the Xanthos trilingual: p. 45 above.

Eleuthera: Frei, "Götterkulte Lykiens," 1787–91 (cf. p. 89).

Ouarathos (?): a priestess of *theos* Ouarathos at Mnara northwest of Phaselis, *SEG* LI 1829; but cf. 1828, a dedication to Men Ouarathos (no provenance); cf G. Petzl, *EpAnat* 33 (2001): 51–53.

Somendes: μεγίστῳ ἐπηκόῳ θεῷ Σομενδε̣[ῖ] Μ(ᾶρκος) · Αὐρή(λιος) · Ἡλιόδωρος Μακεδόνος εὐχαριστήριον, *IArykanda* 82 (3rd c. A.D.).

Tobaloas: θεῷ ἐπηκόῳ Τοβαλόᾳ Ζωσίμη Ἀπολλωνίου εὐχήν, *TAM* II.758, from between Arneai and Arykanda.

Trosobios: one of the Lycian "hard" gods (σκιροί or σκιρροί: Plut. *De def. or.* 21, 421D, with important variants in the app. crit.: cf. p. 89), doubtless a Hellenised form of Trzzubi known from a text in Lycian from Limyra, *TAM* I.111; cf. G. Neumann, in *Florilegium Anatolicum: Mélanges . . . Laroche* (Paris, 1979), 261; Frei, "Götterkulte Lykiens," 1836.

LYCAONIA

Yoe: On Υοη in Υοη Ὀρονδίω εὐχήν, MAMA I 11 (Laodikea Katakekaumene), see Zgusta, *Ortsnamen*, ¶ 985.

ISAURIA

Stallos: p. 79 above.

CILICIA

Kasakarbas (?): in *Westkilikische Inschriften*, Ayasofya 3, Λεωνίδης Τειμοθέου ἀρχιερεὺς ΚΑ̣Ϲ̣ΑΚΑΡΒΑΟΥΩΝΙ εὐχήν; Zgusta, *Ortsnamen*, ¶ 455–4, detects two gods, Κασακαρβας and Ουω(ς).

Olybris: dedications to Zeus/*theos* Olybris, *IAnazarbos* 44–47 (= Ehling et al., *Kulturbegegnung*, 232–33, nos. 20–23; 233, no. 24 and *SEG* XLIII 993 are new instances from the region); cf. *OGIS* 577, Rome, again linked with Anazarbos, and instances from Palestine, *SEG* XLVII 2061–62 (plus an unpublished instance from Caesarea Philippi mentioned ibid.), possibly dedicated by Cilicians. Olymbros and Adanos (eponym of the Cilician city Adana) are among seven children of Heaven and Earth in Steph. Byz. s.v. Adana: cf. Zgusta, *Ortsnamen*, ¶ 931. An Aemilius Valens (origin unknown) dedicated Γῇ Μετρ[ὶ] Ὀλυβρι θεᾷ δεσποίνᾳ in Armenia ca. 150–200: *SEG* XLII 1322. See too Galatia and Cappadocia.

Ouos: see Kasakarbas.

Perasia: p. 39n20.

CAPPADOCIA

Astarte: decree to be displayed in *pronaos* of temple of Astarte, Robert, *Noms indigènes,* 458–59, line 31.

Olybris: Δὶ Ὀλυβρε[ι], *AnatSt* 19 (1969): 27, no. 3.9 (= *IKomana* 99); cf. s.v. Cilicia.

COMMAGENE

Soumanas (?): θεῷ ἐπηκόῳ Σουμανας, *SEG* XXX 1660, Doliche, 2nd/3rd A.D. (on interpretations, see the note in *SEG*).

NO PROVENANCE

Dedications of rider reliefs to Ordas, Ourdas: *SEG* LIX 1918–19; Tybout in *SEG* LIX 1918 compares *SEG* LIV 1384 from Attaleia, a rider relief inscribed Τροκονδας Ουροδοβα, but Ουροδοβα might be an anthroponym; cf. too Delemen, *Rider-Gods,* no. 153, possibly to Ορδω (Attaleia).

APPENDIX E

Thasian Herakles

Herakles twice in early sources receives the epithet "Thasian," and in both cases it is problematic. Reporting on his researches into the history of figures called Herakles, Herodotus tells how he travelled to Tyre and there saw two shrines of the god/hero, one ancient, venerable, and magnificent, another where he bore the epithet "Thasian" (2.44.1–3). What meaning we can attach to a shrine of "Thasian Herakles" in a Phoenician city in the fifth century is hard to see. The postulate that he was so called, in Greek, by Greek traders residing in Tyre is rather desperate.[1] But it is unlikely that the Tyrians will have adopted such a Greek cult, under the Greek title. Possibly "Thasian Herakles" is Herodotus's *interpretatio* of a Phoenician name. But what name it might render we have no way of knowing.

The other instance of the epithet is a sacred law from Thasos itself (5th c. B.C.?) prescribing sacrificial norms in a cult of Thasian Herakles (*IG* XII Supp. 414; *LSS* 63). The problem here is the opposite: if it is obscure why Tyrians might have wished to dub a "Herakles" "Thasian," it is also unclear why the Thasians should have felt any need to do so.[2] A specific problem relating to that text is to know whether the epithet distinguished a particular sanctuary of Herakles on the island from others, not carrying that epithet, or whether the hero was Thasian throughout Thasos. The sacred law was found in the so-called passage of the *theoroi* in the agora, away from the excavated Herakleion, a Herakleion which is both the largest sanctuary on the island and the largest of Herakles known anywhere. Some therefore suppose that the law attests a

1. J. Pouilloux, *Recherches sur l'histoire et les cultes de Thasos* (Paris, 1954), 1:54; J. Teixidor, *RHR* 200 (1983): 247; *Guide de Thasos*², 177.

2. Cf., however, *IG* XII.8.361, a *horos* of Zeus Agoraios Thasios from Thasos: a dull conclusion might just be that Thasian epithet use was idiosyncratic. Herakles has a few other toponymic epithets: Rheginos (*LSAG*² 244, 248 no. 11), Bouraikos (Paus. 7.25.10); Mekisteus/Makistios (Lycoph. *Alex.* 652 with Σ, Strabo 8.2.21, C 348).

second sanctuary in the agora region.[3] But several similar sacred laws were found in the same region, and it looks as if this public space was used to publish laws regulating cults of the island located elsewhere. If that conclusion is sound, Thasian Herakles at this date was not one Herakles among many, but the general name for the great god of the island. He never receives it again, however; on coins he becomes "Herakles Saviour" of the Thasians.[4]

In the passage just cited about Tyre, Herodotus goes on to explain how he travelled also to Thasos and saw there a shrine of Herakles set up by Phoenicians who sailed out in search of Europe and founded Thasos, five generations before the birth of the Greek Herakles. He then expresses approval of those Greeks who maintain two cults of Herakles, as both god and hero. From the juxtaposition some suppose that Herodotus, in speaking of a double cult of Herakles, is referring precisely to Thasos; and attempts have repeatedly been made to detect both forms of cult on Thasos, whether in the same sanctuary (which is not what Herodotus says), or at different sites. But neither as a reading of Herodotus nor of the archaeological evidence is that compelling.[5]

The relevant data from Thasos are rather that the cult of Herakles there was of unusual importance, and that it was believed by Herodotus, and very plausibly by the islanders themselves, to have been a Phoenician foundation. It can accordingly be supposed that "Thasian Herakles" is the title chosen by the islanders to distinguish their Herakles, who is by origin Phoenician Melqart, from the Herakles of the rest of the Greek world.[6] Nothing in the architecture or ritual confirms the postulate of Phoenician origin: the sacred law mentioned above forbids sacrifice of "goat and pig" to Thasian Herakles, a ban comprehensible in a Phoenician cult;[7] but just the same ban is found in the cults of the Charites and Peitho on the island, and so need not be given an external explanation. All the same, "Thasian Herakles" ought to indicate such a sense on the part of the islanders that their Herakles differed from some other. We can allow then that the expression, characterising the Thasian cult as in some way unusual, might derive from the islanders' belief, right or wrong, that it was a Phoenician foundation. But this theory, plausible for the Thasian instance of the title, does not help with Herodotus's claim of a Tyrian cult under that name: if Thasian Herakles is Melqart, what need is there to reimport him to Tyre?

3. So notably J. Pouilloux, "L'Héracles thasien," *RÉA* 76 (1974): 305–16, but already rejected by Van Berchem, "Hercule-Melqart," 91–92, whom I follow; rejected also by J. des Courtils and A. Pariente, "Problèmes topographiques et religieux à l'Hérakleion de Thasos," in R. Étienne and M. T. le Dinahet, *L'espace sacrificiel* (Paris, 1991), 67–73, but unnecessarily (see n. 5 below) returned to by them in *BCH* 120 (1996): 799–800. *Guide de Thasos*², 142–45, registers only one Herakleion.

4. See *Guide de Thasos*², 311.

5. As shown, e.g., by B. Birgquist, *Herakles on Thasos* (Uppsala, 1973), a conclusion not altered by her changed view of the sacred law, based on new evidence for the verb ἐνατεύειν, in R. Hägg and B. Alroth, *Greek Sacrificial Ritual, Olympian and Chthonian* (Stockholm, 2005), 61–70. The 1996 position of des Courtils and Pariente (n. 3 above) can be adjusted in the light of that new evidence.

6. So, e.g., Van Berchem, "Hercule-Melqart," 88–109, who believed the connection to be architecturally visible (a position no longer accepted); and, tentatively, *Guide de Thasos*², 230. On the islanders' belief, H. Seyrig, *Syria* 51 (1927): 187.

7. So Van Berchem, "Hercule-Melqart," 99; but see *IG* XII.8.358b; *IG* XII Supp. 394 (*LSCG* 114b; *LSS* 71). Bonnet, *Melqart*, 346–71, argues that nothing in the cult is demonstrably Phoenician, but stresses (371) the unusual civic prominence of Thasian Herakles as a possible trace of Melqart.

APPENDIX F

Some Epithets in Bilingual Texts

Three bilinguals from Cyprus show epithets of the god called RŠP (Resheph) in Phoenician, Apollo in the Cypriot syllabic rendering of Greek, crossing between the two languages in simple transliteration.[1] In two cases the epithets concerned are toponymic, where transliteration was inevitable, and are taken by Masson to be originally Greek; but in the third the Phoenician epithet MKL must be primary, since it is independently attested, and the Greek rendering Apollo Amyklos is derivative. A monolingual dedication (in Greek script) to Apollo Amyklaios from the same site warns us of the hazards involved in interpreting epithets: seen in isolation, it might encourage speculation about Spartan influence, but in context we can see that the obscure MKL/Amyklos has merely been assimilated to the famous Spartan god. Another instance of a Phoenician epithet transcribed into Greek comes in a trilingual of the second century B.C. from Sardinia (*KAI* 66), which gives the equivalences to Lord 'Eśhmun M'RḤ—Aesculapio Merre—Ἀσληπίῳ Μηρρη. The theonym is translated, the epithet, though it has a meaning—something like "Leader"—merely transcribed.

But another Cypriot bilingual treats the epithets differently (*KAI* 42, early 3rd c.?). The Greek text gives Athena Soteira Nike, "Athena Saviour, of Victory"; the Phoenician Anat ('NT) with an epithet, the reading and interpretation of which are disputed: M'Z ḤYM, "refuge of living creatures," according to some; 'Z ḤYM, "force of life/living creatures" for

1. On Phoenician-Greek bilinguals in general, see F. Briquel Chatonnet, *CRAI*, 2012, 619–38. The texts are O. Masson, *Les inscriptions chypriotes syllabiques*[2] (Paris, 1983), nos. 215 (*KAI* 41), 216, 220 (*KAI* 39; on MKL, see Masson, p. 248). Note that the linguistic origin of the epithet is a distinct issue from that of the language in which the dedication was first conceived, in each of these cases apparently Phoenician (Burkert, "Mein Gott," 5). Apollo Amyklaios: Masson, 235 n. 2.

others.[2] Here, for the divine name, an *interpretatio* also attested in other Cypriot contexts, based on similarity of function aided perhaps by a vague phonic similarity, is accepted; as for the epithet, simple transcription has clearly been avoided, though doubt about interpretation of the Phoenician makes it unclear how close a correspondence in sense was sought. Soteira is a standard epithet for Athena, while commentators are silent about parallels for (M)'Z ḤYM associated with Anat; possibly the phrase has been created ad hoc to render the Greek epithet Soteira. A bilingual from Malta[3] similarly combines *interpretatio* of a theonym with the attempt to create partial equivalence in the epithet: the Phoenician has "our lord Melqart, lord of Tyre," the Greek Herakles Archegetes. Archegetes, literally something like "First Leader" and applied to deities intimately associated with a place (cf. Athena Archegetis at Athens), is not a regular epithet of Herakles;[4] it seems to have been ingeniously reapplied to Herakles to calque "lord of Tyre," which is predictable for Melqart.

In this admittedly small sample of cases, we find two attempts to translate epithets (one apparently going in either direction), two Greek toponymic epithets transcribed into Phoenician, two nontoponymic Phoenician epithets merely transcribed into Greek; in these last cases it is hard to say whether the ground for this failure to translate lies in linguistic weakness or indifference on the part of the dedicator, or actual obscurity of the epithets.

In a quite different context, a Greek-Aramaic bilingual dedicated by a veteran on the upper Tigris in the third century A.D.(?) appears to add the Aramaic divine title MRLH', "Lord of the Gods," in simple transcription (Μαρη[α]λλη) to Zeus Olympios; the Aramaic is fragmentary (no theonym survives) but MRLH' can probably be read there too.[5]

A bilingual foundation plaque of Ptolemy IV (W. C. Hayes, *JEA* 34, 1948: 114–15)[6] is made in Greek to Aphrodite Ourania, in hieroglyphic to "Hathor who- is- (in-) Heaven" (Hayes's rendering of his reading *r nty (m) pt*).[7] Hayes declares the epithet to be unique for Hathor and thus probably a translation of Ourania, a conclusion accepted by P. M. Fraser (*JEA* 42, 1956: 97–98, who draws attention to Aelian's mention, *NA* 10.27 of a temple of Aphrodite Ourania at Kusae in Upper Egypt). But such foundation plaques are an Egyptian tradition and in other cases concern Egyptian gods;[8] "Master/Mistress of Heaven," little different

2. For the former, see *KAI;* for the latter, M. Bianco and C. Bonnet, "Sur les traces d' Athéna chez les Pheniciens," *Pallas* 100 (2016): 155–80, at 172, citing a forthcoming study by M. G. Amadasi Guzzo; for dedications to Anat from (certainly or probably) the temple of Athena at Idalion, see Bianco and Bonnet, 161, no. 2.2; 164, no. 2.4.

3. *KAI* 47 (3rd/2nd c. B.C.); cf. Bonnet, *Enfants de Cadmos,* 289–90.

4. Though now apparently found in the much later *SEG* LII 894.

5. C. S. Lightfoot and J. F. Healey, "A Roman Veteran on the Tigris," *EpigAnat* 17 (1991): 1–7. They write that MRLH' (on which, cf. p. 123n49) originated as a title of Sîn, the moon-god of Harran; it has apparently been reapplied here.

6. The hieroglyphic text of the twin (E. Breccia, *BSRAA* 26, 1931: 276, no. 1; *SB* 5 7782; *SEG* VIII 360) is apparently unpublished.

7. John Baines (whom I consulted) reads rather *r nty s't pt,* which gives the bizarre "daughter of the sky."

8. J. M. Weinstein, "Foundation Deposits in Ancient Egypt" (PhD diss., University of Pennyslvania, 1973).

from "who is in Heaven," is a stock epithet of Horus and Hathor:[9] it seems likely that the translation went in the other direction. If so, a Greek equivalent was readily available. But in another bilingual, "Koptite Osiris, Foremost of the Gold House," becomes "Sarapis, the Great God," in Greek (Vleeming, "Short Texts in Demotic," no. 250: "Ptolemaic"), while in another, "Pan Euodos," "of Fair Paths," renders "Koptite Min, the Great God" (Vleeming, no. 118: "probably Ptolemaic"). Here correspondence between the epithets in the two languages was not sought.[10]

9. See the index to Vleeming, "Short Texts in Demotic," 319, s.v. nb-p(.t).
10. Despite a possible Egyptian original for Pan's epithet Euodos: chapter 3, n. 120 above.

APPENDIX G

Divine and Human Names Juxtaposed

Seven cases have been identified of a god accompanied by an adjectival form of a human's name or what may be such. Only in two is the identity of the human not in doubt. Herakles Diomedonteios (ca. 200 B.C.) is the Herakles of a private cult on Cos founded by one Diomedon: this is therefore a certain instance of a cult named after its founder, the relation often postulated but never demonstrated for the Anatolian "Men of Artemidorus" type. The context in which an altar of Zeus Philippios is mentioned at Eresos on Lesbos in the late fourth century, one in which Alexander is of central importance, shows that the Philip must be the great Macedonian king, Alexander's father. Then follow the problematic cases. The museum at Bursa contains a dedication of unknown provenance to Apollo Leonteios. The best guess is that Leon is a founder like Diomedon—if indeed the epithet derives from a man's name, which is not inevitable. If the Aphrodite Stratonikis who possessed what seems to have been Smyrna's most important precinct was named after a mortal, this was Stratonike, the wife of Antiochus I Soter. But some doubt whether a cult newly established in honour of a queen could have acquired central civic prominence, and take Stratonikis instead as an ordinary cult epithet "of the victorious army"; even on this view, it is allowed that the epithet will have contained a secondary reference to Stratonike, who certainly received cult in the city. Morphologically the former view is easier, since Stratonikis relates straightforwardly to the name Stratonike but not to any attested epithet or compound noun. Apollo Pasparios, attested for Paros and Pergamum by Hesychius, raises a similar problem: some suppose that it evokes the great Pergamene benefactor Diodoros Pasparos, who ultimately received cultic honours; others, that a probably pre-Greek word of unknown meaning generated both the epithet and the benefactor's name.[1]

1. Diomedonteios: see p. 113 above. Philippios: *Chiron* 42 (2012): 206, line 5. Leonteios: *IPrusa* II 1017 (late Hell./early imp.). Stratonikis: see Rigsby, *Asylia*, 95–99. Pasparios (Hesych Π 1064, Πασπάριος· ὁ Ἀπόλλων· παρὰ Παρίοις καὶ Περγαμηνοῖς): contrast Habicht, *Gottmenschentum*,

The other two instances are attested only in the Roman period: Zeus Antigonios /Antigoneos/Antigona(i)os, known from numerous private dedications from two different sites in Lydia and Phrygia, and Zeus Seleukeios, known from a dedication by the village of Nisyra (followed by the Nymphai Karpodoteirai, "Givers of Produce") and a private dedication (preceded by an obscure Μητὴρ θεοῦ, "Mother of the god") from separate sites in Lydia.[2] The most obvious view of these epithets is that they originated like Zeus Philippios, as forms of honour to monarchs. But there exists an oblique argument for the view that Seleukeios was an old Macedonian epithet that preexisted the Seleucid monarchy, even if it acquired new resonance in Seleucid-controlled Lydia. There is, it is pointed out, an obscure festival Seleukeios in Alexandria, in order to explain which we need (since the Ptolemies would scarcely have instituted such a festival for their rivals) to postulate an older Macedonian cult from which it would have been named.[3] Both views of the epithet face difficulties. The primeval Macedonian Zeus Seleukeios is a deity who has to be postulated ad hoc; on the other hand, desperate measures are needed to find any other explanation for the Alexandrian festival. As for Zeus Antigonios, if he is named for a monarch, it would have to be Monophthalmos, since later members of the dynasty did not control Lydia; perhaps the cult was established by an ordinary Antigonos, like the Herakles Diomedonteios of Diomedon.[4] On any view the persistence of the cults under these names deep into the Roman period is striking. Perhaps they served for a time as badges of allegiance worn by settlers proud of their Macedonian origins. But by the time we meet them they are indistinguishable in clientele and functions from the great amorphous mass of Lydo-Phrygian cults of Zeus.

Where in these cases the human is a monarch, it is sometimes claimed that he or she is identified with the god. But identification is precisely what the adjectival form given to the human's name avoids. It leaves the relation between the god and the mortal completely undefined. Explicit verbal[5] identification, commonplace for Roman emperors and members of their families,[6] is less so in the Hellenistic world. The obvious but isolated example is the designation, sometimes perhaps casual and jocular but which could be formalised in cult, of a woman as Aphrodite; Arsinoe II, wife of Ptolemy Philadelphos, is the most conspicuous

---

14 n. 2; and Ohlemutz, *Kulte in Pergamon,* 5–6. The type is much commoner in Latin: see, e.g., Carter, *De deorum romanorum cognominibus,* 35; *ILS* 3080 with note.

2. Antigonios: *Manisa Museum,* nos. 70–74 (Paz de Hoz, *Die lydischen Kulte,* no. 61.9–13); and *SEG* XLVIII 1437 from İmrenler (Saittai); *SEG* LI 1806–7 and LII 1347–49 (some certainly, and perhaps all, from the region of Kotiaion). Seleukeios: *TAM* V.1.426 (Nisyra: Seleukios) and V.2.1306 (Hyrcanis: Seleukeos). On the interpretation, cf. Schuler, *Ländliche Siedlungen,* 40.

3. So J. Keil and A. v. Premerstein, *Bericht über eine zweite Reise in Lydien* (Vienna, 1911), 101–2, no. 200 (Σελεύκιος? = "der Glänzende"), in the first publication of *TAM* V.1.426, followed by Robert in publishing *TAM* V.2.1306 (*Hellenica* VI, 24–26, with the evidence for the Alexandrian festival in 25 n. 4; cf. *BÉ* 1951, no. 46). For the other view, see A. D. Nock, *Essays,* 1:157 (from *JHS* 48, 1928: 41–42), but conceding his explanation for the festival to be "rather desperate."

4. See the brief comments of C. Habicht, *Gnomon* 69 (1997): 270 (if a monarch, then Monophthalmos); and K. Rigsby, *AJP* 117 (1996): 169 (probably not a monarch).

5. Iconography is a different story.

6. Cf. A. D. Nock, *JHS* 48 (1928): 34 = *Essays,* 1:148; more fully, P. Riewald, *De imperatorum romanorum cum certis dis et comparatione et aequatione* (Diss. phil., Hall., 20.3, 1912).

case.[7] It is tempting to contrast the brashness of Arsinoe Aphrodite with the more evasive Aphrodite Stratonikis (if indeed the latter is named for Stratonike); but we lack contemporary evidence to show how strongly such distinctions were felt, and Aphrodite Laodike of Iasos is a Seleucid instance of identification.[8] An earlier case of simple juxtaposition of the name of the ruler with that of the god has been detected in a decree again of Iasos which mentions a (priesthood?) of Zeus Idrieus and of a figure only one letter of whose name remains, but who could be supplemented as Hera Ada (Ada being the wife of Idrieus). But an earlier view that this is Zeus of Idrias, a region of Caria (Hdt. 5.118.1), has been strongly revived.[9] A clear case from the other end of the Greek world is the altar from Syracuse (small, but probably implying a larger public model) inscribed "Of Zeus Soter Hieron" (Διὸς Σωτῆρος Ἱέρωνος); this will relate to Hieron II, tyrant in the third century.[10]

Another modality is for the god to be given an epithet particularly associated with a mortal. Zeus Nikator is now known from Arykanda in Lycia, Tymandos in northern Pisidia, and two sites in Pamphylia.[11] In contrast to the problematic Zeus Seleukeios there is no obstacle here to the view that Seleukos I was evoked by his defining epithet Nikator even when not named;[12] all the places in question were under Seleucid control before the treaty of Apameia, and three of the documents relate to priesthoods, that is, to an organised cult. In the Seleucid heartland at Seleucia in Pieria there was in fact a cult of Seleukos Zeus Nikator (i.e., Seleukos I) and Antiochos Apollo Soter: two further instances, though posthumous, of simple juxtaposition of god and mortal. The preference for Zeus Nikator elsewhere might have been designed, like Aphrodite Stratonikis, to evoke an association while avoiding an outright identification. The evidence for these Anatolian cults of Nikator postdates the period of Seleucid control, sometimes by centuries. Perhaps the openness of the epithet, which could be understood simply as "Conqueror," helped it to outlast the dynasty.

What may be a parallel case is the epithet Eumenes, "Well Disposed," given to Zeus in a private foundation from Philadelpheia and as a second epithet, with Larasios, on two

7. See Posidippus XII.7, in Gow-Page *HE*, with the note; for other assimilations by the Ptolemies, see W. F. Otto and H. Bengtson, *Zur Geschichte des Niederganges des Ptolemäerreiches* (Munich, 1938), 74 n. 1; Fraser, *Ptolemaic Alexandria*, 1:240; for Aphrodite Laodike (wife of Antiochus III), *IIasos* 4.80. For the Aphroditai Leaina, Lamia, and Phila, see Parker, *Athenian Religion*, 258 n. 7, 259 n. 14. Note U. von Wilamowitz-Moellendorff, *Hellenistische Dichtung* (Berlin, 1924), 1:74, on such expressions: they are not true identifications, but merely indicate "the sphere in which the godlike activity of the human thus elevated takes place."

8. See previous note. Later we encounter Antiochos (VI) Epiphanes Dionysus, so named while alive on coins from some parts of the empire (Houghton et al., *Seleucid Coins*, II.1:318), a title taken on later by Antiochos XII.

9. See R. Fabiani, *SCO* 61.2 (2015): 174–81, on *IIasos* 52.8.

10. Habicht, *Gottmenschentum*, 260–62. R. Fabiani, *SCO* 61.2 (2015): 178 n. 51, reports the doubts (not compelling as so far published) of A. Dimartino of the text's authenticity.

11. *SEG* XLVI 1704 (late 1st c. B.C.); *MAMA* IV 426 (cf. Cohen, *Hellenistic Settlements in Europe, the Islands, and Asia Minor*, 286; 2nd c. A.D.?); *ISide* 90 (2nd/1st c.); Swoboda, *Denkmäler*, no. 109.I.5 (cf. *SEG* XLII 1228; late imperial).

12. Cf. J. Nollé's note on *ISide* 90; he stresses the vocalisation Nikator, not Niketor. Seleucia in Pieria: *SEG* XXXV A 11–13. There was also a cult of "King Seleukos Nikator" at Dura-Europos in the second century A.D. (*PDura* 25.19–20).

inscriptions of Tralles; Hellenistic coins of Tralles similarly bear the legend Διὸς Λαρασίου καὶ Διὸς Εὐμένου. Both places were under the sway of Eumenes II of Pergamum, and one of the inscriptions containing the epithet speaks of musical competitions in his honour and in his lifetime. On the other hand, the hope that Zeus and other gods would be well disposed to mortals long antedated the accession of Eumenes; we do not need the king to understand the epithet.[13] If it was in fact introduced to honour the king, it did so obliquely, at the level of a secondary connotation.

13. *TAM* V.3.1539.6–7 (2nd/1st c. B.C.): *ITralles* 23.29 (reign of Eumenes II) and 8.1–3 (imperial); L. Robert, *RPhil* 8 (1934): 287–91 (*OMS*, 2:1186–90), with references for εὐμένης as a divine epithet and for the coins; for Zeus Eumenes, see now *SEG* XLIII 630.8. Robert, following Imhoof-Blumer, raises the possibility that the coins are too early to be associated with Eumenes (so too F. Daubner, *EpAnat* 41, 2008: 178), but Thonemann, *Maeander Valley*, 40 n. 100, leaves them in his reign.

APPENDIX H

Exported Gods

The Cults of Hellenistic Colonies

THE PROBLEM

This book has been about cultural interaction in religion, predominantly in the Hellenistic period. Much of the emphasis has lain on how non-Greek gods were treated in the Greek language, the extent to which they were assimilated to known Greek gods through *interpretatio,* and the implications of that partial assimilation. But a phenomenon of obvious interest to the broader subject is the creation of new Greek settlements away from old Greece in and after the time of Alexander. Arrian describes Alexander's arrival at the site of the future Alexandria as follows:

> It struck him that the position was admirable for founding a city and that it would prosper. A longing for the work therefore seized him; he himself marked out where the city"s marketplace was to be built, how many temples there were to be and the gods, some Greek, and Isis the Egyptian, for whom they were to be erected, and where the wall was to be built around it. With this in view he offered sacrifice, and the sacrifice proved favourable.[1]

Arrian goes on to report a further detail which he finds "not incredible": lacking any other material with which to mark out the city's circuit, Alexander used barley helpfully offered by his troops, and the seer Aristander of Telmessus predicted that the city would flourish particularly with the fruits of the earth.[2] (According to Curtius Rufus this use of barley was, not an improvisation, but the Macedonian custom.[3])

1. Arr. *Anab.* 3.1.5 trans. Brunt.

2. Arr. *Anab.* 3.2.1–2.

3. Curt. Ruf. 4.8.6; for elaborations found in other sources (birds devour the barley) see A. B. Bosworth's commentary on Arr. *Anab.* 3.2.1.

Alexander's action as described by Arrian is foundation in the ideal mode: it recalls Homer's account of how Nausithoos brought the Phaeacians to Scheria, "and drew a wall around the city, and built houses, and made temples for the gods, and divided out fields."[4] Such accounts bring out the special interest of city foundations for the student of cults. They are moments of incision: with the new beginning, religion ceases to be a matter of tradition; choices have to be made, amid a myriad possibilities, as to "how many temples there are to be and the gods . . . for whom they are to be erected." They are moments, we can suppose, at which cults of fading allure can be quietly left aside in favour of the most popular gods of the moment. The Hellenistic age was the second great age of new foundations, most of them outside the confines of mainland Greece: the first volume of G. M. Cohen's indispensable compendium of "Hellenistic settlements," which listed "over 180"[5] in "Europe, the islands and Asia Minor," was followed by those for "Syria, the Red Sea Basin, and North Africa," with 135 entries, and "the East from Armenia and Mesopotamia to Bactria and India," with 198. The new foundations ought to provide the best of all laboratories for the study of the state and trends of early postclassical Greek religion. It is remarkable how little attention has been paid to the religious aspect of these foundations, whether by those interested in colonisation or in cults.

Different Types of Settlement

There are, however, reasons for this neglect. Very varied phenomena are covered by the usefully neutral term "settlement."[6] Many were not or were not originally cities but smaller and less formal settlements based on a nucleus of soldiers or ex-soldiers and their families (and ultimately descendants); it is convenient to term them *katoikiai*, even though the term is barely attested before the Roman period.[7] Dura-Europos on the Euphrates, the remains of which used so to impress the modern visitor, is the paradigm case of a city of the Roman period that may have begun its life as a modest *katoikia*.[8] Where a city was in fact founded, it was very seldom implanted on virgin soil. In cases where the evidence admits an opinion, Cohen in his first volume tentatively enumerates seventy-five "re-foundations" as against eighteen "new foundations"; in the second volume the proportion is fifty-one to four, in the third twenty-two to one. New cities almost always incorporated one or more smaller existing settlements (this was probably true even of Alexandria), or were renewals of destroyed

4. Hom. *Od.* 6.9–10.

5. Cohen, *Hellenistic Settlements in Europe, the Islands, and Asia Minor* (p. 1 for the figure quoted); Cohen, *Hellenistic Settlements in Syria, the Red Sea Basin, and North Africa*; and Cohen, *Hellenistic Settlements in the East*; hereafter *Hellenistic Settlements,* I, II, and III, respectively.

6. Cf. Tscherikower, *Städtegründungen,* 113–29, underlined by Briant, *Rois, tributs et paysans,* 229–30; cf. Mileta, "Charakter und Entwicklung der Neuen Poleis."

7. Related words such as *katoikoi* and *katoikountes* are common. See in particular Schuler, *Ländliche Siedlungen,* 34–41; Schuler, *ZPE* 128 (1998): 33–34; on the debate also Capdetrey, *Pouvoir Séleucide,* 159; Bencivenni, *Progetti di riforme,* 339–41, with references. F. Daubner argues, in "Seleukidische und Attalidische Gründungen in Westkleinasien—Datierung, Funktion und Status," in Daubner, *Militärsiedlungen,* 41–63, that the concept of military colony (on the Roman model) is inapplicable to the Greek cases.

8. Fraser, *Cities of Alexander,* 188, wonders whether even Alexander's Far Eastern settlements were closer to *katoikiai* than true poleis.

or derelict earlier cities;[9] an existing city might be renamed as an Antiocheia or Ptolemais or whatever after receiving an injection of new inhabitants or being restored or revivified in some other way. Louis Robert has warned severely against the view that new cities were ever created by the mere stroke of a pen;[10] but the kind of change or benefit that a monarch brought to a city that changed its name to honour him or his family must have varied greatly. Where an old settlement acquired a new, nondynastic name—as, for instance, Castabala, renamed Hierapolis, in flat Cilicia—one may wonder whether any initiative on the part of a monarch needs to be postulated. The question whether characteristic Greek institutions such as tribes were introduced in a given settlement is often unanswerable.[11]

Different Types of Settler

To variations in types of settlement correspond variations in settlers. The literary sources speak of Alexander's own foundations in the Far East as being peopled by a mixture of Macedonian veterans, Greek mercenary veterans, and (whether as volunteers, or more often under compulsion) natives of the region; the natives were perhaps attached to the city, not citizens of it.[12] The circumstances of those foundations, made in some haste while on the move with whatever materials came to hand, were doubtless somewhat special. For later foundations, we can suppose that Greeks and Macedonians normally provided the politically dominant element, on the basis, for instance, of the decrees in good Greek form proposed by individuals with Greek names and patronyms that survive from many of them.[13] Occasionally a little more precision is possible. Strabo reports that Antioch near Pisidia was settled by Magnesians from the Maeander; the same Magnesians "after making prayers and sacrifices" dispatched "men sufficient in number and outstanding in courage" when Antiochus I appealed for reinforcements for another Antioch (probably that on the Persian Gulf, not the Pisidian one again);[14] and plausible theories can sometime be offered from tribe

9. Tscherikower, *Städtegründungen*, 119–20. Alexandria: this was certainly the ancient belief, though the supposed original village "Rhakotis" has apparently been unmasked as a misunderstanding of an Egyptian term meaning "building site" (i.e., Alexandria itself under construction): see Cohen, *Hellenistic Settlements*, II, 355. Fraser, *Cities of Alexander*, 186 (my emphasis), speaks of a "*largely* unpopulated site." For the probability of Achaemenid-era settlement at Ai-Khanum, see Mairs, "Founder's Shrine," 108–9.

10. *Hellenica* XI-XII, 155–57, with powerful counterexamples. For the opposite view in relation to Phoenicia, see H. I. Macadam, *Topoi* 3 (1993): 332, 341–44; in relation to cities receiving Ptolemaic dynastic names, see E. Winter, in Daubner, *Militärsiedlungen*, 65; in relation to foundations of Antiochus IV, see Cohen, *Hellenistic Settlements*, I, 22; M. H. Sayar, *AMS* 34 (1999): 129.

11. On the importance of tribes, see Briant, *Rois, tributs et paysans*, 261; for those known, see N. F. Jones, *Public Organization in Ancient Greece* (Philadelphia, 1987), 348–84; and the appendices "Civic Institutions" to Cohen, *Hellenistic Settlements*, I, II, and III.

12. Briant, *Rois, tributs, paysans*, 253; Fraser, *Cities of Alexander*, 186–88.

13. But, as Robert, *Noms indigènes*, 487, showed from the example of Hanisa in Cappadocia, and as the decree of the Kiddiokometai (n. 101 below) has confirmed, the mere ability to draft a decree in good Greek form proves nothing about ethnicity.

14. Strabo 12.8.14, C 577; *OGIS* 233.12–20 (= Rigsby, *Asylia*, no. 111: on the attribution see Rigsby, 260): the latter text shows that the Antioch in question preexisted the Magnesian assistance and cannot have been originally populated thence.

names or from onomastics about the origins of particular settlers or groups of settlers, as Robert, for instance, detected Ionians and Macedonians in Laodicea on the Lykos in the southern extremes of Phrygia.[15] Ptolemy Soter is said to have summoned settlers from Argos, two further cities lost through a break in the stone, and Thessaly to his foundation of Ptolemais Hermiou in Upper Egypt, a city spoken of in another text as a "Nile-born sanctuary of the Greeks."[16]

But the number of foundations is such that non-Greeks must often have had a place, as in those of Alexander, even if not with full citizen rights. Pausanias says that Seleucus transferred population from Babylon to Seleucia on the Tigris.[17] We shall see below how "the locals," that is, Phrygians, were included in the citizen body when the *katoikia* of Toriaion became a polis in the second century; when the "settlers at Magnesia (ad Sipylum)" were given Smyrnaean citizenship under Seleucus II, the "Persians under Omanes" were included along with the rest of the garrison.[18] Where a new city was, as so often, an expansion of a nucleus of existing villages, the original villagers were still there. When a preexisting town took a Greek dynastic name, it is unclear what influx, if any, of Greek settlers needs to be postulated; even in cases where a significant intervention by the monarch occurred, this was not necessarily one that affected the composition of the population. Where new inhabitants were needed, they may often have been recruited from Hellenised subjects of the king in question rather than from Hellenes by origin.[19] As for settlements below polis level, though the men of *katoikiai* typically describe themselves as "Macedonians," a famous passage of Josephus quotes a letter of Antiochus III in which he orders the resettlement of two thousand Jewish families from Mesopotamia to Phrygia and Lydia as a check on disorders in those regions; Mysians too form *katoikia*-like settlements, or blend with Macedonians as

15. Robert, in Gagniers, *Laodicée,* 328–33. Antioch on the Maeander's supposed kinship ties with Samos may suggest a Samian component there: *IG* XII.6.1.6.19–20. For a possible role of Nagidos in Cilicia in providing settlers for an Antioch, see C. P. Jones and J. Russell, *Phoenix* 47 (1993): 297–304 (*SEG* XLIII 998). The late sources for the population of Antioch on the Orontes quoted by Cohen, *Hellenistic Settlements,* II, 81, do not inspire great trust, given the elaborate mythologisation of that foundation (T. S. Scheer, *Mythische Vorväter,* Munich 1993, 329–36); Strabo 16.2.5, C 750 gives no details about the settlers. But we can accept veterans of Seleucus and settlers from the nearby and now-destroyed Antigoneia. Fraser, *Ptolemaic Alexandria,* 1:62–66, regards the origins of the first Greek settlers of Alexandria as unknowable.

16. *SEG* XX 665 (cf. Fraser, *Ptolemaic Alexandria,* 2:146 n. 189; 2nd c. A.D.); *IGPhilae* II 166; on the population of Ptolemaic settlements, see Mueller, *Settlements of the Ptolemies,* 166–74.

17. Paus. 1.16.3; but the cuneiform evidence is now read as attesting only transfer of "Macedonians" (Cohen, *Hellenistic Settlements,* III, 133 n. 2), though this does not exclude transfer of locals also. Josephus, *AJ* 18.372 later speaks of Seleuceia as a city inhabited by "many Macedonians, a Greek majority, and not a few Syrians."

18. *IMagnSip* 1 (*OGIS* 229).104–6. Kuhrt and Sherwin-White, *From Samarkhand to Sardis,* 168–70, oppose the apartheid model of Hellenistic colonisation; but note the exclusion of non-Greeks earlier in the same decree, *IMagnSip* 1.50–51.

19. On the case, based on cults, for Hellenised Tyrians as settlers in Philadelpheia Rabbat Amman, see Cohen, *Hellenistic Settlements,* II, 268, who cautions that they were not necessarily brought there at the foundation.

"Mysomakedones." One of the three representatives of the *katoikia* of Toriaion early in the second century has a Gallic name.[20]

The Evidential Problem

The classic model of a city foundation, then—settlers of homogeneous origin introduced in a single act to an unoccupied site—can almost never have applied. The phenomena we are concerned with are extended in time (from Alexander to at least the mid-second century), and, to a dizzying extent, in space. Every permutation is possible.[21] One regularity is that, alas, the crucial early stages are almost never available for close inspection. The gloomy verdict of P.M. Fraser on Egyptian Alexandria, famous city though it became, is that "no significant conclusions regarding the original cults of the inhabitants can be drawn."[22] The most spectacular foundation programme was surely that by which Seleucus I planted four major cities in northern Syria in a short space of time. From one of them, Seleucia in Pieria, there survives part of a list of annual priesthoods and priestly offices from the reign of Seleucus IV (187–175); we have the beginnings of the list for two years, which present the offices in the following order (I omit the names of the actual holders of the office):

> Priests: of Zeus Olympios (and the Saviour Gods and Zeus Kasios) and Zeus Koryphaios; [of Apollo at Daphne]; of Apollo; of Seleukos Zeus Nikator and Antiochos Apollo Soter and Antiochos God and Seleukos Kallinikos and Seleukos Soter and Antiochos and Antiochus the Great; of king Seleukos; Sceptre-bearer; Thunder-bearers.[23]

This is a precious fragment, but even this postdates the foundation by more than a century. Much later still is our evidence for the priesthoods of Dura-Europos.[24] At Ptolemais

20. "Macedonians": Schuler, *Ländliche Siedlungen*, 307, list F. Josephus, *AJ* 12.147–53; for the debate on its authenticity, see J. Ma, *Antiochos III and the Cities of Western Asia Minor* (Oxford, 1999), 267. Mysians: *SEG* XL 1062. Mysomakedones: Pliny, *HN* 5.120; and see the close relations between a Mysian group and the military settlement at Apollonioucharax attested in the new letter of Eumenes II, *New Documents from Lydia*, no. 32, as reinterpreted by M. Wörrle, *Chiron* 39 (2009): 427 n. 76; and P. Thonemann, "Eumenes and Apollonioucharax," *Gephyra* 8 (2011): 19–30. Note too the Persians at Palaiomagnesia, n. 18 above (but on the "village of Kardakes," see Schuler, *Ländliche Siedlungen*, 149 n. 63). Daubner (in *Militärsiedlungen*, 52–54), however, argues for renewed Macedonian emigration to Asia Minor in the second century B.C. Toriaion: *SEG* XLVII 1745 (now *ISultan Dağı* I, 393) line 3 (Brennos); on this south Phrygian settlement, see p. 213 below.

21. Motives for foundation varied greatly too: on Seleucid (strategic) v. Attalid (economic) aims, see Thonemann, "Attalid State," 27. On Seleucid colonies, see Kosmin, *Elephant Kings*, 183–251.

22. Fraser, *Ptolemaic Alexandria*, 1:193.

23. *SEG* XXXV 1521. The items in round brackets are listed only for the second year, that in square brackets only for the first.

24. *PDura* 25.18–20, 180 A.D., priests of Zeus, Apollo, Forebears (Πρόγονοι), King Seleucus Nicator. *SEG* VIII 33 (Scythopolis in Palestine, 2nd c. B.C.?), as rather heavily restored, gives priesthoods of Zeus Olympios, Saviour Gods, Forebears of the king, the king, thus blending elements from Seleuceia and Doura rather strikingly. (Later at Scythopolis there are dedications to Zeus Akraios and Zeus Bak- : *SEG* XX 456–57; Cohen, *Hellenistic Settlements*, II, 298 n. 19; cf. *BÉ* 1964, no. 516). But even Scythopolis's status as a Hellenistic foundation, still less a Seleucid one, is conjectural. *SEG* VIII 96, Samaria Sebaste, has been still more heavily restored to give a similar list of priests; a reference to "Forebears" is trustworthy.

Hermiou in Upper Egypt, founded by Ptolemy Soter, we have third-century evidence for Dionysia and Hellenistic evidence for cults of Zeus, Isis, and a cluster of Egyptian gods, as well as Ptolemy Soter himself.[25] From Kalindoia in the Chalcidice in northern Greece there survives a dedication made in the late fourth century by "Agathanor son of Agathon to Apollo after serving as priest of Asclepius"; it lists those who served as priests (of Asclepius) "from the time when King Alexander gave the Macedonians Kalindoia and the places around Kalindoia, Thamiskia, Kamakaia, Tripoatis."[26] This is apparently the record of the creation of a Macedonian city at Kalindoia, and if so attests a cult certainly established at the foundation; the years were probably dated by priests of Asclepius as in other Macedonian cities (whence the need for the list). (We can note incidentally that the choice of Asclepius as the god to provide the eponymous priest was not a nod to Macedonian heritage, not at least to ancient Macedonian heritage, but something rather up-to-date.) But as unshakable near-contemporary evidence for a cult established at the time of foundation the Kalindoia text is unique. Nothing survives comparable to the law governing a Roman colonial foundation, the Lex Coloniae Genetivae from Urso in Spain.[27]

KATOIKIAI

Katoikia *to City—Two Cases, and the Role of "Founders"*

These no doubt are the reasons why this chapter in the history of Hellenistic religion has never been written: the phenomenon itself fragments under investigation, and, above all, the kind of documents that inspire curiosity are absent. In these circumstances, an investigation can only be tentative and inconclusive; and for that reason this attempt has been consigned to an appendix. But the issue is too interesting to be neglected without a serious investigation of problems and possibilities. And in fact not all is gloom. Two recently discovered texts provide strong side-lighting on the process of foundation. A letter from the 230s (?—certainly after 238) of Thraseas, Ptolemaic general of the region, to Arsinoe in the far west of Cilicia "is the first document to provide something more than a fragmentary picture of the organization of a Ptolemaic overseas settlement."[28] The crucial information comes from a decree passed by the nearby city of Nagidos subjoined by Thraseas to his

25. *OGIS* 51, 52, 103; *SEG* XX 665 (the mix recalls Arrian's remark, cited on p. 205, on Alexandria); for the cult of Soter himself, see Cohen, *Hellenistic Settlements,* II, 351 n. 1 (the attestations are of the Roman period, but the cult must be much older).

26. *SEG* XXXVI 626: see the references and discussion in Cohen, *Hellenistic Settlements,* I, 94–95. Asclepius priest as eponym: Hatzopoulos, *Macedonian Institutions,* 1:384.

27. M. Crawford, *Roman Statutes* (London, 1996), 1: no. 25, esp. secs. LXIV–LXVIII: cf. J. Scheid, "Aspects religieux de la municipalisation," in M. Dondi-Payre and M. T. Rapsaet-Charlier, *Cités, municipes, colonies* (Paris, 1999), 381–423, at 388–97; J. Rüpke, "Religion in the Lex Ursonensis," in C. Ando and Rüpke, eds., *Religion and Law in Classical and Christian Rome* (Stuttgart, 2006), 34–46; G. Woolf, "Roman Diaspora," 246–47.

28. Cohen, *Hellenistic Settlements,* I, 3; cf. 363–64, on *SEG* XXXIX 1426; some crucial improvements to the text by G. Petzl, *ZPE* 139 (2002): 83–88 (*SEG* LII 1462), are incorporated in Austin, *Hellenistic World,* no. 272; the main historical study is by C. P. Jones and C. Habicht, *Phoenix* 43 (1989): 317–46. On Nagidos, see C. P. Jones and J. Russell, *Phoenix* 47 (1993): 297–304.

letter. We learn that Arsinoe had been founded (ca. 270–260) by the then general of Cilicia, Aetos of Aspendos (father of the current general Thraseas), on territory that according to the Nagidians belonged to Nagidos but was under attack by "barbarians." Now Thraseas is keen to enhance his father's foundation and equip it with magistracies and laws of its own; and he has persuaded the Nagidians to yield (by the present decree) the "public land" to "the settlers" and "any others whom Thraseas may settle." Henceforth the Arsinoeis are to be *apoikoi* (i.e., colonists originating from a place) of the Nagideans, and, though establishing their own laws, are to enjoy equal political rights (*isopoliteia*) in Nagidos and be invited to the festivals of "the king and Arsinoe and Berenice" and of Concord (Homonoia); the Nagideans in turn are to attend sacrifices of the Arsinoeis to the Theoi Adelphoi. The key element in the decree is the concession by Nagidos, probably made very unwillingly, of "the public land" to Arsinoe; commentators infer that, during the interruption in Ptolemaic control of Cilicia caused by the Second Syrian War, the Nagidians had reasserted their claim to some of the land ("the "public land" of the decree) assigned by Aetos to Arsinoe, and are now being called back to order by his son.

A first point that emerges is the role of the *ktistes*, "founder." It had always been obvious that, however many cities kings founded in name, they could not have been involved in the practical details very often.[29] Here we find a general explicitly credited as having founded a settlement, admittedly in a document designed for his son's eyes, and his son energetically involved in enhancing the foundation. A later Ptolemaic general, Boethos, is spoken of as *ktistes* of three "cities" in Egypt in contemporary documents, in one case the title being clearly used to honour him, and two further such foundations by Ptolemaic officers are mentioned by Strabo.[30] The role of subordinate founders was clearly publicly acknowledged under the Ptolemies; it has been noted that the title of *ktistes* is never applied explicitly to any member of that dynasty.[31] But we hear in Pliny of an Arabian Antioch too founded by "Nicanor prefect of Mesopotamia"; this Seleucid officer may also be responsible for the description of Dura-Europos in Isidore of Charax as a "city of Nikanor, foundation of the Macedonians."[32] The discrepancy between sources that speak of Gerasa and Samareia as

29. Tarn, *Greeks in Bactria and India*, 6. *TAM* V.2.1187 gives credit for the synoecism of Apollinis in Lydia to one of the brothers of Eumenes II.

30. Boethos: see B. Kramer and H. Heinen, "Der Ktistes Boethos und die Einrichtung einer neuen Stadt I and II," *ArchPF* 43 (1997): 315–39, 340–63, esp. 350–63, on two Trier papyri of 132 B.C. (also Heinen in L. Mooren, ed., *Politics, Administration, and Society in the Hellenistic and Roman World*, Leuven 2000, 123–53); *OGIS* 111.9–10 (between 152 and 149); in brief Cohen, *Hellenistic Settlements*, II, 347–48. Strabo: 16.4.5, C 769; 16.4.7, C 770, confirmed by the hieroglyphic Pithom stele ¶ 23–24 (p. 70, 2.14 in the new edition and translation by Thiers, *Stèle de Pithom*), translated in Mueller, *Settlements of the Ptolemies*, 198.

31. Leschhorn, *Gründer der Stadt*, 223, who, however, is not convinced that the silence is significant: the Pithom stele ¶ 20–21 (Mueller, *Settlements of the Ptolemies*, 197; p. 65, 2.12 in Thiers, *Stèle de Pithom)* in fact seems to credit Ptolemy II with the foundation of an Arsinoe.

32. Pliny, *HN* 6. 117; Isidore *FGrH* 781 F 2 (1) (p. 779.26 Jacoby): see Cohen, *Hellenistic Settlements*, II, 157, with 163 n. 6. Cohen, *Hellenistic Settlements*, II, 347, mentions as other possible cases Dokimeion and Themisonium; if those cities were named for Dokimos and Themison when acting in royal service the fact is noteworthy, but has been doubted (see Cohen, *Hellenistic Settlements*, I, 296–97, 325–26).

foundations of both Alexander and Perdiccas can be reconciled if Perdiccas refounded these cities on Alexander's instructions.[33] But there is no hint that any of these subordinate founders received the heroic honours traditionally granted to a colony"s founder. At Arsinoe in Cilicia, we hear of a precinct of Arsinoe in which decrees are displayed, and of sacrifices to the Theoi Adelphoi; there is no trace of Aetos the founder.[34] Modest separate rites for Aetos can perhaps be envisaged, but it is more plausible that the role of founder could be, so to speak, desacralised; the royal eponym was the symbolic centre of the settlement. As a consequence, when as often happened a king received cult in a city which he had nominally founded, it is unclear whether we should label this "founder-cult";[35] the distinction between this and other forms of cult paid to monarchs is blurred. A puzzle here is the Kineas who had a *temenos*, "sacred precinct," at Ai Khanum in Bactria and whom Robert argued to be a Thessalian and the founder-hero of the settlement. If this is indeed an instance of a traditional cult of a nonroyal founder, it looks like an exception to the trend.[36]

The decree states that the Arsinoeis are to be *apoikoi* of the Nagideans; here we have another point of confrontation with earlier norms, but one where on the surface nothing has changed. As a general rule, Hellenistic settlements had no mother-cities; the role played by the mother-city in archaic and classical Greek colonisation was taken over by the royal sponsor.[37] But here we find a mother-city/colony relationship. The exception, however, is one only superficially: the relationship is apparently being created at the time of the decree in the 230s or 220s, and had not been recognised when the foundation occurred a generation earlier, plausible though it is that some Nagideans may have migrated to the new settlement. It was an aspect of Thraseas's attempt to generate warm feelings between the two communities. Thraseas wishes to give to Arsinoe its own magistracies and laws. When founded, therefore, though it had a good dynastic name, it seems to have lacked the institutions that in normal Greek understanding were prerequisites for recognition as a polis.[38] It became a full polis only at this second stage. A similar transition from *katoikia* to polis is what is dramatically illustrated by a second new discovery, a text of cardinal interest, an

33. Cohen, *Hellenistic Settlements*, II, 248.

34. Boethos, the Ptolemaic founder, was highly honoured along with the reigning monarchs by an association in *OGIS* 111, but what he there receives is not founder-cult. Kramer and Heinen, *ArchPF* 43 (1997): 357–59, consider that Boethos might have been honoured within the dynastic cult in the cities founded by him.

35. Leschhorn, *Gründer der Stadt*, is constantly faced by this dilemma. Note too the refoundation of Mantinea when captured by the Achaeans, aided by Antigonus Doson, in 223. Aratus was chosen oikist, and renamed the town Antigoneia (Plut. *Arat.* 45.8); an inscription honours Antigonus as Saviour and Benefactor (*IG* V.2.299), but there is no mention of cult honours for the oikist Aratus. He was by contrast so honoured as a refounder at Sicyon, where there was no royal competitor (Plut. *Arat.* 53.1–7).

36. L. Robert, *CRAI* 112 (1968): 431–38; cf. Kosmin, *Elephant Kings*, 237–38 (with 348 n. 86 on the evidence for actual cult); Mairs, "Founder's Shrine."

37. Leschhorn, *Gründer der Stadt*, 215–16.

38. On the ambiguities of the vocabulary used in the decree, see A. Chaniotis, *EpigAnat* 21 (1993): 33–42. On laws, cf. E. Bickerman, *RPhil* 65 (1939): 343–45.

exchange of letters between Eumenes II and a southeastern Phrygian settlement, Toriaion.[39] The community has petitioned Eumenes through three representatives, one bearing the Galatian name Brennos, to be granted "a constitution, laws of its own, a gymnasium and all that follows from these things." Eumenes grants the request, stressing that with firm Roman backing he is in a position to confer real and lasting favours (possibly an allusion to earlier dealings between Toriaion and Antiochus III on the same subject, leading to nothing); he expands the idea of "constitution" as council, magistracies, and a tribal structure, and specifies that "the natives who share your settlement" are to be included in the new entity. Showily, he emphasises the change by subjoining to his first letter, to "the settlers of the Toriaitai," a second addressed to "the council and people of the Toriaitai" which thereby recognised their new status. Toriaion, then, is another city which has grown out of what probably originated as a Seleucid Macedonian military settlement; it has been given a persuasive location at an impressive Hellenistic fortress at Kale Tepesi.[40] The men of our decree will be descendants in the second or third generation of veteran settlers. Arsinoe was a colony promoted through intervention of a local official; Toriaion sought its own elevation.[41] What matters most for the present argument is an absence from the new text. The defining marks of a city are very clearly laid out: laws, a council, magistracies, a tribal structure, a gymnasium. But not a word is said about religion, neither about the choice of gods nor about how cults are to be financed;[42] and this although Eumenes makes special arrangements to ensure that oil can be provided for the gymnasium until his agent creates a permanent source of funding. It is hard to avoid the conclusion that the religious life of the settlement will be unaffected by its new status.

Religion in Military Settlements, Failaka in Particular

Such need not have been the universal pattern: lands were donated to Zeus and the city of Aizanoi in northern Phrygia by Prusias and Attalos, perhaps at a moment of transition for that community between *katoikia* and colony.[43] But in cases such as Toriaion's, enquiry

39. *SEG* XLVII 1745 (now *ISultan Daği* I, 393), translated in Austin, *Hellenistic World*, no. 236, but the translation needs adjusting in the light of a crucial correction in 26–27 (see the note in *SEG*) which introduces the "natives who share your settlement." On Attalid dealings with inherited Seleucid *katoikiai*, cf. R. E. Allen, *The Attalid Kingdom* (Oxford, 1983), 95–96.

40. See P. J. Thonemann, *NC* 168 (2008): 43–53, who also associates it with Attalid cistophoric coins countermarked TOR.

41. "How the change-over from military colony to *polis* took place is not known. There were cases where the king of set purpose enlarged some colony and formally made it a *polis*, but usually the settlement itself must have outgrown the *koinon* form and petitioned for the change: mere permission from the king probably sufficed, but was certainly needed": prescient words of Tarn, *Greeks in Bactria and India*, 9.

42. Contrast the explicit mention of *temenē* and money for the finance of sacrifices, along with laws and the ancestral constitution, in the list of things the "restoration" of which to an uncertain city in Hellespontine Phrygia was sought from the king by the Attalid governor Korragos: *IPrusa* II, 1001 (Austin, *Hellenistic World*, no. 235).

43. See M. Wörrle, *Chiron* 39 (2009): 430–31, who cites a study on the Hellenisation of the cult image.

moves back a stage, to ask about the organisation of religious life in a military settlement. A settlement intended to have any permanence—something more rooted than a garrison—necessarily had its origin in an administrative act, a distribution of land-lots to the settlers; at Arsinoe we can name the administrator concerned, Aetos. In principle then we could transfer back the classic model of colony foundation to the *katoikia,* and imagine cults being chosen, and land being set aside to fund them, by the founder. But it is also conceivable that all the land went to the settlers, and they were left in religious terms to shift for themselves.[44] There are, however, some traces of *katoikiai* which had authority over cult buildings and even, faintly, of plots set aside for the gods. A decree found in the region of Apollonia on the Rhyndakos in Hellespontine Phrygia is a resolution of "the settlers at Daphnous" to crown a general and another soldier, and place a record in the shrine of Apollo; a series of dedications found with it were made by individual Greeks to an "Apollo Daphnousios" depicted in familiar Greek guise.[45] They must have had some right to use the sanctuary as a place of display. The Arsinoeis had a shrine of Arsinoe; a sizeable Doric temple (ca. 13 m by 20 m) housing a large cult statue has been found in the Hellenistic Greek settlement, very likely a Seleucid *katoikia,* on an unoccupied site at Jebel Khalid on the middle Euphrates.[46] As for plots set aside, a letter of an Attalid king which treats the assignation of lots within a military settlement mentions, in a very fragmentary context, "*temenē* [to support the cult of the gods?] . . ."[47] What seems to be another such settlement at Apollonioucharax in Lydia petitions Eumenes II for help in recovering a village which can henceforth (as before?) be "sacred and tax-free" and provide revenues for the "sacrifices of Zeus Stratios and those for you." The settlement also asks that an already existing right of asylum at the precinct of Zeus Stratios be extended; the sanctuary was evidently very important in the life of the settlement, and one can imagine that the cult and the settlement were created together.[48] Much the most vivid evidence, however, comes from an excavated

44. One hundred ninety-nine names, almost exclusively Greek, appear on an unpublished dedication to Zeus Synetenos from Bucak Köyü (eastern Caria) described by A. Chaniotis in *AJA* 102 (1998): 248–50: these look like settlers, dedicating to a local (?) god.

45. *SEG* XLIII 879–84; cf. Schuler, *Ländliche Siedlungen,* 33, 191, 255. *TAM* V.1.543 is a similar text, but with no place of display indicated. *TAM* V.2.1321, a dedication of "those from Lasnedda" to Papias, might (Cohen, *Hellenistic Settlements,* I, 215) show colonists approaching a local god at a shrine they probably did not control; but we cannot be sure who they were.

46. G. Clarke, in *Religious Identities in the Levant,* 143–55; Wright, *Sacred Spaces,* 81–90; for the *katoikia,* Cohen, *Hellenistic Settlements,* II, 178–80; and Kosmin, *Elephant Kings,* 200–201. The proportions and internal arrangement of the temple are unusual ("fusion of Greek and Mesopotamian influences," Wright, 90); so too is the line of "incense altars" that flanked parts of it and perhaps the whole.

47. *RC,* no. 51 = B. Virgilio, *Lancia, diadema e porpora*², Studia Hellenistica 14 (Pisa, 2003), doc. no. 29 (with bibl.). Welles's comment that "the *temenē* would not have been income-producing" is surely wrong (for income-producing *temenē,* see, e.g., *SEG* XXXV 1170). The classic model for such setting aside of *temenē* is *ML* 49 (Brea), on which see I. Malkin, *Chiron* 14 (1984): 43–48; cf. Thuc. 3.50.2.

48. *New Documents from Lydia,* 32 B 15–17 (cf. n. 20 above); for the suggestion that the settlers introduced Zeus Stratios, see M. Wörrle, *Chiron* 39 (2009): 431. The *katoikoi* associated with a high priest of ApolloTarsenos in Mysia in *RC,* no. 47 (*SEG* XXXIX 1337) are mysterious (Schuler, *Ländliche Siedlungen,* 193).

Seleucid settlement on the island (modern Failaka) at the head of the Persian Gulf to which, according to Arrian,[49] Alexander gave the name Ikaros in reminiscence of the island of that name in the Aegean. It was already reported to Alexander as being sacred to Artemis; a man from Alexander's entourage, Androsthenes of Thasos, spoke of a very sacred (*hagion*) temple of Apollo there and an "oracle" of (Artemis) Tauropolos. These identifications must be products of *interpretatio*. The core of the settlement is a fortress, to judge by archaeological finds populated by both Greeks and non-Greeks; two sanctuaries have been found within the fortress (F5), one housing a temple of Greek style (though with Persian column bases), and another sanctuary outside it, the "seaside shrine" (B6) (argued to be Iranian in plan). Further inland (Tell Khazneh) an earlier Neo-Babylonian or Achaemenid sanctuary appears to have been taken over by the Greek settlers, just as the native cult of "Artemis" of which Alexander was told seems to have encouraged the settlers to take Artemis Soteira as their chief goddess.[50] Dedications by "Soteles the Athenian and his troops to Zeus Soter, Poseidon, Artemis Soteira" and by Soteles to "Poseidon Asphaleios" apparently come from the fortress, and one to Artemis from the sanctuary at B6. Sculptures and terracottas suggest worship of Herakles, possibly too Dionysus, Pan, and Aphrodite, in domestic contexts.[51]

What is most important for our purposes is that, according to a letter by a Seleucid functionary Ikadion of 238/7 (?),

> The king is concerned about the island Ikaros, because his forebears consecrated fields (?) and attempted to relocate the shrine of Soteira and wrote to the officers in charge to relocate it. But they, whether because it was impossible for them or for some other reason, did not relocate it. But when the king wrote to me I quickly relocated it and set aside money for an athletic and musical competition, wishing to carry out the intentions of the king and his forebears. And concerning those who live on the island, the temple attendants (*neōkoroi*) and the others . . . [52]

49. *Anab.* 7.20.5—the claim has been contested (see the discussion in Cohen, *Hellenistic Settlements*, III, 144–45), but is supported by the presence of a Tauropolion (Strabo 14.1.19, C 639) on the Aegean Ikaros. Androsthenes: ap. Strabo 16.3.2, C 766 (*FGrH* 711 F 2); cf. J. Gachet and J. F. Salles, *Mesopotamia* 25 (1990): 206–7; Dionys. *Per.* 608–11 also speaks of Tauropolos receiving worship on the Persian Gulf island.

50. On all this, see now Cohen, *Hellenistic Settlements*, III, 140–54, with references; on Soteles, P. L. Gatier, *Arabian Archaeology and Epigraphy* 18 (2007): 75–79 (with a dating to the early 3rd c. B.C.; but J. Marcillet-Jaubert, *Failaka: Fouilles Françaises, 1984–85*, Lyon 1986, 265–67, contemplates the second half of the 3rd c. for *IEstrOriente* 417, also possibly mentioning a Soteles); for the broader context, P. Kosmin, "Reconsidering the Hellenistic Gulf: The New Greek Inscription from Bahrain," *JHS* 133 (2013): 61–79. Native cult: Arr. *Anab.* 7.20.3–4. But J. Naveh, *BASOR* 297 (1995): 1–4, downdates an Aramaic inscription found at Tell Khazneh to the third century and thus denies that it proves in itself an Achaemenid presence; he reads it as a dedication to "Bel of Ikaros."

51. See Cohen, *Hellenistic Settlements*, III, 152–53 n. 25, citing studies by J. B. Connelly.

52. *IEstrOriente* 421–22; for the other Greek inscriptions from Failaka, see 416–20, 423; *SEG* LV 1590. The date of *IEstrOriente* 421–22 is very controversial (to references in Cohen, *Hellenistic Settlements*, III, 152 n. 23, add M. Z. Petropoulou, *EpigAnat* 39, 2006: 139–47: proposals range from 243/2 to 203/2, Cohen, 143). Evidence on the findspot of *IEstrOriente* 416, the dedication to Zeus Soter, Poseidon, and Artemis Soteira, is conflicting (fortress or Tell Khazneh?—see J. P. Salles, *Failaka: Fouilles*

At this crucial point, where the rights of the natives were perhaps being discussed and protected against abuses, the text becomes frustratingly fragmentary and obscure. We cannot correlate the "relocation" very confidently with the sanctuaries on the ground. But from a combination of the inscriptions and the remains some conclusions emerge reliably. The settlers took up a preexisting cult of "Artemis," which may have remained to some degree in the control of indigenous *neōkoroi*. But they built further sanctuaries, presumably to other gods. And the kings themselves took an interest in the cultic life (though surely not in this only) of the colony, an interest which could lead to subvention.[53] The settlers were not, then, left completely to themselves. But one can imagine a scenario whereby only rudimentary provision for cults was made at the time of foundation; the settlers then developed others, partly by frequenting existing native sites; and the conglomerate that so emerged was taken over if (as was not the case with Ikaros) the *katoikia* eventually became a city.[54] For what it is worth, a process of gradual accumulation of cults, though for a city, not a colony, is how Libanius of Antioch envisages the growth of the pantheon of his native place. The Cypriot gods forced Antiochus II, son of the founder, to fetch them to Antioch; they persuaded Apollo to issue an oracular instruction to that effect. Isis forced Seleucus IV to do the same, through dreams. Libanius's purpose is eulogistic, to stress the gods' own yearning for the blessed city. But the pinning to specific reigns might suggest a historical basis.[55]

CITIES FOUNDED/REFOUNDED IN DIFFERENT DEGREES

Cities Significantly Refounded

The case illustrated by Arsinoe and Toriaion, the *katoikia* that mutates into a city with limited royal involvement, is only one possibility. Some cities were founded through decisive royal intervention.[56] Legends developed around the royal role: Alexander and the barley-grain at

---

Françaises, 1984–5, Lyon 1986, 133–35), but *IEstrOriente* 418 (Soteles to Poseidon Asphaleios) was from the fortress (as was *IEstrOriente* 420) and probably should bring *IEstrOriente* 416 with it. *IEstrOriente* 419 (Artemis) is from B6, and 417 attests sacrifice by Greek-speakers at Tell Khazneh.

53. On such direct involvement at local level by Seleucid kings, see Sherwin-White, "Seleucid Babylonia," 22–23.

54. Tscherikower, *Städtegründungen*, 121–22, suggests Peltai and Blaundos as cases of *katoikiai* that became *poleis*. To the arguments suggesting that Phrygian Metropolis may have started as a garrison or veteran settlement (Cohen, *Hellenistic Settlements*, I, 313–14), Daubner, *Militärsiedlungen*, 45–46, adds the late-attested, but thoroughly Macedonian and military, cult of Artemis Tauropolos: *MAMA* IV 122; on this goddess in Macedonia, see Hatzopoulos, *Macedonian Institutions*, index, p. 516.

55. Libanius, *Or.*11.109–17 (cf. T.S. Scheer, *Die Gottheit und ihr Bild*, Munich 2000, 266–68); possible historical base: J. Aliquot, in Bricault and Versluys, *Cults of Isis*, 135–46. The stories are interwoven with two others, about attempts by foreign rulers (Ptolemy and the Romans) to steal Antiochene gods, again frustrated by the gods themselves. Nothing that is very reliable can be said about the early cultic history of Antioch, except that a thundering Zeus seems to have been central and was joined by the cults of Apollo, and apparently Artemis, established by Seleucus at Daphne. See B. Cabouret, "Les cultes grecs d' Antioche," *Topoi* 7 (1977): 1005–22.

56. So L. Robert and J. Robert, *BÉ* 1980, no. 108, dissenting from G.M. Cohen, *The Seleucid Colonies* (Wiesbaden, 1978). Another possibility is the *katoikia* implanted within an existing settlement,

Alexandria; instructions given to the sleeping Alexander by the two Nemeseis of Smyrna to refound their city (whether he did so is very unclear);[57] the eagles that snatched meat from sacrifices made by Seleucus I and carried it to the destined sites of Seleucia in Pieria and Antioch on the Orontes (but a probably earlier tradition spoke of a thunderbolt guiding him to Seleucia);[58] the frustrated attempt of the magi, consulted by Seleucus about the best hour to found Seleucia by the Tigris, to trick him and avert destiny; the wooden statue of Dionysus found by "Attalos and Eumenes" at the site of Dionysopolis in western Phrygia; the arrowhead inscribed "of Phoebus" and shot by the god himself which Seleucus found at the site of Daphne by Antioch.[59] The point of all such stories is to confirm that the foundation in question was well-omened, god-approved. Those concerning Syrian Seleucia and Antioch explicitly showed Seleucus sacrificing to check whether he was founding his city in an appropriate place (though in the event the eagle intervened to provide proof of a different kind); and it would be surprising if such consultative sacrifice did not take place at a time of foundation. Robert supposed that the Seleucids followed the good old practice and consulted the oracle at Delphi about their plans.[60] But the only text that he quotes in direct support is an instruction concerning Laodicea on the Lykos in southern Phrygia supposedly conveyed via the wife of Antiochus II in a dream:

> To King Antiochus Phoibos Apollo gives this instruction
> to found a glorious city, as ordered
> High-Thundering Zeus, sending Hermes Eriounios.[61]

---

which it comes to dominate: *IMagnSip* 1 (*OGIS* 229), with Bencivenni, *Progetti di riforme*, 223–26. The case is too singular to be helpful here. Mileta, "Charakter und Entwicklung der Neuen Poleis," distinguishes three types: refashioned old cities (such as Thyateira); promoted military settlements (such as Toriaion); synoecisms of indigenous villages (such as Laodicea on the Lykos).

57. Paus. 7.5.1–3; for more on the tradition of Alexander as founder (and its possible validity), see Cohen, *Hellenistic Settlements*, I, 181 n. 1; Fraser, *Cities of Alexander*, does not mention Smyrna. The legends were collected by Tscherikower, *Städtegründungen*, 129–31. For dreams leading to foundations, cf. Steph. Byz. s.v. Antiocheia; Plut. *Alex.* 26.4–7.

58. Eagles: Malalas, *Chronographia* 8.11 Thurn (8.199–200 Dindorf) (for the legends of Laodikeia by the Sea and Apameia, with variations on the eagle theme, see 8.17–18 Thurn, 8.202–3 Dindorf); thunderbolt: Appian *Syr.* 58 (299). For the numismatic and artistic evidence for the Antioch legend (also found in Libanius *Or.* 11.85–88), see H. Seyrig, *RÉA* 42 (1940): 340–44; for the history of the legends, Chuvin, "Fondations syriennes." On the (surviving) "Charonion" mentioned by Malalas 8.21 (8.205 Dindorf), see Wright, *Sacred Spaces*, 71–73.

59. Seleuceia by the Tigris: App. *Syr.* 58 (300–307). Part of a more extended "Seleucus romance," according to Fraser, *Cities of Alexander*, 37. Dionysopolis: Steph. Byz. s.v. Dionysou Polis. Daphne: Libanius, *Or.* 11.95–96.

60. L. Robert, in Gagniers, *Laodicée*, 295; cf. *CRAI*, 1968, 442. Apollo's title Archegetes at Phrygian Hierapolis (Ritti, *Hierapolis*, 108, 114) and at Attaleia in Pamphylia (*IGR* III 780–81) is also adduced, but need not refer to an oracular role. For Ai-Khanum, Mairs, "Founder's Shrine," 111–16, argues that an association with Delphi is created some time after the foundation.

61. Steph. Byz. s.v. Laodikeia; Meineke ad loc. supposes Stephanus's source to be the *Chronica* of Phlegon of Tralles. The oracle is, rightly, acknowledged neither by H. W. Parke and D. W. Wormell, *The Delphic Oracle* (Oxford, 1956) nor in J. Fontenrose's book of the same title (Berkeley, 1978). Zeus and Hermes have an association in Laodicean cult: see *ILaodLyk* 64 and the relief that accompanies *ILaodLyk* 63.

Even if we take this at face value—but the dream instruction is a common motif of foundation legends—it is not a response to a question put to an oracle. Favourable oracular responses ought to have stuck in collective memory: the absence of the motif from the various legends listed above suggests that none were sought. Seleucus was, by contrast, said to have consulted "the magi" about the right moment to found Seleucia on the Tigris. The detail forms part of the "Seleucus romance," but it is very plausible that where prestigious local divinatory methods existed, the kings made use of them.[62]

Gods from Home With cities that experienced a definite act of foundation, we are back with the problem of choice of gods. As was noted earlier, no founding charters or early calendars exist; we must scavenge amid a miscellaneous debris of information. A working assumption of many scholars is that most cults were imported from the settlers' place of origin: Antioch in Persis (precise site unknown), which had been reinforced by an important influx of settlers from Magnesia on the Maeander, agreed to a request of the Magnesians partly "from respect for our common gods."[63] Plausible instances of such importations can certainly be found. Both the altar of Zeus Bottiaios in Antioch on the Orontes and the shrine of Athena Kyrrhestike northeast of there, by their epithets, evoke places in Macedonia;[64] they are the cultic equivalent of the Macedonian toponyms scattered through Syria, or the Macedonian institutional term Peliganes, "counsellors," now attested even in Susa. The sanctuary of Apollo Komaios at Seleucia on the Tigris, first mentioned though it is by Ammianus Marcellinus, must go back to early settlers: two dedications to the same god are now known from early Hellenistic Philippi.[65] The late-attested cult of Artemis Tauropolos at Metropolis in Phrygia (*MAMA* IV 122) should similarly be an early Macedonian introduction. Apollo and Artemis Daittai came to Seleucia on the Eulaios (the refounded Susa) from Daphne near Antioch: here we have movement within the Seleucid realm.[66]

But a set of gods can never have been transferred wholesale from one place to another. A whole pantheon with all its local particularities could not be replicated; very few new cities will have been populated with settlers all from one place anyway. Selections and combi-

62. App. *Syr.* 58 (300–307). Kosmin, *Elephant Kings,* 213, notes that concern with the right moment is distinctively Mesopotamian. But Ramsay's statement that "at the foundation of any Greek colony, the sanction of the native God was obtained in order to give legitimacy to the new city in the eyes of the natives" (*The Cities and Bishoprics of Phrygia,* Oxford 1895, 1:53) goes well beyond the evidence.

63. *OGIS* 233 (Rigsby, *Asylia,* no. 111) 40–41.

64. Zeus Bottiaios: Lib. *Or.* 11.76, 88; Malalas, *Chronographia,* 8.12 Thurn (8.200 Dindorf). Athena Kyrrhestike: in Macedonia, Diod. Sic. 18.4.5; *SEG* XXVII 258; in the Kyrrhestike region of north Syria: Steph. Byz. s.v. Kyrrhos; Strabo 16.2.7, C 751 (cf. Cohen, *Hellenistic Settlements,* II, 181–82; see Cohen, 407–8, for imported toponyms, with the subtle discussions of P. Bernard, *Topoi* 5, 1995: 382–408; and Kosmin, *Elephant Kings,* 106–10). Peliganes: P. L. Gatier, *ZPE* 184 (2013): 205–10 (*SEG* LIX 1654), with the other extra-Macedonian attestations.

65. Amm. Marc. 23.6.24; *SEG* XXIV 620, LIX 691 (with evidence for its possibly Ionian origins).

66. *SEG* VII 17 (*IEstrOriente* 190); *RC* 44.21–22. "Daittai" is not understood (Macedonian?): see Welles, in *RC,* p. 183. Zeus Seleukeios is for Robert another Macedonian importation (see p. 202 above). Metropolis: cf. n. 54 above.

nations would certainly have had to be made.[67] A possibility was to give gods nonspecific epithets found in many old Greek cities, such as the pairing Zeus Polieus/Athena Polias found at Aigeai and Antioch on the Pyramos in Cilicia or the cult of Athena Polias in Attaleia in western Pamphylia; that these cults arrived at the time of the foundation is, however, never certain.[68] Some gods came in with particular institutions, Herakles and Hermes with the gymnasium, Dionysus with the theatre; but again we cannot be sure precisely when.[69] New foundations perhaps provided an opportunity for new gods: Sarapis appears quite often, though probably in many cases through private initiative in the postfoundation phases rather than founder's fiat.[70] By late in the third century, if epigraphic dating can be trusted, Stratonicea in Caria, founded soon after 268 apparently by synoecism of existing villages, contained a Samothrakion and a Sarapieion, as well as a Herakleion.[71]

Gods Favoured by Dynasties Gods particularly favoured by the royal sponsors of the settlement certainly also entered the mix, perhaps even as a central ingredient. From the time of Antiochus I, Apollo was honoured as the ancestor of the Seleucid dynasty; Seleucus I himself, though emphasising Zeus on his coinage, built a spectacular new sanctuary for Apollo at "Daphne" by Antioch in Syria. Wherever, therefore, Apollo appears in a Seleucid foundation or refoundation, influence direct or indirect of that dynastic claim is regularly suspected.[72] Given the god's universal popularity the case can scarcely be proved, but sometimes acquires circumstantial support. At Amyzon, a Carian town centred on a sanctuary of "Artemis," Apollo starts to be mentioned in first place before the goddess at the time when the sanctuary comes under the protection of Antiochus III.[73] At Alabanda/Antioch in Caria,

67. Cf. Bickerman, *Institutions*, 159, noting the institutional diversity of Seleucid foundations: "They [the kings] created international centres, open to all the Greeks, even to all adepts of Hellenic culture, and allowed the new cities to organise themsleves as they chose. In other words, the Seleucid foundations are not 'colonies' reproducing the type of a metropolis, but authentic new cities." And on the general point, cf. Woolf, "Roman Diaspora."

68. Aigeai: *ICilicie* 74 (*SEG* XXXVII 1245: 39/40 A.D.); note that Athena perhaps bore a second epithet. Antioch on the Pyramos: *SEG* LVI 1797–98, ca. 200–180 (cf. n. 92 below). Attaleia: *SEG* VI 647 (also Apollo Archegetes, 651). For more on all this, see R. Parker, "Athena in Anatolia," *Pallas* 100 (2016): 73–90, at 74.

69. Hermes/Herakles: e.g., at Tyana/Eusebeia near Taurus, *ITyana* 29; at Apamea Kelainai, *MAMA* VI 173.16; at Ai Khanum, *IEstrOriente* 381. On gymnasia in the East, see F. Daubner's list in *Religious Identities in the Levant*, 42. Dionysus: e.g., honours proclaimed at the Dionysia of Antioch on the Maeander, *IMM* 90.28.

70. See *IEstrOriente* 280 (Hyrcania, reign of Antiochos I); *SEG* VIII 95 (*RICIS* 403/0501; Samareia, 3rd c. B.C.?); *IG* X.2.1.3 (187 B.C.); and more in Cohen, *Hellenistic Settlements*, I, 105 n. 8 (Thessalonike); *IGLS* IV.1261 (*RICIS* 402/0301, Laodicea by the Sea, 174 B.C.—certainly private; cf. J. Sosin, *CQ* 55, 2005: 130–39).

71. *IStrat* 1004; *SEG* XLIV 917 (*IStrat* 1531); cf. Debord, "Questions stratonicéennes," 161.

72. See, e.g., on Seleuceia/Tralles, Boffo, *Centri religiosi*, 311–15, following Rehm's note on *Milet* I.3.143, where (lines 8–9) Apollo is "origin of the kinship" between Seleuceia and Miletus; on Nysa, Boffo, 287–89; on Apollonia Salbake, L. Robert, in Gagniers, *Laodicée*, 288, 303 n. 136. Cf. Robert, in Gagniers, 295 n. 12: "The Seleucid Apollo and Artemis are worshipped at Dura-Europos [*SEG* VII 352, 2/3 A.D.] and at Susa [*IEstrOriente* 190]."

73. See Robert and Robert, *Amyzon*, 140–41, with further evidence.

the god bears the title Isotimos, "equal in honour." He acquired the unique epithet, it has been suggested, when his cult was introduced by a Seleucid alongside that of the hitherto dominant Zeus.[74]

Attalid influence is a little easier to trace, since the Attalid special gods bore distinctive epithets: Athena was Nikephoros, "Victory-Bringer," Dionysus Kathegemon, "Leader." It is not a coincidence that a dedication at Panion on the north shore of the Propontis, which acclaimed Eumenes II as founder, was made "for Eumenes, saviour and benefactor and founder of the city, to Zeus Soter, Athena Nikephoros, and Apollo," nor that Dionysus Kathegemon is attested at Philadelpheia in Lydia, an Attalid foundation, and at Thyateira in Lydia, a Seleucid foundation taken over by the Attalids. The presence of this god at Acmonia, and nearby Sebaste, in Phrygia may hint at an otherwise unattested Attalid involvement with Acmonia; the same undoubtedly applies to the priest of Athena Nikephoros at Blaundos on the Lydo-Phrygian border.[75] Near Thyateira "the Heracleasts settled in Mernouphyta from (the time of?) King Attalos and Eumenes" survive as late as the second century A.D.; Heracleasts too are no surprise given the Attalids' claim to descent from that hero.[76] A Lydian settlement Pantheotai has been tentatively identified as Attalid because of the Attalid cult of "all the gods";[77] and the Attalid refoundation Dionysopolis in western Phrygia clearly reflects their enthusiasm for Dionysus. But it is never certain in such cases, except the last named, that the "Attalid" cult was introduced precisely at the moment of foundation/refoundation. "Founder's preference" does appear, in bizarre form, in the city Ouranopolis consecrated to Aphrodite Ourania on Mount Athos by Cassander's eccentric brother Alex-

74. So L. Robert, in *Études Déliennes*, BCH Suppl. 1 (Athens and Paris, 1973), 460, following Dittenberger's note 20 on *OGIS* 234; followed by Boffo, *Centri religiosi*, 131, 309–31; rejected by Rigsby, *Asylia*, 327.

75. Panion: *OGIS* 301: "These are the gods of the Pergamenes," comments Dittenberger ad loc. Dionysus Kathegemon in Thyateira: *TAM* V.2.976.8, 979.5–6; in Philadelpheia: *TAM* V.3.1462.10–12 (Jaccottet, *Choisir Dionysos*, no. 113); 1497.9–11; note too K. Buresch, *Aus Lydien* (Leipzig, 1898), 11–13, no. 8 (Jaccottet, no. 114), from Tepe Köi. Acmonia: *MAMA* VI 239 (Jaccottet, no. 84); Sebaste: Drew-Bear and Naour, "Divinités de Phrygie," 1947–48 (Jaccottet, no. 85), with, 1949 n. 143, the cautious suggestion about Attalids and Acmonia. Cf. D. Musti, "Il Dionisismo degli Attalidi," in *L'Association Dionysiaque dans les sociétés anciennes* (Rome, 1986), 105–26. Blaundos: *SEG* XLVI 1491 ("late Hellenistic") = A. Filges, *Blaundos* (Tübingen, 2006), 321, no. 1. Blaundos is commonly taken, without reference to this cult, to be a possibly Seleucid military colony by origin (so, e.g., Filges, 18–21). P. Weiss, *Chiron* 30 (2000): 628–30, shows connections between the attested tribe names of Eumeneia in Phrygia and cults of Pergamum, and concludes that comparable cults were introduced at the time of foundation. This is possible, but the tribe names are of very common Greek gods without epithet; they lack a distinctive Pergamene imprint.

76. Heracleasts: *TAM* V.2.959; Attalid descent from Herakles: Nicander fr. 104 Gow; cf. L. Robert, "Héraclès à Pergame," *RPhil* 58 (1984): 7–18 (*OMS*, 6:457–68). Robert mentions too (15) the coins showing Herakles from (Attalid) Apollonis in Lydia (Cohen, *Hellenistic Settlements*, I, 203). The group of "young men around Herakles" at Thyateira (*TAM* V.2.949, etc.) is less diagnostic, given the universal association of Herakles with the young. For possible Attalid influence on the cults of Attaleia in Pamphylia (n. 82 below), see Cohen, *Hellenistic Settlements*, I, 338 n. 3 (but the attestations are late, and lack the diagnostic epithets); on Philadelpheia, see L. Robert, *RPhil* 8 (1934): 287–91, citing O. Weinreich.

77. C. Habicht, *JRS* 65 (1975): 79.

archos.[78] What to say in this regard about Antiochus IV is very controversial: can the sudden return to prominence of Zeus on his coinage be combined with known cases where he promoted cults of Zeus Olympios to give a general policy? And, if so, can such cults not associated with him by any source be nonetheless ascribed to his initiative?[79] Decisive arguments have yet to be found.[80]

Traditions of the Territory There were also, of course, local gods to consider; for, though in one perspective piety was to worship one's ancestral gods, in another every territory belonged to the gods who "held" or "occupied" it. Hellenistic monarchs, it is now well recognised, supported and participated in existing cults; the "Borsippa cylinder," in which Antiochus I describes in cuneiform how "his heart urged" him "to [re]build Esagil and Ezida" (sanctuaries at Babylon and Borsippa),[81] is a key exhibit. Even where, amid our scanty evidence for the cults of a given settlement, we encounter none but old Greek gods,[82] we should be cautious about supposing that this was, in cultic terms, a "wholly Hellenic" centre.[83] Arrian states explicitly that, at the foundation of Alexandria, Alexander established a shrine for "Isis the Egyptian," as well as the Greek gods. The importance of Alexander's decision should not be reduced by the suggestion that Isis was, by now, virtually domesticated as a Greek goddess; for, though it is true that her name had long been familiar to Greeks and there was already a shrine of Isis in the Piraeus at this date, that shrine served an immigrant Egyptian clientèle: Alexandria was the first Greek city to offer the goddess public cult. It was so, that is, if Arrian's notice is reliable; Alexander's shrine of Isis seems to have left no other trace of itself.[84] The pseudo-Aristotelian *Economica* tells how Cleomenes, Alexander's future governor of Egypt, extorted money from the priests and property-owners of Canopus to be allowed to stay where they were, and then forced them to move to Alexandria all the same; the priests

78. See Cohen, *Hellenistic Settlements,* I, 105–6; M. Winiarczyk, *Eos* 89 (2002): 269–85.

79. As, e.g., Zeus Olympios at Gerasa, Zeus Nikephoros (?) at Gadara (Cohen, *Hellenistic Settlements,* II, 248–49, 282–83). Cf. n. 113 on Castabala.

80. See pro (the idea goes back to Bevan), above all, Seyrig, "Zeus à Séleucie," with Dupont-Sommer and Robert, *Déesse de Hiérapolis Castabala,* 97–98, and the works he cites; contra, Bickerman, *Institutions,* 231–35; and (too robust) P. F. Mittag, *Antiochos IV Epiphanes* (Berlin, 2006), 139–45; intermediate, O. Mørkholm, *Antiochos IV of Syria* (Copenhagen, 1966), 122–32; Houghton, *Seleucid Coins,* II.1, p. 48.

81. On the cylinder, see A. Kuhrt and S. M. Sherwin-White, *JHS* 111 (1991): 71–86; K. Stevens, *JHS* 134 (2014): 66–88; on Mesopotamia, Linssen, *Uruk and Babylon,* 18–19; more generally, Kuhrt and Sherwin-White, *From Samarkhand to Sardis,* 36–37, 132, 144, 154–55; Boffo, *Centri religiosi,* passim; strong claims for Antiochos IV in this regard in J. D. Grainger, *Hellenistic Phoenicia* (Oxford, 1991), 117.

82. As, e.g., at Attaleia in Pamphylia, where *IGR* III 780 attests priests of Apollo Archegetes, "great Dionysus," and Ares; and *SEG* VI 647.12–13 (1st c. A.D.), a priestess of Athena Polias. (But even at Attaleia contrast from the Roman period, *SEG* XVII 590, 592.) The many "Greek" gods seen on the coins of Ptolemais/Antioch Ake in Phoenicia (Cohen, *Hellenistic Settlements,* II, 215) are Syro-Phoenician gods in disguise (H. Seyrig, *Syria* 39, 1962: 193–207).

83. L. Robert and J. Robert, *Carie,* 312, on Apollonia Salbake in Caria: "a purely Hellenic town in an indigenous milieu which was Hellenising itself"; cf. 270: "These cults [those revealed by the coins] have an entirely Hellenic character."

84. Fraser, *Ptolemaic Alexandria,* 1:267–68.

presumably were meant to continue plying their trade.[85] However all that may be, Ptolemy Soter was soon to create for the city what was in effect the new cult of Sarapis; and, whatever his motives, whatever worshippers he had in view,[86] the decision to create or promote a figure so closely linked to Isis was unquestionably a response to the Egyptian environment.

The cult of "the thunderbolt" instituted, according to Appian (*Syr.* 58 [299]), by Seleucus in Seleucia in Pieria on the coast of northern Syria is perhaps an exemplary case of cultural compromise. The Jebel Aqra south of Seleucia had been associated with a thunder-god since the Bronze Age, and such a figure was characteristic of the whole region. There was, it is true, almost complete conceptual compatibility between a Syrian thunder-god and Homer's cloud-gathering Zeus, and the Zeus of Seleucia is Olympios and Koryphaios (from Mount Koryphe near Seleucia) as well as Kasios (the god of Jebel Aqra) in the first attestation. But the same early second-century text that attests those titles attests also an office of "thunderbolt-bearers"; Appian speaks of a cult of "thunderbolt," not of Zeus of Thunderbolts, and coins of the city (from the late 2nd c. onward) depict a garlanded thunderbolt resting on a stool. It looks as if Zeus was depicted aniconically as a thunderbolt, and this aniconic image was periodically paraded, in accord with north Syrian traditions. But nothing in this was so alien as to be difficult for Greek or Macedonian settlers.[87] A somewhat parallel case may be that of the Apollo Propylaios who is prominent in the cultic life of Eumeneia in Phrygia. He bears a good Greek epithet, and his cult might well go back to the Attalid foundation; but dedications to him are regularly accompanied by a double ax, the typical Anatolian attribute of a male god, and coins show him wielding one.[88] Either then

85. Arist. [*Oec.*] 2.1352a28-b4.

86. Fraser, *Ptolemaic Alexandria,* 1:252, sees it as designed to "give the Greek population of Egypt, and especially that of Alexandria, an overriding patron deity," but adds that, given the context, "Egyptian elements, in some form or another, were fundamental to the nature of the new god." For very varying views on date and motivation of the creation of the cult, see various authors in *Cults of Isis,* 9 n. 23, 77–79, 102, 135. Note the cult of Egyptian gods in Ptolemais Hermiou, n. 25 above.

87. See Seyrig, "Zeus à Séleucie"; and on the Jebel Aqra and Zeus Kasios, Lane Fox, *Travelling Heroes,* 256–64. First attestation: n. 23 above. Thunderbolt on stool: conveniently visible in Cook, *Zeus,* 2:809; cf. K. Butcher, *Coinage in Roman Syria* (London, 2004), 413–14. The many coins of Seleuceia that merely show a thunderbolt prove little, the type being common on many coinages; it has even been claimed (unnecessarily) as a Macedonian inheritance (Chuvin, "Fondations syriennes," 104). Seyrig pointed to the appearance of a thunderbolt in the hand of Nike in images of Zeus Nikephoros on rare coins of Alexander Balas (1799 and 1841, in Houghton, *Seleucid Coins*—but 1841 is there reassigned away from Seleuceia to Ptolemais/Ake) to argue that Zeus Olympios had combined in Seleuceia with Keraunios. Zeus Keraunios is not attested in the city *illo nomine* until the second century A.D. (Seyrig, "Zeus à Séleucie," 296) but noted as distinctive of Seleuceia by Hesych. s.v. κεραύνιος. On Syrian parade-processions, see Seyrig, *Syria* 18 (1937): 373–74. If the colossal goddess figure carved into the rock near Antioch on the Orontes (the "Charonion," n. 58 above) is Hellenistic, it would represent a comparable cultural compromise (Wright, *Sacred Spaces,* 73).

88. See, e.g., *GRBS* 17 (1976): 260–61, no. 15 with plate 9, no. 4; more in Cohen, *Hellenistic Settlements,* I, 304 n. 6; and fully G. Labarre, "L'Apollon *Propylaios* d'Eumeneia et les *theoi propylaioi,*" in P. Brun, ed., *Scripta Anatolica: Hommages à Pierre Debord (*Bordeaux, 2007), 283–96 (who thinks, 295, that the cult was certainly introduced at the time of the Pergamene foundation). The double ax appears already on the first coins of Eumeneia.

the familiar Greek epithet has been applied to an indigenous figure, or, which is perhaps more plausible, the imported god has gone iconographically native.

These are cases where an imported cult may have assumed a new shape under the influence of local traditions. There must also have been many cases where "new towns," since very few started from nothing, incorporated existing sanctuaries and cults.[89] (The possibility by contrast that some existing sanctuaries were not incorporated, were neglected through policy or mere indifference,[90] is intriguing but for the moment beyond our observation.) Almost all that we knew until 2006 of the religious life of Antioch on the Pyramos in Cilicia (a city with good Greek institutions and tribes and decrees) is that it centred on the shrine of Athena Magarsia, named from the place Magarsos, which was this Antioch's original name. "Athena" Magarsia was a goddess to whom Alexander the Great had already paid his respects, a goddess who predated Greek influence therefore. As she appears on the coins of neighbouring Mallos, she is a compromise between an Anatolian/north Syrian goddess and an aegis-wearing Greek Athena.[91] Athena Magarsia stayed on, therefore, when Magarsos was elevated to city status. In 2006, however, it emerged that this Antioch also hosted a joint cult of Zeus Polieus and Athena Polias important enough for a decree honouring a benefactor to be displayed in "the sanctuary of Athena Polias" (we infer she was dominant in the combination).[92] Indigenous and imported Athenas in parallel, therefore?

Nysa in Caria was a city founded by a Seleucid, apparently on an empty site, by synoecism of three existing villages; the eponym of one of the villages, Athymbros, was regarded by the Nysians as their "archegete." Nysa was famous for the Ploutonion in the nearby village of Acharaka, where the festival Theogamia celebrated the wedding of Plouton and Kore, and the sick came to be healed through dreams in "Charon's cave."[93] Though the names are Greek, the functions are not; it is no normal business of Plouton and Kore to heal, and this looks like an older cult redescribed in Greek terms, a process which led to rich elaborations in the Roman period. It was probably the role in the healing process of a cave interpreted as a point of access to the underworld that drew Plouton (and with him Kore) into the cult.

89. On existing towns incorporated into colonies, see Kosmin, *Elephant Kings,* 203–4.

90. Milena Melfi in forthcoming work studies this issue in relation to the Roman occupation of Greece.

91. *SEG* XII 511.8 and 15 (new ed. by I. Savalli-Lestrade in *Studi Ellenistici* 19, 2006: 119–247 [*SEG* LVI 1799]; on the history of this city, see too Savalli-Lestrade, in P. Fröhlich and C. Müller, eds., *Citoyenneté et participation à la basse époque hellénistique,* Geneva 2005, 22–35: it was in existence by 209 or 204; whether founded by Antiochos III or a predecessor is unclear, as is Mallos's attitude to the foundation); Arr. *Anab.* 2.5.9; the goddess is still mentioned in *IGR* III 889. Coins: A. Houghton, in Houghton et al., eds., *Festschrift für/Studies in Honor of Leo Mildenberg* (Wetteren, 1984), 91–110.

92. *SEG* LVI 1797–98. I. Savalli-Lestrade, *Studi Ellenistici* 19 (2006): 229, suggests that this cult derives from Mallos and might go back to the "Argive origins" of Mallos; but these origins are doubtless fictional.

93. Strabo 14.1.44–45, C 649–50. Athymbros as archegete: Strabo 14.1.46, C 650; and Roman period coins (Leschhorn, *Gründer der Stadt,* 234–35); whether that proves actual cult of Athymbros, and if so from what date, is unclear. On a comparable local archegete (Kranaos) on the coins of Antioch on the Maeander, see L. Robert, *OMS,* 1:359–61; Kosmin, *Elephant Kings,* 231. On Nysa, see especially Robert, *BCH* 101 (1997): 64–77 = *Documents d' Asie Mineure,* 22–35.

Control of the great sanctuary passed from the villagers of Athymbros to Nysa[94] at a moment which is strictly unidentified, but which is tempting to take as the foundation of the city.

These are merely two clear instances of what must have been a very common phenomenon.[95] The startling discovery has recently been made that an Attalid settlement (perhaps ultimately polis) at Kleonnaeion was apparently embedded in the ancient city of Pessinous in Galatia and issued coins featuring Cybele, goddess of Pessinous; the new settlement (admittedly one chiefly populated by Galatians, not Macedonians) chose the local goddess as its emblem.[96] Another plausible though disputed case concerns Thyateira in northern Lydia. Thyateira is a Seleucid military colony, perhaps implanted in an older Lydian town, that grew into an important city. In the second century A.D., when relevant evidence begins, there emerges as its chief god one Tyrimnos, soon also known as Apollo Tyrimnos; this figure is initially indicated on coins by a double ax, then appears as a clothed horseman, and finally becomes a naked standing Apollo in the Greek style. At first sight then he is an indigenous god who has preserved both his image, as a riding ax-bearing horseman, and also his name, and only quite late coalesces with Apollo.[97] But a doubt has arisen: should not the name of this god of Macedonian colonists relate in some way to that of the legendary Macedonian king Tyrimmas?[98] In that event Tyrimnos would be a Macedonian who, iconographically, was depicted in the local idiom for a period, though eventually reemerging in Greek garb as Apollo.[99] But it is hard to see how a legendary Macedonian king, not demonstrably worshipped in his home country, could have been elevated even in a Macedonian settlement to the dominant place in a pantheon. More probably the Tyrimmas/Tyrimnos similarity is a coincidence sent to delude us, and the simpler explanation should be preferred. Tyrimnos if so is a rare instance of a non-Greek god who has lived on in a Macedonian colony/city with native guise and name.

Laodicea on the Lykos in southern Phrygia may also show continuity, though in this case of gods already given a Greek *interpretatio*. It probably received that name from Antiochus II; according to Pliny it had previously been called first Diospolis, then Rhoas. Per-

94. This emerges from *RC* 9 taken with 64 (Rigsby, *Asylia*, no. 185; cf. no. 186). The question of the precise privileges conferred in *RC* 9 (contrast Rigsby, *Asylia*, 44; and Van Bremen, "Leon Son of Chrysaor," 221–22) need not concern us here. The almost universal assumption that *RC* 9 must antedate the foundation of Nysa is not completely secure, given what we know about the occasional retention of vested interests in cults (cf. Van Bremen, "Leon Son of Chrysaor," 227–35), but nonetheless plausible.

95. One would like to know more of Zeus Larasios the chief god of Tralles, whose iconography on coins reveals little, but who in late antiquity emerges as having barefoot concubines of high status associated with him: *ITralles* 6, etc. (cf. Cook, *Zeus*, 2:957–62; *BÉ* 1940, no. 147).

96. See P. Thonemann's brilliant paper, "Pessinous and the Attalids," *ZPE* 194 (2015): 117–28.

97. For the many inscriptions, see *TAM* V.2 p. 314; for iconography, see P. Herrmann, in *LIMC* VIII.1.152–53 s.v. Tyrimnos. Tyrimnos on the whole precedes Apollo Tyrimnos, but *TAM* V.2.882 (cf. the note on 884) may be an early (2nd c. A.D.?) instance of the latter. The dedication from the region by a Greek to an unknown Apollo Pituaenos (*TAM* V.2.881, 276/5 B.C.) does not help; nor does the pairing of ApolloTyrimnos with an unknown Artemis Boreitene in *TAM* V.2.882 with 884.

98. L. Robert, *RPhil* 52 (1978): 247 n. 36 (*OMS*, 5:443). He mentions also a probably Macedonian name Torrimas attested in two papyri.

99. Cf. p. 222 above on the Apollo Propylaios of Eumeneia.

sonal names suggest that it attracted a good quotient of Greek settlers. One tribe was Apollonis; a figure of a standing Zeus with a sceptre in his left hand, an eagle perched on his right, is so common on the coins of the city that numismatists call the type "Laodikean Zeus" wherever it appears.[100] In 1970 or 1971 was found, about four kilometres from the ruins of Laodicea, an honorary decree full of perfect Greek formulae issued in 267 by the "Neoteichitai" and "Kiddiokomitai"; the honorands are to receive permanent privileges at "the publicly financed festivals," and copies of the decree are to be displayed in the sanctuary of Zeus in Babakome and the sanctuary of Apollo in Kiddioukome.[101] So here we find two communities, of which one at least (the Kiddiokomitai) is clearly Phrygian,[102] conducting an organised religious life, centred on shrines of "Zeus" and "Apollo," very close to the future Laodicea. These are the same gods which, we have seen, were especially prominent in Laodicea itself. The bold suggestion has even been made that Babakome might underlie Pliny's former name for Laodicea, Diospolis: Babas is not so distant phonologically from Pap(i)as, a Phrygian god sometimes identified with Zeus.[103] Antiochus could well, it is true, have introduced cults of Zeus and Apollo without regard to local tradition; Apollo is Pythios, not Kiddiokomites, in the later attestations. But the coincidence is at the least suggestive.

The important issue of architecture can only be touched on here. It is no longer believed that Seleucia on the Tigris, a city the institutions of which were "loyal to its founder Nicanor" in Tacitus's phrase, housed a ziggurat.[104] But a visitor to Ai Khanum from Greece, if any there were,[105] would have been surprised to see almost square temples of "Mesopotamian" broad-room type. The question, which will apply also in several like cases, is whether Greek gods could be honoured with Greek rituals amid such unfamiliar architecture. A dedication to Hermes and Herakles has been found, but that came predictably from what has been identified as a gymnasium of familiar Greek design. No temples in the Greek style are

100. On all this, see Cohen, *Hellenistic Settlements,* I, 309–11. Pliny: *HN* 5.105. Greek settlers: L. Robert, in Gagniers, *Laodicée,* 328–33. "Laodikean Zeus" is not to my knowledge attested as a cult title. For Apollo in Laodicea, see further Robert, 290–91, 294–95.

101. The text is now *ILaodLyk* 1 (*SEG* XLVII 1739); cf. Corsten, "Foundation of Laodikeia."

102. What the Neoteichitai are is disputed. For the first editor, M. Wörrle, followed by Schuler, *Ländliche Siedlungen,* 51–52, 161, 188–89, they were a second village; the Roberts argued that Teichos implies "an establishment of military origin" (*BÉ* 1976, no. 667), but Cohen observed (*Seleucid Colonies,* 25–28) that their subjection, along with a native village, to a royal *epimeletes* does not suggest a military colony. Corsten, "Foundation of Laodikeia," notes that they lack a sanctuary of their own, using that of Babakome, and suggests "a fortified settlement or manor, probably established . . . for the protection of the population"; he also points out that the Greek word proves the ethnicity of the founder, not of the occupants.

103. Corsten, "Foundation of Laodikeia." A myth concerning "Zeus Aseis," a goat, and a child leaves traces on coins (and on *ILaodLyk* 65) from the second century A.D.: what the myth is, and how Zeus Aseis relates to other Zeuses at Laodicea, cannot at present be determined (L. Robert, in Gagniers, *Laodicée,* 333, 359–60).

104. Tacitus: *Ann.* 6.42 (cf. Briant, *Rois, tributs et paysans,* 268, for the few such judgements in ancient texts); no ziggurat: Cohen, *Hellenistic Settlements,* III, 158 (citing Invernizzi).

105. The general belief, following Robert, that Klearchos of Soloi was such a visitor, coming from Delphi, is contested by Mairs, "Founder's Shrine," 113–14.

known. The main temple ("temple à redans," "temple with niches") contained fragments of a large statue in the Greek manner which may have depicted Zeus; but this might, it has been suggested, have been a kind of bilingual god, Zeus to some worshippers, Mithra or Ahura Mazda to others. That suggestion is based more on general considerations than any specific feature of the very fragmentary remains of the statue. In the same temple was found, however (but may have been brought there from elsewhere), a most remarkable gilded silver plate showing Cybele and Nike in a large-wheeled chariot drawn by lions; a priest standing behind shades them with a parasol; they approach another priest who is making a libation on a high, stepped altar. There is clearly a confluence of iconographies here. There are also signs that the ritual palette has been extended: vases buried upside down along the shady wall of the "temple with niches" are taken to attest libations of unfamiliar type; more drastically, a stepped podium in an enclosure on the acropolis recalls Herodotus's remarks on Persian open-air devotions addressed to the heavens (1.131.1–2). Inscriptions that might have clarified these possibilities are unfortunately lacking.[106]

Ancient Cities Renamed or Receiving Settlers

Where what occurred was the refoundation with a Greek name of an ancient city with established cults, the situation was reversed: existing cults would necessarily remain central to the life of the city. Thus, most conspicuously, when a whole series of ancient cities in Phoenicia and Cilicia that had received new Seleucid names issued quasi-municipal coins in and after the reign of Antiochus IV, they very regularly showed on the reverse gods certainly not imposed on them by the Seleucids, figures such as Kronos-El, Sandan, Atargatis, and Hadad.[107] In such cases the question (usually unanswerable) becomes whether changes or additions were made in the context of the refoundation. Hierapolis, "Sacred City," was a name that seems often to have become attached to the kind of "city" that was dominated by a sanctuary.[108] Bambyke in Syria with its rich sanctuary of the Dea Syria/Atargatis probably became Hierapolis under Seleucus I,[109] but the cult described by Lucian (*Syr. D.* passim) of

106. On the temples, see especially P. Bernard, "L'architecture religieuse de l'Asie centrale à l'époque hellénistique," in *Akten des XIII Internationalen Kongresses für Klassische Archäologie* (Mainz, 1990), 51–59; also now (with further references) Cohen, *Hellenistic Settlements,* III, 236 n. 23; and Mairs, *Hellenistic Far East,* 85–89, who stresses possible Achaemenid-period antecedents such as the podium at Pačmak-tepe. Hermes and Herakles: *IEstrOriente* 38; note too the figurine of Herakles, *Afghanistan: Crossroads,* 113. Takht-i-Sangin is another site with Greek material but a temple of non-Greek form (Boyce and Grenet, *Zoroastrianism,* 3:173–79; R. Mairs, *The Archaeology of the Hellenistic Far East,* BAR, Oxford 2011, 25), but the main users of the sanctuary were not to my knowledge demonstrably Greek. Contrast the Greek or at least mixed-style temples at Failaka and Jebel Khalid. Bilingual god: see, e.g., Boyce and Grenet, *Zoroastrianism,* 3:165–72; and references in Cohen, *Hellenistic Settlements,* III, 233 n. 12. Cybele plate: *Afghanistan: Crossroads,* 118–19.

107. See most recently Kosmin, *Elephant Kings,* 238–42 (cf. Bonnet, *Enfants de Cadmos,* 157 n. 16; Wright, *Sacred Spaces,* 32–37); the captions in some cases were in Phoenician. "Seleukid adoption or appropriation" of pre-Hellenistic cult sites, and a resulting ritual amalgam, are the central theme of Wright, *Sacred Spaces,* 71–142 (for the phrase quoted, see p. 142, with a summary).

108. See P. Debord, "Hiérapolis: Du sanctuaire-état à la cité," *RÉA* 99 (1997): 415–26.

109. See Lightfoot, *Syrian Goddess,* 38–42.

what he quaintly calls "the Assyrian Hera" (i.e., Atargatis) is spectacularly un-Greek. When coinage begins at Castabala in flat Cilicia under Antiochus IV, the inhabitants have become Hieropolitai; but the cultic life is dominated by the "goddess Perasia" (probably attested under that name already in an Aramaic inscription from the region) with her fire-treading, wild-roaming priestesses.[110] Hierapolis (original name unknown) near Laodicea on the Lykos on the margins of Caria and southern Phrygia contained a cave sanctuary, surely indigenous by origin, remarkable enough to attract almost a page of description in Strabo.[111] Strabo calls it in one place a Ploutonion, in another a Charonion, but the role of Galloi suggests it should rather belong to Cybele. The vapours issuing from the cave were so deadly as to kill any living creature entering even the enclosure in front of it—any living creature except castrated Galloi, who, whether through divine favour or certain antidotes, alone could go within. The royal footprint is much clearer here than in the two other "sacred cities" just mentioned: this Hierapolis was given a tribal structure (which received Attalid additions) by the Seleucids, and a decree votes posthumous honours for Apollonis the mother of Eumenes II.[112] But in none of the three cases can we be sure what changes, if any, the monarchs brought to the cultic life of the city.

Suggestions of some plausibility have been made: that the Apollo Archegetes much mentioned from the second century A.D. in Phrygian Hierapolis presided over the Seleucid refoundation; that the eagle who accompanies the images and emblems of Perasia on the first coins of Hierapolis/Castabala attests a cult of Zeus Olympios introduced to the city by Antiochus IV, Zeus perhaps slipping into the role of a preexisting male consort of Perasia; that the representation of Apollo/Nebu at Hierapolis/Bambyke was influenced by the Seleucid enthusiasm for Apollo.[113] But the problem (chronological imprecision aside) is to know whether these phenomena can only be understood in terms of direct intervention from outside; they might rather show internal development under influence from the ambient Greek culture.[114] It is indeed doubtful whether the concept of "refoundation" is applicable at all in these cases, except for Phrygian Hierapolis with its new tribes; the scale of actual remodelling that went with the change of name is unclear. As far as direct evidence in

110. See Dupont-Sommer and Robert, *Déesse de Hiérapolis Castabala,* passim: for the Aramaic inscription, 11; for the priestesses, Strabo 12.2.7, C 537; and Iambl. *Myst.* 3.4 (110), with Robert"s brilliant discussion, 53–64; for the goddess, see still in the second century A.D. *SGO* 4, 19/18/01.

111. Strabo 13.4.14, C 629–30; cf. 12.8.17, C 579. For the other literary sources on this cave cult, see Ritti, *Hierapolis,* 7–15. Cybele and caverns: Nic. *Alex.* 7–8, with schol.

112. *OGIS* 308. Tribes: F. Kolb, *ZPE* 15 (1974): 255–70, resolving a long-standing debate; cf. P. Debord, *RÉA* 99 (1997): 421–22.

113. Apollo Archegetes: Cohen, *Hellenistic Settlements,* I, 307 n. 4. There was a tribe name Apollonias at Hierapolis, but this does not prove that the god was unknown in the place before the tribal structure was established. Eagle: so Robert, in Dupont-Sommer and Robert, *Déesse de Hiérapolis Castabala,* 97–98; cf. the cautious comments of Boffo, *Centri religiosi,* 58–59. Contrast Bickerman, *Institutions,* 232: "The eagle of Castabala is the bird of Hadad." Robert also notes (93) the assimilation of Perasia on coins to the type of city-protecting Tyche: a phenomenon of representation, no doubt, not of cult. Apollo/Nebu: Lightfoot, *Syrian Goddess,* 41–42, 462–64.

114. Note, e.g., H. Seyrig, *Syria* 13 (1932): 355: "The Eleusinian cults penetrated these regions profoundly in the time of the Lagid domination."

relation to religion goes, all we have is a picturesque story in Lucian which shows Stratonike building a new temple at Hierapolis/Bambyke for the Syrian goddess, promoting, not opposing, a cult of the place.[115] Certainly to be rejected is the old suggestion that Apollo was brought into Phrygian Hierapolis to drive out the evil earth-power Cybele as he had once expelled Pytho from Delphi;[116] such a fantasised divine war drastically misrepresents the situation.

Babylon received a Greek garrison in the third century, and in the second hosted a Greek community known in cuneiform texts as *pu-li-te-e/pu-li-ta-nu* (i.e., *politai,* "citizens"); whether it was actually refounded as a Greek polis (or received one within itself) is contested.[117] At all events there was a significant Greek presence. Houses of Greek and Babylonian design are found in the same areas; there are a theatre, and a gymnasium. An astronomical diary of 169/8 says that "the *politai* [established] a *pompē* and a ritual (?) according to a Greek design," probably to celebrate achievements of Antiochus IV. But, this text aside, there is no trace of Greek cults; it is not even quite certain that this *pompē* was performed in Babylon.[118] One can scarcely doubt that Herakles and Hermes were present in the gymnasium. The intriguing question[119] is whether, for their other needs, the *politai* went to Babylonian shrines (in which we know their monarchs carried on traditional Babylonian rituals). Seleucia on the Eulaios was another Greek city superimposed on one much older, Susa. A cult of Apollo and Artemis Daittai was, we saw, imported from the Seleucid capital Antioch; and the early coins all show traditional Greek gods. But almost all the "sacred manumissions" were made to "Nanaia goddess," whose sanctuary must have retained considerable importance.[120]

115. Lightfoot, *Syrian Goddess,* 388–91, 427–31, argues that the temple with a podium described by Lucian should be of the Roman period, though Stratonike might have financed an earlier construction; Wright, *Sacred Spaces,* 92–93, cautiously defends a Hellenistic date for it.

116. L. Weber, *Philologus* 69 (1910): 178–251, on the basis of Hierapolitan coins showing Apollo Pythoktonos, the statement of Dam. *Isid.* (ap. Phot. *Bibl.* 344b) that the vaporous cave was under the temple of Apollo (their proximity is now confirmed archaeologically—but not for the Hellenistic period), and the legend of Apostle Philip's fight with snakes in the city (on which, see now Ogden, *Drakōn,* 422–25).

117. See Cohen, *Hellenistic Settlements,* III, 130–38, citing the now abundant bibliography. On the loanword *pu-li-te-e/pu-li-ta-nu,* see van der Spek, "Babylonian City," 68.

118. G. F. del Monte, *Testi dalla Babilonia ellenistica* (Pisa, 1997), 1:76 (cf. van der Spek, "Babylonian City," 67). For the doubt, see Cohen, *Hellenistic Settlements,* III, 136 n. 10. *BCHP* 6 obv. 5–8 is another cuneiform reference to an offering made in the reign of Antiochus I "in the Greek fashion." On the persistence of traditional rituals perfomed by natives (with participation too by Seleucid monarchs), see Linssen, *Uruk and Babylon,* passim; striking is the new claim by R. Sciandra, reported by Cohen, *Hellenistic Settlements,* III, 131, that a Greek was appointed as šatammu of Esagila in 125 B.C. Linssen tentatively allows (124–28; cf. Cohen, *Hellenistic Settlements,* III, 131 n. 12) the possibility that cult for the Seleucids may have been added.

119. Cf. R. van der Spek, in T. Derks, *Ethnic Constructs in Antiquity* (Amsterdam, 2009), 110–11, who leaves it unresolved.

120. So Tarn, *Greeks in Bactria and India,* 29, 463–66; on the goddess, see *DDD,* 612–14. Coins: G. Le Rider, *Suse sous les Séleucides et les Parthes* (Paris, 1965), 288, 293. Manumissions: *IestrOriente* 189, 192, 193, 197; just one to Apollo and Artemis Daittai (190; on them cf. n. 66).

Cities Founded Close to Ancient Sanctuaries

The first subsection above, "Cities Significantly Refounded," discussed existing cults perhaps adopted directly into the pantheon of a foundation in whose territory they lay; the second, "Ancient Cities Renamed or Receiving Settlers," was about ancient cities which were dominated by ancient sanctuaries and received an influx of settlers or at least adopted a Greek name. To be distinguished from such cases[121] are those of settlements established close to, but not encompassing, prestigious ancient sanctuaries. The old theory that military colonies were implanted on land seized from such sanctuaries has long been rejected,[122] but the question of relations between city and sanctuary remains. Two cases are particularly interesting.[123] On a hill just three kilometres from Antioch in Pisidia stood an important sacred place of Men Askaenos. Its organisation as described by Strabo (12.8.14, C 577) was, until the death of Amyntas in 25 B.C., of an un-Greek type, with numerous priests and sacred slaves; but the first thorough survey of the remains has revealed a Greek temple and associated buildings of the early second century B.C., with nothing of an earlier date.[124] The temple resembles two at Magnesia on the Maeander sufficiently closely to bring to mind the tradition that Antioch was in part settled from that city. Ramsay, noting with surprise the absence of early material, argued that the sanctuary near the city was founded specifically to serve the third-century colonists, as an offshoot from a second sanctuary of Men somewhere further off in the territory of Antioch also attested by Strabo (Strabo 12.3.31, C 557); the colonists, he argued, felt it necessary to honour the powerful local god, and the clergy of the original sanctuary succeeded in imposing the traditional sacerdotal structure.[125] Men Askaenos in Antioch near Pisidia would then be the equivalent of Isis in Alexandria according to Arrian, the indispensable native deliberately imported into or at least near to the Greek settlement. But Strabo's supposed second sanctuary of Men has been impugned as a gloss in his text, while study of the architectural remains has put them in the second century.[126] They are perhaps a product of Attalid patronage of sanctuaries,

121. They are combined in the lists of Cohen, *Hellenistic Settlements,* I, 432; II, 426. Aizanoi may be another relevant case, but its early history is too obscure to make it useful (cf. M. Wörrle, *Chiron* 39, 2009: 426–31).

122. See, e.g., Robert and Robert, *Carie,* 295–96; Cohen, *Seleucid Colonies,* 21–22; Boffo, *Centri religiosi,* 295–96; note, e.g., *SEG* XXXV 1170. How the newly discovered relation between Kleonnaeion and Pessinus (n. 96 above) worked is not clear.

123. Seleucus Nicator built at the sanctuary of Zeus at Olba, twenty kilometres from Seleucia on the Kalykadnos: Heberdey-Wilhelm, *Reisen in Kilikien* (Vienna, 1896), 85, no. 166 (*Westkilikische Inschriften,* Olba-Diokaisareia, no. 36); on the cult, see T. S. Mackay, *ANRW* 18.3 (1990): 2087–2103. But the city never acquired a formal link with the sanctuary; this was simply a case of monarchic subvention to an established cult. The relations of Laodicea on the Lykos to the nearby sanctuary of Men Karou are not known: L. Robert, in Gagniers, *Laodicée,* 291–92.

124. Mitchell and Waelkens, *Pisidian Antioch,* 37–90.

125. See, e.g., *JHS* 8 (1918): 112–13.

126. Gloss: R. Syme, *Anatolica* (Oxford, 1995), 344–47; Mitchell and Waelkens, *Pisidian Antioch,* 16 n. 29. (But S. Radt in his ed. of Strabo retains the disputed words, citing B. Levick, *JRS* 81, 1971: 82–83.) Remains: Mitchell and Waelkens, 67–68, suggesting Attalid investment. That cities in the region respected this Men is shown by late Hellenistic coinage of Philomelion and Antioch itself

unconnected with the original Seleucid foundation. In that event, all that we can say about the attitude of the founders of Antioch to the important native cult is that they left the traditional structures and priesthood unaltered.[127]

The problem of getting back to the foundation arises also with Stratonicea in Caria. We noted above (n. 71) that it contained within its urban centre, probably by the late third century, a Samothrakion, a Sarapieion, and a Herakleion. According to Strabo (14.2.25, C 660), it also had in its territory "two sanctuaries, the famous one of Hecate at Lagina holding great festivals annually, and near the city the shrine of Chrysaorean Zeus shared by all the Carians, where they come together to sacrifice and debate common concerns"; some twelve kilometres southeast of the town there was also the sanctuary of "Carian Zeus" at Panamara. The city in due course certainly achieved control of the shrines at Panamara and Lagina, incorporating them in its own cultic system, and acquired membership in the Chrysaorean league.[128] But this does not prove that, at the moment of foundation, Stratonicea simply gobbled up the old cults, in defiance of existing vested interests. What happened in detail is for the moment very obscure. A *koinon* of the Panamareis still ran the affairs of that cult a hundred years or so after the foundation of Stratonicea (admittedly the priest now came from the city). But a plausibly redated inscription now attests a demarch of the Panamareis near the date of the foundation, which should indicate that the place Panamara had already been absorbed, though perhaps with a privileged status, into the civic organisation of Stratonicea.[129] As for Hecate at Lagina, her cult itself is not attested until the early second century B.C., but can scarcely have been created at that date. Koranza, site of the sanctuary, is probably one of the locales out of which Stratonicea was brought together; here too then there might have been immediate absorption at the time of the foundation. But it remains possible that (though unattested) an organizing body comparable to the Panamareis retained certain traditional privileges.[130]

(F. Imhoof-Blumer, *Kleinasiatische Münzen,* Vienna 1901–2, 1:285; 2:356–58; *BMC Phrygia,* 353 with n. 1), but this does not prove who controlled the sanctuary.

127. But Labarre, *Le dieu Mèn,* 109, argues that Strabo has abbreviated the sanctuary's history in implying that nothing changed till 25. He points to the Seleucid use of the sanctuary as a place of document display (*SEG* LIV 1353.13–14)—but this does not prove control from the Seleucid colony—coinage (see previous note), and the building programme. This could be right. For control by the later Roman colony, see Labarre, 113–17.

128. Debord, "Questions stratonicéennes," 172: "Stratonicea built its fortune on its sanctuaries."

129. On the continuing existence of the *koinon,* see the fundamental study of Van Bremen, "Leon Son of Chrysaor." She there suggested (234) that it was perhaps Philip V who attached Panamara to Stratonicea. But her own redating of *IStrat* 6, with its reference to a demarch, to the first half of the third century (*CRAI,* 2008, 1404–20) tends to imply (but note her caution, 1419) at least partial incorporation earlier.

130. First evidence for Hecate: *IStrat* 504 and coins (A. Meadows, *NC* 162, 2002: 98–101). For Lagina as part of Koranza, see *IStrat* 501–3. The cult of Apollo and Artemis at Koranza attested by *IStrat* 502 will have been inherited by Stratonicea by the same argument. On "pre-Stratonicea," cf. now M. Ç. Şahin's introduction to *Inschriften von Stratonikeia,* vol. 3 (Bonn, 2010).

ROYAL CULTS

One prediction that could, however, be made with great confidence is that Stratonicea had a cult of Stratonice; and that prediction can be extended mutatis mutandis to almost all the royal foundations even where nothing of the kind is attested.[131] Amid so much that is uncertain, the role in new foundations of cults of monarchs is clear, and of fundamental importance. It is not necessary to go through the list of attested cases here. What is important is not just the ubiquity of these cults but their prominence. At Arsinoe in Cilicia, we have seen, the only attested cults are of Arsinoe, in whose shrine the city's documents were displayed, and of the Theoi Adelphoi. The hieroglyphic Pithom stele describes the foundation of a colony on the Red Sea as follows: "After this his Majesty (Ptolemy Philadelphos) arrived at Kem-Our. He founded a great port for the king, in the great name of the daughter of the king (Ptolemy). A temple was built for the love of his sister; he set up (statues of) the Theoi Adelphoi (and) all the rites of foundation of the temple were performed by the prophets and pure-priests . . ."[132]

We lack comparable early evidence for the Seleucid realm; it is generally supposed that a centrally promoted dynastic cult was not established before the reign of Antiochus III. But the festival Antiocheia attested at Laodicea on the Lykos ca. 200–189 is a specimen of the kind of honour paid to the founder that was probably typical. It was held in the like-named month Antiocheon, contained an athletic contest, and was evidently the chief event in the city's calendar; at Apollonia Salbake in eastern Caria and Ilion in the Troad, festivals of Seleucus had a similar role.[133] Once the Seleucid dynastic cult was established, it naturally penetrated the royal foundations. At Seleucia in Pieria in the early second century B.C. are attested, as we saw, priests of Zeus (in several aspects), of Apollo, of the dead Seleucids, of the reigning Seleucus. At Dura-Europos in the second century A.D. the attested priesthoods are still of Zeus, Apollo, "the Forebears," and King Seleucus Nicator.[134] It might seem that the "minimum kit" for a Seleucid foundation consisted of cults of Zeus, Apollo, and a member or members of the dynasty.[135] When one recalls how emphatically the Seleucids insisted on their descent from Apollo, and through him from Zeus, and notes also the assimilations expressed in the cult titles at Seleucia in Pieria "Seleukos Zeus Nikator" and "Antiochos Apollo Soter," it becomes clear how central the figures of the monarchs were in the cultic life

131. For attested cases, see Cohen, *Seleucid Colonies,* 82–83; Cohen, *Hellenistic Settlements,* I, 28, 438–39; II, 430; III, 344–45; Kosmin, *Elephant Kings,* 343 n. 189. On the application to colonies of Bickerman's distinction between municipal and centralised ruler-cults, see Robert and Robert, *Carie,* 302 n. 6.

132. For the Pithom stele (¶ 20–21; Thiers, *Stèle de Pithom,* p. 65, 2.12), see n. 30. I translate here the French of Thiers, which differs in salient points from the English of Naville cited by Cohen, *Hellenistic Settlements,* II, 310 and 339 (note that "second abode" there is a miscopying from Naville's "sacred abode"); the translation in Mueller, *Settlements of the Ptolemies,* 197, is close to that of Thiers.

133. *ILaodLyk* 5.19–21; cf. 2.16–17, 4 a 16 -17. Apollonia Salbake: Robert and Robert, *Carie,* 285–86, no. 166.33–34; Ilion: *IIlion* 31.10–12.

134. See nn. 23 and 24 above. *OGIS* 233.1–5 (Rigsby, *Asylia,* 111), of Antioch in Persis, is dated by a similar priest, who was presumably local; contrast the dating by the "high-priest" of the cult at Amyzon and Xanthos (Cohen, *Hellenistic Settlements,* I, 247–48). For the possibility that the "Doric temple" at Seleucia-Pieria is the Nikatoreion, see Wright, *Sacred Spaces,* 71–77.

135. On the problematic cults of Zeus Seleukeios and Zeus Antigoneios, see appendix D, p. 202.

of their foundations. (For the Attalids there is less evidence; whether that reflects the more modest self-presentation of that dynasty,[136] the many fewer foundations attributable to them than to the Seleucids, or merely a quirk of what survives to us is unclear.) As was noted earlier, they took over the symbolic centralising function that in earlier Greek colonies had fallen to the heroized "founder."

CONCLUSION

The results of this long survey have not been startling. The cults of new settlements have emerged, to the very limited extent that we can observe or make plausible guesses about them, as a mixture in various proportions of particular cults imported from home, of broader and less specific cults of standard Greek gods (including gods newly popular), of already existing cults of gods who possessed the region, sometimes carrying with them traditional cult practices, and finally, and very importantly, of the monarchs who ordered the establishment of the new settlement. Anybody asked to predict what the results of such an enquiry would be might have produced a similar list; but it seemed worthwhile to try to test the prediction empirically, and draw attention to the problem. What is frustrating is that it is never possible to discover the proportions of the various ingredients with anything like precision. One may wonder why the evidential situation is so very disappointing. Robert explained it by the ordinary principles of archaeological stratification: where a city flourishes, the later history effaces the earlier; paradoxically, only short-lived foundations such as Ai-Khanum survive for posterity.[137] If that is so, we can hope that chance discoveries of inscriptions that have somehow escaped the normal destructive processes will one day lighten our darkness. One might certainly think that the foundation or refoundation of a city was the kind of event which, in a world dominated by the epigraphic habit, would have been commemorated on stone. But, though royal letters are numerous, there is none with instructions for a new settlement; nor is there any trace of a calendar of sacrifices as performed in such a city. Not even fragments survive. It is doubtful whether texts such as we need were ever inscribed on stone.

136. Cf. Thonemann, "Attalid State," 34–44. For such evidence as there is, see Leschhorn, *Gründer der Stadt,* 246–52.

137. Robert, in Gagniers, *Laodicée,* 248–49.

BIBLIOGRAPHY

Abbreviations of journals are those used in the American Journal of Archaeology, available at http://www.ajaonline.org/submissions/abbreviations. Abbreviations of ancient authors and standard reference works are taken from S. Hornblower and A. Spawforth, eds., The Oxford Classical Dictionary[4] (2012); of epigraphical and numismatic corpora from T. Corsten, ed., A Lexicon of Greek Personal Names, VA (Oxford, 2010); of papyrological publications from H. G. Liddell, R. Scott, and H. S. Jones, A Greek-English Lexicon[9], with a revised supplement (Oxford, 1996). Some additions to these abbreviations appear below. What follows is not a list of all the works mentioned in the notes, but of those cited in an abbreviated form (author's name plus short title for monographs and journal articles, short title only for collections of inscriptions and essays).

Afghanistan: Crossroads: Hiebert, F., and P. Cambon, eds. *Afghanistan: Crossroads of the Ancient World.* London, 2011.

Aliquot, J. *La vie religieuse au Liban sous l'Empire romain.* Beirut, 2010.

Allen, S. L. *The Splintered Divine: A Study of Ištar, Baal, and Yahweh: Divine Names and Divine Multiplicity in the Ancient Near East.* Boston, 2015.

Allouche-Le Page, M. T. *L' art monétaire des royaumes bactriens.* Paris, 1956.

Amarna Letters: Moran, W. L., ed. and trans. *The Amarna Letters.* Baltimore, 1992.

AMS: Asia Minor Studien.

ANET: Pritchard, J. B., ed. *Ancient Near Eastern Texts Relating to the Old Testament.* 3rd ed. Princeton, N.J., 1969.

ANRW: Temporini, H., ed. *Aufstieg und Niedergang der römischen Welt.* Berlin, 1972–.

Asie mineure dans l'antiquité: Bru, H., et al., eds. *L'Asie mineure dans l'antiquité: Échanges, populations et territoires.* Rennes, 2009.

Assmann, J. *Moses the Egyptian.* Cambridge, Mass., 1997.

———. *The Search for God in Ancient Egypt.* Trans. D. Lorton. Ithaca, N.Y., 2001.

Austin, M. M. *The Hellenistic World from Alexander to the Roman Conquest: A Selection of Ancient Sources in Translation.* 2nd ed. Cambridge, 2006.

Baslez, M. F. "Les associations à Délos." In P. Fröhlich and P. Hamon, eds., *Groupes et associations dans les cités grecques*, 227–49. Geneva, 2013.

———. *Rechérches sur les conditions de pénétration et de diffusion des religions orientales à Délos (IIe-Ier s. avant notre ère).* Paris, 1977.

Baudissin, W. W. von. *Kyrios als Gottesname im Judentum und seine Stelle in der Religionsgeschichte.* 4 vols. Giessen, 1926–29.

BCHP: Finkel, I., and R. J. van der Spek. *Babylonian Chronicles of the Hellenistic Period.* http://www.livius.org.html.

BÉ: Bulletin Épigraphique, in *Revue des Études Grecques.*

Bean, G. E., and T. B. Mitford. *Journeys in Rough Cilicia, 1964–8.* Vienna, 1970.

Belayche, N. "*Angeloi* in Religious Practices of the Imperial Roman East." *Henoch* 32.1 (2010): 45–65.

———. "De la polysémie des épiclèses: ΥΨΙΣΤΟΣ dans le monde Gréco-Romain." In *Nommer les dieux*, 427–42.

———. "*Deus deum . . . summorum maximus* (Apuleius): Ritual Expressions of Distinction in the Divine World in the Imperial Period." In *One God*, 141–66.

———. "*Hypsistos:* A Way of Exalting the Gods in Graeco-Roman Paganism." In *Religious History of the Roman Empire*, 139–74. Translated with addenda from *Archiv f. Religionsgeschichte* 7 (2005): 34–55.

———. *Iudaea-Palaestina: The Pagan Cults in Roman Palestine (Second to Fourth Century).* Tübingen, 2001.

Bencivenni, A. *Progetti di riforme costituzionali.* Bologna, 2003.

Bernand, A. *De Thèbes à Syène.* Paris, 1989.

———. *Le delta égyptien d'après les textes grecs.* Cairo, 1970.

———. *Les portes du désert: Recueil des inscriptions grecques d'Antinooupolis, Tentyris, Koptos, Apollonopolis Parva et Apollonopolis Magna.* Paris, 1984.

———. *Pan du désert.* Leiden, 1977.

———. *Recueil des inscriptions grecques du Fayoum.* Vol. 1. Leiden, 1975.

Bernand, E. *Inscriptions grecques d'Alexandrie ptolémaïque.* Cairo, 2001.

Bettini, M. *Elogio del politeismo.* Bologna, 2014.

Bickerman, E. *The God of the Maccabees.* Trans. H. R. Moehring. In Bickerman, *Studies in Jewish and Christian History*, 2:1025–1149. German original, Berlin, 1937.

———. *Institutions des Séleucides.* Paris, 1938.

———. *Studies in Jewish and Christian History.* 2 vols. Leiden, 2007.

Bikerman, E. "Anonymous Gods." *Journal of the Warburg Institute* 1.3 (Jan. 1938): 187–96 (= Bickerman, *Studies in Jewish and Christian History*, 2:947–60).

Boffo, L. *I re ellenistici e i centri religiosi dell'Asia Minore.* Florence, 1985.

Bonnet, C. "Héraclès en orient." In C. Bonnet and C. Jourdain-Annequin, *Héraclès d' une rive à l'autre de la Méditerranée*, 165–98. Brussels/Rome, 1992.

———. *Les enfants de Cadmos: Le paysage religieux de la Phénicie hellénistique.* Paris, 2015.

———. *Melqart: Cultes et mythes de l'Héraclès Tyrien en Méditerranée.* Studia Phoenicia 8. Leuven, 1988.

Bopearachchi, O. *Monnaies gréco-bactriennes et indo-grecques.* Paris, 1991.

Bosworth, A. B. *A Historical Commentary on Arrian's "History of Alexander."* Vol. 2, *Commentary on Books IV–V*. Oxford, 1995.
Boyce, M., and F. Grenet. *A History of Zoroastrianism*. Vol. 3. Leiden, 1991.
Briant, P. *Rois, tributs et paysans: Études sur les formations tributaires du Moyen-Orient ancien*. Paris, 1982.
Bricault, L. "Isis Myrionyme." In C. Berger et al., eds., *Hommages à Jean Leclant*, 3:67–86. Cairo, 1994.
———. *Myrionymi: Les épiclèses grecques et latines d' Isis, de Sarapis et d' Anubis*. Stuttgart, 1996.
Bruchmann, C. F. H. *Epitheta deorum quae apud poetas graecos leguntur*. Leipzig, 1893.
Bruneau, P. *Études d'archéologie délienne*. Athens, 2006.
———. *Recherches sur les cultes de Délos à l'époque hellénistique et à l'époque impériale*. Paris, 1970.
Burdur Museum: Horsley, G. *The Greek and Latin Inscriptions in the Burdur Archaeological Museum*. London, 2007.
Burkert, W. *Greek Religion: Archaic and Classical*. Trans. J. Raffan. Oxford, 1985. German original, 1977.
———. "'Mein Gott'? Persönliche Frömmigkeit und unverfügbare Götter." In H. Cancik, ed., *Geschichte—Tradition—Reflexion, FS Martin Hengel*, 2:3–14. Tübingen, 1996.
Cadotte, A. *La romanisation des dieux: L'interpretatio romana en Afrique du Nord sous le Haut-Empire*. Leiden, 2006.
Capdetrey, L. *Le pouvoir séleucide: Territoire, administration, finances d'un royaume hellénistique, 312–129 avant J.-C.* Rennes, 2007.
Carter, J. B. *De deorum romanorum cognominibus quaestiones selectae*. Leipzig, 1898.
CCET: Corpus cultus equitis Thracii. 5 vols. Leiden, 1979–84.
Celtic Religion: Arenas-Esteban, J. A., ed. *Celtic Religion across Time and Space*. Toledo, 2010.
Chaniotis, A. "Acclamations as a Form of Religious Communication." In *Religion des Imperium Romanum*, 199–218.
———. "Megatheism: The Search for the Almighty God and the Competition of Cults." In *One God*, 112–40.
Chantraine, P. *Dictionnaire étymologique de la langue grecque: Histoire des mots*. Paris, 1968–80.
Chiai, G. F. "Königliche Götter und gehorsame Untertanen im Kleinasien der Kaiserzeit: Zur Funktion der Machtepitheta in religiöser Kommunikation." In *Religion des Imperium Romanum*, 219–47.
———. "Medien religiöser Kommunikation im ländlichen Kleinasien." In G. Schörner and D. Šterbenc Erker, *Medien religiöser Kommunikation im Imperium Romanum*, 67–91. Stuttgart, 2008.
Chiekova, D. *Cultes et vie religieuse des cités grecques du Pont gauche*. Bern, 2008.
Chrysostomou, P. Η Θεσσαλική θεά Εν(ν)οδία η Φεραία θεά. Athens, 1998.
Chuvin, P. "Les fondations syriennes de Séleucos Nicator dans la Chronique de Jean Malalas." In P. L. Gatier et al., eds., *Géographie historique au Proche Orient*, 99–110. Paris, 1988.
CIIP: Ameling, W., and H. Cotton. *Corpus inscriptionum Iudaeae/Palaestinae*. Berlin, 2010–.
CIRB: Struve, V. V. *Corpus inscriptionum regni Bosporani*. Moscow, 1965.

CIS: Corpus inscriptionum semiticarum. Paris, 1881–.

Clinton, K. *Myth and Cult: The Iconography of the Eleusinian Mysteries.* Stockholm, 1992.

Cohen, G. M. *The Hellenistic Settlements in the East from Armenia and Mesopotamia to Bactria and India.* Berkeley, 2012.

———. *The Hellenistic Settlements in Europe, the Islands, and Asia Minor.* Berkeley, 1995.

———. *The Hellenistic Settlements in Syria, the Red Sea Basin, and North Africa.* Berkeley, 2006.

———. *The Seleucid Colonies: Studies in Founding, Administration, and Organization.* Wiesbaden, 1978.

Cook, A. B. *Zeus: A Study in Ancient Religion.* 3 vols. Cambridge, 1914–40.

Corsten, T. "The Foundation of Laodikeia on the Lykos." In H. Elton and G. Reger, eds., *Regionalism in Hellenistic and Roman Asia Minor*, 131–36. Bordeaux, 2007.

Cowley, A. *Aramaic Papyri of the Fifth Century B.C.* Oxford, 1923.

Crossroads of Asia: Errington, E., and J. Cribb, eds. *The Crossroads of Asia.* Cambridge, 1992.

Cults of Isis: Bricault, L., and M. J. Versluys. *Power, Politics, and the Cults of Isis.* Leiden, 2014.

Dahlquist, A. *Megasthenes and Indian Religion.* Uppsala, 1962.

Day, J. *Yahweh and the Gods and Goddesses of Canaan.* Sheffield, 2000.

DDD: Van der Toorn, K., et al. *Dictionary of Deities and Demons in the Bible.* 2nd ed. Leiden, 1999.

Debord, P. "Questions stratonicéennes." In A. Bresson and R. Descat, eds., *Les cités d'Asie mineure occidentale au IIe siècle a.C.*, 157–79. Bordeaux, 2001.

———. "Sur quelques Zeus cariens: Religion et politique." *Studi Ellenistici* 13 (2001): 19–37.

Delemen, I. *Anatolian Rider-Gods: A Study on Stone Finds from the Regions of Lycia, Pisidia, Isauria, Lycaonia, Phrygia, Lydia, and Caria in the Late Roman Period.* AMS 35. Bonn, 1999.

DGE: Schwyzer, E. *Dialectorum Graecarum exempla epigraphica potiora.* Leipzig, 1923.

Dickey, E. *Greek Forms of Address.* Oxford, 1996.

Dieux des autres: Bonnet, C., et al., eds. *Les représentations des dieux des autres.* Supplement to *Mythos* n.s. 2 (2011). Caltanissetta, 2012.

Doniger O'Flaherty, W. *Śiva: The Erotic Ascetic.* Oxford, 1981.

Drew-Bear, T., and C. Naour. "Divinités de Phrygie." In *ANRW* 18.3, ed. W. Haase, 1907–2044. Berlin, 1990.

Drew-Bear, T., C. M. Thomas, and M. Yildizturan. *Phrygian Votive Steles.* Ankara, 1999.

Du Mesnil du Buisson, R. *Inventaire des inscriptions palmyréniennes de Doura-Europos.* Paris, 1939.

Dumezil, G. *La courtisane et les seigneurs colorés.* Paris, 1983.

Dupont-Sommer, A., and L. Robert. *La déesse de Hiérapolis Castabala, Cilicie.* Paris, 1964.

Dura V: Rostovtzeff, M., ed. *The Excavations at Dura-Europos: Preliminary Reports,* V. New Haven, Conn., 1934.

Dura IX: Rostovtzeff, M., et al., eds. *The Excavations at Dura-Europos: Preliminary Reports,* IX.3. New Haven, Conn., 1952.

Dussaud, R. *Les Arabes en Syrie avant l'Islam.* Paris, 1907.

EAM: Rizakis, A., and G. Touratsoglou. Ἐπιγραφὲς ἄνω Μακεδονίας. Athens, 1985.

EBGR: Chaniotis, A., et al. "Epigraphic Bulletin for Greek Religion." Published annually in *Kernos.*

Ehling, K., D. Pohl, and M. H. Sayar. *Kulturbegegnung in einem Brückenland: Gottheiten und Kulte als Indikatoren von Akkulturationsprozessen im Ebenen Kilikien.* AMS 53. Bonn, 2004.

Faith, Hope, and Worship: Versnel, H. S., ed. *Faith, Hope, and Worship: Aspects of Religious Mentality in the Ancient World.* Leiden, 1981.

Fraser, P. M. *Cities of Alexander the Great.* Oxford, 1996.

———. *Ptolemaic Alexandria.* 2 vols. Oxford, 1972.

Frei, P. "Die Götterkulte Lykiens in der Kaiserzeit." In *ANRW* 18.3, ed. W. Haase, 1729–1864. Berlin, 1990.

Gagniers, J. des. *Laodicée du Lycos: Le nymphée.* Quebec City/Paris, 1969.

Gaifman, M. *Aniconism in Greek Antiquity.* Oxford, 2012.

Gladigow, B. "Gottesnamen (Gottesepitheta) I." In *Reallexikon f. Antike u. Christentum* 11:1202–38. Stuttgart, 1981.

Goukowsky, P. *Essai sur les origines du mythe d' Alexandre.* Vol. 2, *Alexandre et Dionysos.* Nancy, 1981.

Graf, F. "Gods in Greek Inscriptions." In J. N. Bremmer and A. Erskine, eds., *The Gods of Ancient Greece*, 55–80. Edinburgh, 2010.

———. *Nordionische Kulte.* Rome, 1985.

Gschnitzer, F. "Eine persische Kultstiftung in Sardeis und die 'Sippengötter' Vorderasiens." In W. Meid and H. Trenkwalder, eds., *Im Bannkreis des Alten Orients*, 45–54. Innsbruck, 1986.

Guide de Délos[4]: Bruneau, P., and J. Ducat, eds. *Guide de Délos*[4]. Athens, 2005.

Guide de Thasos[2]: Grandjean, Y., and F. Salviat. *Guide de Thasos*[2]. Athens, 2000.

Haas, V. *Geschichte der hethitischen Religion.* Leiden, 1994.

Habicht, C. *Die Inschriften des Asklepieions. Altertümer von Pergamon* 8.3. Berlin, 1969.

———. *Gottmenschentum und griechische Städte.* 2nd ed. Zetemeta 14. Munich 1970.

Hajjar, Y. *La triade d' Héliopolis-Baalbek.* Leiden, 1977.

Hatzopoulos, M. *Macedonian Institutions under the Kings.* 2 vols. Athens, 1996.

Haute Terres: Debord, P., and E. Varinlioğlu, eds. *Les hautes terres de Carie.* Bordeaux, 2001.

Hawkins, J. D. *Corpus of Hieroglyphic Luwian Inscriptions.* Vol. 1. Berlin, 2000.

Healey, J. F. *Aramaic Inscriptions and Documents of the Roman Period.* Oxford, 2009.

———. *The Religion of the Nabataeans.* Leiden, 2001.

Heinz, M. "Thessalische Votivstelen." PhD diss., Bochum, 1998.

Hellenism in the East: Kuhrt, A., and S. M. Sherwin-White, eds. *Hellenism in the East: The Interaction of Greek and Non-Greek Civilizations from Syria to Central Asia after Alexander.* London, 1987.

Hellenistic Karia: Van Bremen, R., and J. M. Carbon, eds. *Hellenistic Karia.* Bordeaux, 2010.

Hengel, M. *Judaism and Hellenism.* Trans. J. Bowden. 2 vols. London, 1974.

Henrichs, A. "Despoina Kybele." *HSCP* 80 (1976): 253–86.

Hermann Usener filologo della religione: Arrighetti, G., et al. *Aspetti di Hermann Usener filologo della religione.* Pisa, 1982.

Herzog, R. "Namensübersetzungen und Verwandtes." *Philologus* 56 (1897): 33–70.

Hoover, O. D. *Handbook of Coins of Baktria and Ancient India.* Lancaster, Pa./London, 2013.

Hornblower, S. "Lykophron and Epigraphy: The Value and Function of Cult Epithets in the *Alexandra*." *CQ* 64 (2014): 91–120.

———. *Mausolus*. Oxford, 1982.

Hornung, E. *Conceptions of God in Ancient Egypt: The One and the Many*. Trans. J. Baines. Ithaca, N.Y., 1996.

Houghton, A., et al. *Seleucid Coins*. New York, 2002–8.

IAmyzon: Robert, L., and J. Robert. *Fouilles d' Amyzone en Carie*. Paris, 1983.

IAnkyra: Mitchell, S., and D. French. *The Greek and Latin Inscriptions of Ankara (Ancyra)*. Vol. 1, *From Augustus to the End of the Third Century AD*. Munich, 2012.

IAph: Reynolds, J., C. Roueché, and G. Bodard. *The Inscriptions of Aphrodisias*. http://insaph.kcl.ac.uk/iaph2007/.

ICam: Segre, M., and I. Pugliese Carratelli. "Tituli Camirenses." *ASAtene* 27–29 (1949–51): 141–318.

ICentral Pisidia: Horsley, G. H. R., and S. Mitchell. *The Inscriptions of Central Pisidia*. IK 57. Bonn, 2000.

ICilicie: Dagron, G., and D. Feissel. *Inscriptions de Cilicie*. Paris, 1987.

IGDS: Dubois, L. *Inscriptions grecques dialectales de Sicile*. Rome, 1989.

IGerasa: Welles, C. B. "The Inscriptions." In C. H. Kraeling, *Gerasa, City of the Decapolis*, 355–496. New Haven, Conn., 1938.

IGLJ: Inscriptions de la Jordanie. IGLS XXI. Paris, 1986–.

IGLS: Inscriptions grecques et latines de la Syrie. Beirut/Paris, 1929–.

IGPhilae: Bernand, A., and E. Bernand. *Les inscriptions grecques de Philae*. Paris, 1969.

IJO: Ameling, W., and D. Noy et al. *Inscriptiones Judaicae Orientis*. 3 vols. Tübingen, 2004.

IK: Inschriften griechischer Städte aus Kleinasien. Bonn, 1972–.

IKeramos: Varinlioğlu, E. *Die Inschriften von Keramos*. IK 30. Bonn, 1986.

IKomana: Baz, F. *Die Inschriften von Komana (Hierapolis) in Kappadokien*. Istanbul, 2007.

Interpretatio Graeca/Romana/Indigena: Chiai, G. F., R. Haüssler, and C. Kunst, eds. *Interpretatio Graeca/Romana/Indigena: Religiöse Kommunikation zwischen Globalisierung und Partikularisierung*. In *Mediterraneo Antico* 15 (2012): 13–273.

IOlbia: Knipovich, T. N., et al. *Nadpisi Ol'vii/Inscriptiones Olbiae (1917–1965)*. Leningrad, 1968.

IOSPE: Latyschev, B. *Inscriptiones antiquae orae septentrionalis Ponti Euxini graecae et latinae*. Petersburg, 1885–1901. Vol. 1, 2nd ed. (1915).

ISelge: Nollé, J., and F. Schindler. *Die Inschriften von Selge*. IK 37. Bonn, 1991.

ISide: Nollé, J. *Side im Altertum: Geschichte und Zeugnisse*. 2 vols. IK 43–44. Bonn, 1993.

IStrat: Çetin Şahin, M. *Die Inschriften von Stratonikeia* (vol. 3, *The Inscriptions of Stratonikeia*). 3 vols. IK 21, 22, 68. Bonn, 1981–2000.

ITyana: Berges, D., and J. Nollé. *Tyana: Archäologisch-historische Untersuchungen zum südwestlichen Kappadokien*. IK 57. Bonn, 2000.

Jaccottet, A. F. *Choisir Dionysos: Les associations dionysiaques, ou, La face cachée du dionysisme*. Lausanne, 2003.

Jim, T. S. F. *Soteria* (provisional title). Forthcoming.

Jongeling, K. *Handbook of Neo-Punic Inscriptions*. Tübingen, 2008.

Jost, M. *Sanctuaires et cultes d' Arcadie*. Paris, 1985.

KAI: Donner, H., and W. Röllig. *Kanaanäische und aramäische Inschriften.* 5th ed. Wiesbaden, 2002.

Kaizer, T. "Religious Mentality in Palmyrene Documents." *Klio* 86 (2004): 165–84.

Karttunen, K. *India in Early Greek Literature.* Helsinki, 1989.

Kearns, E. *Heroes of Attica.* BICS Suppl. 57. London, 1989.

Keltische Götternamen: Spickermann, W., ed. *Keltische Götternamen als individuelle Option?* Rahden/Westf., 2013.

Konaris, M. D. *The Greek Gods in Modern Scholarship.* Oxford, 2016.

Konya Museum: McLean, B. H. *Greek and Latin Inscriptions in the Konya Archaeological Museum.* London, 2002.

Kuhrt, A., and S. Sherwin-White. *From Samarkhand to Sardis.* London, 1993.

Labarre, G. *Le dieu Mèn et son sanctuaire à Antioche de Pisidie.* Brussels, 2010.

Lane, *CMRDM:* Lane, E. *Corpus monumentorum religionis dei Menis.* 4 vols. Leiden, 1972–78.

Lane Fox, R. *Travelling Heroes.* London, 2008.

Laumonier, A. *Les cultes indigènes en Carie.* Paris, 1958.

Lerat, L. *Les Locriens de l' Ouest.* 2 vols. Paris, 1952.

Leschhorn, W. *Gründer der Stadt.* Stuttgart, 1984.

Lieven, A. von. "Translating Gods, Interpreting Gods: On the Mechanisms behind the *Interpretatio Graeca* of Egyptian Gods." In I. Rutherford, ed., *Greco-Egyptian Interactions: Literature, Translation, and Culture, 500 BCE–300 CE,* 61–82. Oxford, 2016.

Lightfoot, J. L. *Lucian, On the Syrian Goddess.* Oxford, 2003.

Linssen, M. J. H. *The Cults of Uruk and Babylon: The Temple Ritual Texts as Evidence for Hellenistic Cult Practice.* Leiden, 2004.

Macmullen, R. *Paganism in the Roman Empire.* New Haven, Conn., 1981.

Mairs, R. "The Founder's Shrine and the Foundation of Ai-Khanum." In N. MacSweeney, ed., *Foundation Myths in Ancient Societies,* 103–28. Philadelphia, 2014.

———. *The Hellenistic Far East.* Oakland, Calif., 2015.

Malay, M. *Researches in Lydia, Mysia, and Aiolis.* Vienna, 1999.

Manisa Museum: Malay, H. *Greek and Latin Inscriptions in the Manisa Museum.* Vienna, 1994.

Marcadé, J. *Au musée de Délos: Étude sur la sculpture hellénistique en ronde bosse découverte dans l'île.* Paris, 1969.

Marek, C. "Der höchste, beste, grösste, allmächtige Gott." *EpigAnat* 32 (2000): 129–46.

———. *Stadt, Ära und Territorium in Pontus-Bithynia und Nord-Galatia.* Tübingen, 1993.

Mette, H. J. "Nekrolog einer Epoche: Hermann Usener und seine Schule." *Lustrum* 22 (1979–80): 5–106.

Mettinger, T. N. D. *In Search of God.* Trans. F. H. Cryer. Philadelphia, 1988.

Mikalson, J. D. *Religion in Hellenistic Athens.* Berkeley, 1998.

Mileta, C. "Überlegungen zum Charakter und zur Entwicklung der Neuen Poleis im hellenistischen Kleinasien." In A. Matthaei and M. Zimmermann, eds., *Stadtbilder im Hellenismus,* 70–92. Berlin, 2009.

Mili, M. *Religion and Society in Ancient Thessaly.* Oxford, 2015.

Militärsiedlungen: Daubner, F., ed. *Militärsiedlungen und Territorialherrschaft in der Antike.* Berlin, 2011.

Millar, F. *The Roman Near East.* Cambridge, Mass., 1993.
Mitchell, S. "The Cult of Theos Hypsistos." In P. Athanassiadi and M. Frede, eds., *Pagan Monotheism in Late Antiquity*, 81–148. Oxford, 1999.
———. "Further Thoughts on the Cult of Theos Hypsistos." In *One God*, 167–208.
———. "Iranian Names and the Presence of Persians in the Religious Sanctuaries of Asia Minor." In E. Matthews, ed., *Old and New Worlds in Greek Onomastics*, 151–72. Oxford, 2007.
Mitchell, S., and M. Waelkens. *Pisidian Antioch.* London, 1998.
Mon. Anc.: Monumentum Ancyranum.
Moyer, I. S. *Egypt and the Limits of Hellenism.* Cambridge, 2011.
Mueller, B. "Megas Theos." PhD diss., Halle, 1913.
Mueller, K. *Settlements of the Ptolemies.* Leuven, 2006.
Musée Soueida: Dunand, M. *Mission archéologique au Djebel Druze: Le Musée de Soueïda.* Paris, 1934.
Namenforschung: Eichler, E., et al., eds. *Namenforschung.* 3 vols. Berlin, 1995.
New Documents from Lydia: Herrmann, P., and H. Malay. *New Documents from Lydia.* Vienna, 2007.
New Pauly: Cancik, H., et al., eds. *Brill's New Pauly: Encyclopaedia of the Ancient World.* Leiden, 2002–.
NGSL: Lupu, E. *Greek Sacred Law: A Collection of New Documents.* Leiden, 2004.
Niehr, H. *Ba'alšamem: Studien zur Herkunft, Geschichte und Rezeptionsgeschichte eines phönizischen Gottes.* Leuven, 2003.
Nilsson, M. P. *Geschichte der griechischen Religion.* 3rd ed. Vol. 1. Munich, 1967.
Nock, A. D. *Essays on Religion and the Ancient World.* 2 vols. Oxford, 1972.
Nommer les dieux: Belayche, N., et al., eds. *Nommer les dieux: Théonymes, épithètes, épiclèses dans l'antiquité.* Rennes, 2005.
Ogden, D. *Drakōn.* Oxford, 2013.
Ohlemutz, E. *Die Kulte und Heiligtümer der Götter in Pergamon.* Würzburg, 1940.
One God: Mitchell, S., and P. van Nuffelen, eds. *One God: Pagan Monotheism in the Roman Empire.* Cambridge, 2010.
One God or Many? Porter, B. N., ed. *One God or Many? Concepts of Divinity in the Ancient World.* Chebeague Island, Maine, 2000.
Oppermann, M. *Der thrakische Reiter des Ostbalkanraumes im Spannungsfeld von Graecitas, Romanitas und lokalen Traditionen.* Langenweiβbach, 2006.
Otto, W. F. *Priester und Tempel im hellenistischen Ägypten.* 2 vols. Leipzig, 1905, 1908.
PAT: Hillers, D. R., and E. Cussini. *Palmyrene Aramaic Texts.* Baltimore, 1996.
Parker, R. "Artemis Ilithye et autres: Le problème du nom divin utilisé comme épiclèse." In *Nommer les dieux*, 219–26.
———. *Athenian Religion.* Oxford, 1996.
———. *Polytheism and Society in Ancient Athens.* Oxford, 2005.
———. "Theonyms in Northern Greece." In C. Antonetti, A. M. Guimier-Sorbets, M. Kalaitzi, and P. Paschidis, eds., *Βορειοελλαδικά: Tales from the Lands of the Ethne.* Μελετήματα. Athens, forthcoming.
———. "Zeus Plus." In M. Albert-Llorca et al., eds., *Puissances divines à l'épreuve du comparatisme: Constructions, variations et réseaux relationnels*, 327–38. Turnhout, 2017.

Paz de Hoz, M. *Die lydischen Kulte im Lichte der griechischen Inschriften.* AMS 36. Bonn, 1999.

PDura: Welles, C. B., et al., eds. *The Excavations at Dura-Europos.* Final Report 5, pt. 1: *The Parchments and Papyri.* New Haven, Conn., 1959.

Perdrizet, P., and G. Lefebvre. *Les graffites grecs du Memnonion d'Abydos.* Nancy, 1919.

Pernot, L. "Le lieu du nom dans la rhétorique religieuse des Grecs." In *Nommer les dieux,* 29–39.

Peterson, E. *ΕΙΣ ΘΕΟΣ.* Göttingen, 1926. Peterson, *ΕΙΣ ΘΕΟΣ*², refers to the re-edition with unchanged pagination and extensive addenda by C. Markschies et al. (Würzburg, 2012).

Petzl, G. *Die Beichtinschriften Westkleinasiens.* EpigAnat 22. Bonn, 1994.

Plassart, A. *Les sanctuaires et les cultes du Mont Cynthe.* Paris, 1928.

Polinskaya, I. *A Local History of Greek Polytheism.* Leiden, 2013.

———. "Shared Sanctuaries and the Gods of Others: On the Meaning of 'Common' in Herodotus 8. 144." In R. M. Rosen and I. Sluiter, eds., *Valuing Others in Classical Antiquity,* 43–70. Leiden, 2010.

Quaegebeur, J. "Cultes égyptiens et grecs en Égypte." In E. van 'T Dack et al., eds., *Egypt and the Hellenistic World,* 303–24. Leuven, 1983.

Radke, G. *Die Götter Altitaliens.* Münster, 1965.

RC: Welles, C. B. *Royal Correspondence in the Hellenistic Period.* Oxford, 1934.

Reichsreligion und Provinzialreligion: Cancik, H., and J. Rüpke, eds. *Römische Reichsreligion und Provinzialreligion.* Tübingen, 1997.

Religion des Imperium Romanum: Cancik, H., and J. Rüpke, eds. *Die Religion des Imperium Romanum.* Tübingen, 2009.

Religio Phoenicia: Bonnet, C., et al., eds. *Religio Phoenicia.* Studia Phoenicia 4. Namur, 1986.

Religious History of the Roman Empire: North, J. A., and S. R. F. Price, eds. *The Religious History of the Roman Empire.* Oxford, 2011.

Religious Identities in the Levant: Blömer, M., A. Lichtenberger, and R. Raja, eds. *Religious Identities in the Levant from Alexander to Muhammad.* Turnhout, 2015.

RÉS: Répertoire d'épigraphie sémitique. Paris, 1900–.

Rey-Coquais, J. P. *Inscriptions grecques et latines de Tyr.* Bulletin d'Archéologie et d'Architecture Libanaises, hors série III. Beirut, 2006.

RIB: Collingwood, R. G., and R. P. Wright. *The Roman Inscriptions of Britain.* Vol. 1, *Inscriptions on Stone.* New ed. with addenda by R. S. O. Tomlin. Stroud, 1995.

Richter, F. *De deorum barbarorum interpretatione Romana quaestiones selectae.* Halle, 1906.

RICIS: Bricault, L. *Recueil des inscriptions concernant les cultes isiaques.* 2 vols. Paris, 2005.

Ricl, M. "Hosios kai Dikaios: Nouveaux Monuments." *EpigAnat* 20 (1992): 95–100. Cited as Ricl, "Nouveaux Monuments."

———. "Hosios kai Dikaios: Première partie; Catalogue des inscriptions." *EpigAnat* 18 (1991): 1–70; "Hosios kai Dikaios: Deuxième partie; Analyse." *EpigAnat* 19 (1992): 71–103. Cited as Ricl, "Hosios kai Dikaios."

———. "Newly Published and Unpublished Inscriptions for Hosios and Dikaios and Their Contribution to the Study of the Cult." In E. Winter and F. Biller, eds., *Vom Euphrat bis zum Bosporus: Kleinasien in der Antike; FS E. Schwertheim,* AMS 65, 563–79. Bonn, 2008. Cited as Ricl, "New Inscriptions."

Rigsby, K. J. *Asylia: Territorial Inviolability in the Hellenistic World.* Berkeley, 1996.

Ritti, T. *Hierapolis: Fonti letterarie ed epigrafiche.* Rome, 1985.

Ritual Dynamics: Hekster, O., et al., eds. *Ritual Dynamics and Religious Change in the Roman Empire.* Leiden, 2009.

Rives, J. B. *Religion in the Roman Empire.* Malden, Mass., 2007.

RO: Rhodes, P. J., and R. Osborne. *Greek Historical Inscriptions, 403–323.* Oxford, 2003. Reference unless otherwise noted is to inscription number, not page.

Robert, L. *Documents d' Asie Mineure.* Paris, 1987.

———. *Noms indigènes dans l'Asie Mineure gréco-romaine.* Paris, 1963.

———. *Villes d' Asie Mineure.* 2nd ed. Paris, 1962.

Robert, L., and J. Robert. *La Carie.* Vol. 2. Paris, 1954.

Robert, L. sometimes with J. Robert. *Hellenica: Recueil d' épigraphie, de numismatique et d'antiquités grecques,* I–XIII. Limoges/Paris, 1940–85.

Robert, *OMS:* Robert, L. *Opera minora selecta.* 7 vols. Amsterdam, 1969–90.

Ronchi, G. *Lexicon theonymon rerumque sacrarum et divinarum ad Aegyptum pertinentium, quae in papyris ostracis titulis Graecis Latinisque in Aegypto repertis laudantur.* 5 vols. Milan, 1974–77.

Rougemont, F. "Les noms des dieux dans les tablettes inscrites en linéaire B." In *Nommer les dieux,* 323–88.

Roussel, P. *Délos, colonie athénienne.* Paris, 1916.

———. *Les cultes égyptiens à Délos du IIIe au Ier siècle av. J.-C.* Paris, 1915–16.

Rübsam, W. J. R. *Götter und Kulte im Faijum während der griechisch-römisch-byzantinischen Zeit.* Bonn, 1974.

Schachter, A. *Cults of Boiotia.* 4 vols. London, 1981–94.

Schuler, C. "Inscriptions and Identities of Rural Population Groups in Roman Asia Minor." In J. Davies and J. Wilkes, eds., *Epigraphy and the Historical Sciences,* 63–100. Oxford, 2012.

———. *Ländliche Siedlungen und Gemeinden im hellenistischen und römischen Kleinasien.* Vestigia 50. Munich, 1998.

Schwabl, H. "Epiklesen." *RE* X A (1972): 153–376.

———. "Zeus, Nachträge." *RE* Suppl. XV (1978): 1441–81.

SCO: Studi Classici e Orientali.

Seyrig, H. "Antiquités Syriennes 29: À propos du culte de Zeus à Séleucie." *Syria* 20 (1939): 296–301.

SGO: Merkelbach, R., and J. Stauber. *Steinepigramme aus dem griechischen Osten.* Vols. 1–4. Munich, 1998–2002.

Sherwin-White, S. M. "Seleucid Babylonia: A Case Study for the Installation and Development of Greek Rule." In *Hellenism in the East,* 1–31.

Smith, M. S. *God in Translation: Deities in Cross-Cultural Discourse in the Biblical World.* Tübingen, 2008.

Sourdel, D. *Les cultes du Hauran à l'époque romaine.* Paris, 1952.

Stucky, R. A. *Das Eschmun-Heiligtum von Sidon.* Basel, 2005.

Swoboda, H., J. Keil, and F. Knoll. *Denkmäler aus Lykaonien, Pamphylien, und Isaurien.* Vienna, 1935.

Syncrétismes religieux: Motte, A., and C. Bonnet, eds. *Les syncrétismes religieux dans le monde méditerranéen antique.* Brussels, 1999.

TADAE: Porten, B., and A. Yardeni. *Textbook of Aramaic Documents from Ancient Egypt.* Jerusalem, 1986.

Tarn, W. W. *The Greeks in Bactria and India.* 2nd ed. Cambridge, 1950.

Thiers, C. *Ptolémée Philadelphe et les prêtres d'Atoum de Tjékou: Nouvelle édition commentée de la "stèle de Pithom."* Montpellier, 2007.

Thompson, D. J. "Demeter in Graeco-Roman Egypt." In W. Clarysse, A. Schoors, and H. Willems, eds., *Egyptian Religion: The Last Thousand Years; Studies . . . Jan Quaegebeur,* 699–707. Leuven, 1998.

Thonemann, P. "The Attalid State." In P. Thonemann, ed., *Attalid Asia Minor,* 2–47. Oxford, 2013.

———. *The Maeander Valley: A Historical Geography from Antiquity to Byzantium.* Cambridge, 2011.

Totti, M. *Ausgewählte Texte der Isis- und Sarapis-Religion.* Hildesheim, 1985.

Tran Tam Tinh, V. *Le culte des divinités orientales en Campanie.* Leiden, 1972.

Tréheux, J. *Les étrangers.* Vol. 1 of *Inscriptions de Délos, Index* I. Paris, 1992.

Trümper, M. "Negotiating Religious and Ethnic Identity: The Case of Clubhouses in Late Hellenistic Delos." *Hephaistos* 24 (2006): 113–40.

Tscherikower, V. *Die hellenistischen Städtegründungen von Alexander dem Grossen bis auf die Römerzeit.* Leipzig, 1927.

Ustinova, Y. *The Supreme Gods of the Bosporan Kingdom.* Leiden, 1999.

Van Andringa, W. *La religion en Gaule romaine.* Paris, 2002.

Van Berchem, D. "Sanctuaires d' Hercule-Melqart." *Syria* 44 (1967): 73–109.

Van Bremen, R. "Leon Son of Chrysaor and the Religious Identity of Stratonikeia in Caria." In S. Colvin, ed., *The Greco-Roman East: Politics, Culture, and Society,* 207–44. Cambridge, 2004 (= *YCS* 31, 2004).

van der Spek, R. J. "The Babylonian City." In *Hellenism in the East,* 57–74.

Versnel, H. S. *Coping with the Gods: Wayward Readings in Greek Theology.* Leiden, 2011.

Vlassopoulos, K. *Greeks and Barbarians.* Cambridge, 2013.

Vleeming, S. P. *Some Coins of Artaxerxes and Other Short Texts in the Demotic Script.* Leuven, 2001.

Vorländer, H. *Mein Gott: Die Vorstellungen vom persönlichen Gott im alten Orient und im Alten Testament.* Neukirchen-Vluyn, 1975.

Wagner, G. *Les oasis d'Égypte a l'époque grecque, romaine et byzantine d'après les documents grecs.* Cairo, 1987.

Weinreich, O. *Ausgewählte Schriften.* Vols. 1–2. Amsterdam, 1969, 1973.

———. "Θεοὶ ἐπήκοοι." In *Ausgewählte Schriften,* 1:131–95 (from *MdI* 37, 1912).

Westkilikische Inschriften: Hagel, S., and K. Tomaschitz. *Repertorium der westkilikischen Inschriften.* Vienna, 1998.

Wilamowitz-Moellendorff, U. von. *Der Glaube der Hellenen.* 2 vols. Berlin 1931–32. Cited from the 1959 Darmstadt reprint, which has slightly different page numbers.

Will, E. *Le sanctuaire de la déesse syrienne.* Paris, 1985.

Williger, E. *Hagios: Untersuchungen zur Terminologie des Heiligen in den hellenisch-hellenistischen Religionen.* Gieβen, 1922.

Wischmeyer, W. "ΘΕΟΣ ΥΨΙΣΤΟΣ: Neues zu einer alten Debatte." *Zeitschrift für antikes Christentum* 9 (2005): 149–68.

Wissowa, G. "Interpretatio Romana: Römische Götter im Barbarenlande." *ArchRW* 19 (1916–19): 1–49.

Woolf, G. *Becoming Roman.* Cambridge, 1998.

———. "Found in Translation: The Religion of the Roman Diaspora." In *Ritual Dynamics,* 239–52.

Wright, N. L. *Divine Kings and Sacred Spaces: Power and Religion in Hellenistic Syria (301–64 BC).* BAR International Series 2450. Oxford, 2012.

Zgusta, L. *Kleinasiatische Ortsnamen.* Heidelberg, 1984.

Zoll, A. L. "Patterns of Worship in Roman Britain: Double-Named Deities in Context." In S. Cottam et al., eds., *TRAC* 94 (1994): 32–44.

INDEX